Natural Additives in Foods

Natural Additives in Foods

Maya Lal
A. J. Anne
Purab Bhatti

2025
Agri Horti Press
New Delhi-110 084

ISBN: 978-93-92969-55-3 (Hardbound)

Published By

Agri Horti Press

New Delhi

E-mail: agrihortipress@gmail.com

Digitally Printed at : Replika Press Pvt. Ltd.

Preface

In the modern food industry, consumers are increasingly seeking natural and healthy alternatives to synthetic additives, driving a growing interest in incorporating natural ingredients into food products. From enhancing flavor and texture to extending shelf life and improving nutritional value, natural additives play a crucial role in meeting consumer demands for clean label products while ensuring food safety and quality.

This book, titled "Natural Additives in Foods," serves as a comprehensive guide to the diverse array of natural ingredients used in food formulation and processing. Drawing upon the expertise of food scientists, nutritionists, and industry professionals, this volume explores the multifaceted landscape of natural additives, providing insights into their sources, functionalities, and applications in various food products.

From plant-derived antioxidants and antimicrobials to microbial enzymes and fermentation products, "Natural Additives in Foods" delves into the rich diversity of natural ingredients available to food manufacturers. Through a blend of scientific research, case studies, and practical applications, we examine the role of natural additives in enhancing food safety, quality, and consumer appeal across different product categories.

Moreover, this book addresses the regulatory frameworks, safety considerations, and labeling requirements associated with the use of natural additives in foods. By navigating the complexities of food regulations and industry standards, we empower food producers, regulators, and consumers with the knowledge and tools to make informed decisions about natural additives and their inclusion in food products.

As we embark on this journey of exploration and discovery, it is our hope that "Natural Additives in Foods" will serve as a valuable resource for food industry professionals, researchers, policymakers, and consumers alike. Whether you are seeking to understand the science behind natural additives, evaluate their efficacy in food products, or make informed choices about the foods you consume, this book invites you to delve into the world of natural ingredients and their role in shaping the future of the food industry.

Maya Lal

A. J. Anne

Purab Bhatti

Contents

Chapter-1

The Implications of Replacing Synthetic Antioxidants with Natural ones In The Food Systems

1. Introduction

The major drive of recent developments in food processing and storage activities is, undoubtedly, to produce food products that have the potential of providing required nutrients and bioactive compounds in order to reduce increasing lifestyle diseases like cancers, cardiovascular, diabetes, and others. Bioactive compounds are non-essential biomolecules that have biological values beyond their calorie content found in foods that are capable of modulating metabolic processes resulting in the promotion of better health [1]. Antioxidants are bioactive compounds contained in foods, though not considered as part of nutrients, but by their antioxidative activities, are capable of enhancing the foods' keeping quality or promoting the consumers' health. Sardarodiyan and Sani [2] posit that antioxidants have become an indispensable group of food additives mainly because of their unique properties of extending the shelf-life of food products without any adverse effect on their sensory or nutritional qualities. Studies about antioxidants have shown that they are chemical compounds that are capable of hindering the generation of reactive species and their derivatives, either in the food systems or in the human body. They are categorized into two major groups which include synthetic and natural antioxidants, based on their sources. Natural antioxidants are produced by natural molecular formations in plants, animals, mushrooms, microorganisms like algae and bacteria and are thus extracted directly from organic sources such as fruits, vegetables, grains, and meat. Synthetic antioxidants are artificially synthesized by combinations of some chemical compounds in laboratories for use, mainly, in the preservation of foods. Though both categories of antioxidants are assumed to perform the same function in food systems, they have a distinct effect. According to the literature, in terms of zero tolerance to side effects and contribution to delay or prevention of occurrence of chronic diseases, natural antioxidants are more effective; while in terms of preservation of high lipid food products synthetic antioxidants are more effective. Morton et al. [3] corroborated with this assertion by reporting preference in the use of synthetic antioxidants in preserving foods with high rancidity levels to natural antioxidants usable in preserving hydrogenated oils with lower rancidity levels.

1.1 Definition of Antioxidant

Antioxidants are substances that prevent or retard oxidative activities in foods or body systems. Halliwell [4] and Arun and Abdul Azeez [5] reported that they are usually present in relatively small concentrations but are capable of frustrating oxidative activities in the systems. Tuberoso et al. [6] and Atta et al. [7] mentioned them as resources for use in preventing or greatly retarding the oxidation of easily oxidizable materials such as fats (and or peroxidation of lipids in food products and cells of the body systems. They are also defined as substances that engage harmful forms of oxygen to prevent them from harming the cells of either the food products or those of the body of food consumers. Kebede and Admassu [8] stated that antioxidants are capable of slowing down the autoxidation process of other compounds or neutralize free radicals. Although Becker et al. [9] and Halliwell [10] specified the above definition in the context of the biological system, Atta et al. [7] alluded to it as a broader definition encompassing many vulnerable macromolecules (e.g. DNA, lipids and proteins) that can be affected by oxidation. Such broad definition means that compounds that inhibit specific oxidizing enzymes, react with oxidants before they damage molecules, sequester dangerous metal ions or even repair systems such as iron transport proteins, can fit into the definition [7]. Ihekoronye and Ngoddy [11] defined them as substances that retard the rate of oxidation which serve two principal functions: breaking the oxidation chain by containing free radicals or acting as hydrogen donors and facilitating the breakdown of peroxides into stable substances that inhibit further oxidation. Atta et al. [7] referred to the above description as the mechanistic definition of antioxidants. The definition considers radical scavenging capacity or amount of free radical captured by antioxidant food components [12]. Asimi et al. [13] considers antioxidants as compounds or systems that can safely interact with free radicals generated in the food products or by metabolic activities to prevent them from reacting with the cells and cause damages, in the case of the body. Their affinity with the free radicals facilitates their disposition to mop up the radicals generated by metabolic processes to protect the cells. Antioxidants, indeed, are substances that at low concentrations retard the oxidation of easily oxidizable biomolecules [14] such as lipids and proteins either in food products or in living cells of the body system to discourage adverse effects of oxidation. Antioxidants act at different levels in the oxidative sequence involving lipid molecules [2]. Bontempo et al. [15] reported several ways they function including reducing oxygen concentration, intercepting singlet oxygen ($1O_2$), scavenging initial radicals like hydroxyl radical to avoid initiation of first-chain reaction, binding metallic ion catalysts, decomposing primary products of oxidation to non-radical species and breaking chain reactions to prevent continued hydrogen abstraction from substrates. The necessity to produce healthier foods to discourage occurrences of lifestyle diseases and their associated increasing intake of drugs propels consideration for replacing synthetic antioxidants with the natural ones in the food system.

2. Operational Mechanism of Antioxidants

The reduction or stoppage of oxidative processes by antioxidants, in any system, follows two principal mechanisms of action. Kebede and Admassu [8] reported a chain-breaking mechanism as the first action in which primary antioxidants donate electrons to the free radicals present in the system. The ways they achieve this include stoppage of formation of free radicals, providing electrons to the existing free radicals to stabilize them and checkmating their reactivity. A free radical can be defined as, "any molecular species capable of independent existence that contains an unpaired electron in an atomic orbital and capture electrons from other substances in order to neutralize themselves" [16]. Atta et al. [7] referred to this action as neutralization of free radicals and identified two major pathways through which this is accomplished to include chain-breaking and preventive processes. In chain-breaking free radicals release or abstract electron to form second radical which does the same thing to the third molecule to continue to generate unstable products to propagate the chain of reactivity and oxidation processes. The free radical has the ability to donate or to accept an electron from other molecules [17]. This stabilizes the free radical at the beginning but starts to produce another in the process [14]. The moment a chain reaction begins, thousands of free radical reactions can occur within a few seconds on the primary reaction [18].

Antioxidants readily donate an electron to the free radicals to get them stabilized. This assertion is in agreement with the report of Brewer [19] that the propagation of free radical chain reaction can be minimized by the donation of hydrogen from the antioxidants and the metal chelating agents. The preventive pathway of antioxidants entails the removal or scavenging of free radicals to prevent their interaction with food substrate. The view of Nawar [20] about the mechanisms of antioxidants indicated that antioxidants scavenge species that initiate peroxidation, chelate metal ions, and disable their potential to generate reactive species or decompose lipid peroxides, quench or prevent the formation of peroxides, break the autoxidative chain reaction, and/or reducing localized O_2 concentration. This assertion described by Kebede and Admassu [8] as the second mechanism of action of antioxidants, entails quenching chain initiator mechanisms that incidentally eliminates initiators of reactive oxygen species (ROS) and reactive nitrogen species (RNS). It is worthy to mention here, according to Pisochi and Pop [21] and Perez and Aguilar [22], that free radicals are derived from oxygen, nitrogen, and fsulfur molecules and, hence the free radicals constitute groups of molecules called reactive oxygen species, reactive nitrogen species, and reactive sulfur species. Atta et al. [7] stated that free radicals of ROS include superoxide anion (O_2–•), perhydroxyl radical (HO_2•), hydroxyl radical (·OH), nitric oxide and other species such as hydrogen peroxide (H_2O_2), singlet oxygen (O_2), hypochlorous acid (HOCl) and peroxynitrite (ONOO–). According to them, whereas RNS are products of the reaction of nitric oxide with O_2–• to form ONOO–; RSS is derived from the reaction of thiols with ROS.

$$RH \rightarrow \mathrm{R}* + \mathrm{H}* \quad (1)$$

$$\mathrm{R}* \rightarrow \mathrm{R}* + \mathrm{O}_2 \rightarrow ROO* \quad (2)$$

$$2\mathrm{ROOH} \rightarrow ROO* + RO* + \mathrm{H_2O} \quad (3)$$

The mechanisms of antioxidants are further explicable with identification of three stages of mechanisms of chain reactions which according to the reports of Rosenblat and Aviram [23], Polumbryk et al. [24], and Kebede and Admassu [8] include: initiation, propagation, and termination stages. In the initiation stage, the abstraction of the hydrogen atom from the system generates free radicals to initiate chemical reactions of oxidation activities. The presence of antioxidants inhibits the formation of free radicals to delay or disable the start of initiation or propagation of the chain reaction. Below is a typical initiation stage of a system RH, a free radical R* formed as a result of the abstraction of a hydrogen atom H*.

The propagation of free radical chain reaction is occasioned by the ability of free radicals or the reactive species to react with a molecule of oxygen from the environment, resulting in the formation of peroxides and peroxy radical ROO* in the propagation stage [8] shown below. Also, the presence of antioxidants frustrates the intermediates from propagating free radicals, which according to Brewer [19] and Atta et al. [7] could be by the donation of hydrogen from the antioxidants. The propagation stage is represented below.

$$\mathrm{R}*+\mathrm{O}_2 \rightarrow ROO* \tag{4}$$

$$ROO*+RH \rightarrow \mathrm{ROOH}+\mathrm{R}* \tag{5}$$

$$RO*+RH \rightarrow ROH+\mathrm{R} \tag{6}$$

In the terminal stage shown below, either two free radicals combine to form a stabilized or nonradical species or the antioxidants donate hydrogen atom (H*) to radicals to terminate the chain reaction. Brewer [19] stated that the free radicals of antioxidants may then form a stable peroxy-antioxidant compound.

$$\mathrm{R}*+\mathrm{R}* \rightarrow \mathrm{R}+\mathrm{R} \tag{7}$$

$$\mathrm{R}*+ROO* \rightarrow \mathrm{ROOR} \tag{8}$$

$$ROO*+ROO* \rightarrow \mathrm{ROOR}+\mathrm{O}_2 \tag{9}$$

$$\text{Antioxidants}+\mathrm{O}_2 \rightarrow \text{Oxidized antioxidants} \tag{10}$$

Although based on the mode of production antioxidants are majorly classified into natural and synthetic antioxidants, the cellular level as the targeted site of free radicals' damage and defensive approach of antioxidants, was also mentioned by Anwar et al. [25] as a criterion for further classifying the antioxidants into enzymatic and nonenzymatic ones. However, the classification reported by Kebede and Admassu [8], Akbarirad et al. [26] and Anbudhasan et al. [14] highlighted the mode of the provision of antioxidants to the body system; and mentioned exogenous, endogenous, and dietary antioxidants, as classes of antioxidants, subsumes the forgoing classification. Lastly, classification based on the course of action was equally mentioned by Manessis et al. [27] in their report on the classification of antioxidants. Some of them will be discussed briefly.

2.1 Natural Antioxidants

Natural antioxidants are, at times, considered as extra nutritional components that occur in small quantities in food materials, especially if such food materials contain compounds like vitamins C or E that dually serve as providers of nutrients and bioactive compounds. Grozea [28] stated that they are found in natural sources, such as fruits, vegetables, and meats. They are also found in all plants parts like nuts, seeds, leaves, roots, and barks [26]. Table 1 shows some natural antioxidants that are increasingly applied in food systems. Though natural antioxidants are products of animals, plants, mushrooms, and algae. Kebede and Admassu [8] reported that natural antioxidants that are mainly used in the food system are mostly synthesized by plants (e.g. vitamins and other naturally occurring chemical compounds in food). Yadav et al. [30] corroborated with the foregoing and mentioned antioxidants commonly found in everyday foods to include vitamin C (ascorbic acid), vitamin E (tocopherols), vitamin A (carotenoids), various polyphenols including flavonoids, anthocyanins, lycopene (a type of carotenoid), and coenzyme Q10, also known as Ubiquitin, which is a type of protein. Some of these antioxidants and others highlighted in Table 1 are significantly sourced from plant-based foods. Natural antioxidants are found in most fresh foods [14]; with fruits, vegetables, and medicinal herbs being the richest sources of antioxidant compounds such as vitamins A, C, and E, ß-carotene, and important minerals [31]. Mohdali [32] reported different variations in phenolic contents not only among different fruits or vegetables but also reports of different authors even for the same fruits or vegetables. Also, two major groups, enzymatic antioxidants and non-enzymatic antioxidants constitute the human antioxidant [33, 34].

Table 1. Some dietary antioxidants and examples of their sources.

Exogenous antioxidants	Dietary sources
Vitamin C (ascorbic acid/ascorbate)	Bell peppers, strawberries, kiwi, Brussels sprout, broccoli, most fruits (particularly citrus fruits), some vegetables, tomatoes.
Vitamin E (tocopherol, tocotrienols)	Vegetable oil (olive, sunflower, safflower) and its derivatives (margarine, salad dressing, nuts and seeds, cereal grains, broccoli, Brussels sprouts, cauliflower, almonds, hazelnuts
Carotenoids(carotene, zeaxanthin, lutein, lycopene, β-cryptoxanthin, etc.)	Orange and red vegetables and fruits (carrots, tomatoes, apricots, plums) and green leafy vegetables (spinach and kale), dark leafy vegetables, sweet potatoes, yams, citrus fruits, kale, papaya
Polyphenols (flavonols, flavanols, catechins, anthocyanins, isoflavones, phenolic acids	Fruits (apples, berries, grapes, citrus), vegetables (celery, kale, onions, lettuce, eggplants, peppers, cruciferous vegetables, onions) legumes (beans, soybeans, nuts), wine, tea, cocoa, oilseeds, black tea
Trace elements(selenium, zinc)	Seafood, red meat, chicken, and whole grains

Sources: Da Costa et al. [29] and Kebede and Admassu [8].

2.2 Dietary Antioxidants

Dietary antioxidants are a complex mixture of micronutrients and bioactive phytochemicals in the diets that exhibit a range of antioxidant functions, and also according to Da Costa et al. [29], play an important role in the defense against stress. They are sufficiently supplied to the body system by the consumption of balanced diets, fruits, and vegetable-based diets. Hence, they are as well, part of exogenous antioxidants. Young et al. [35] indicated that the members of the Food and Nutrition Board of the National Research Council in the United States, described dietary antioxidants as components of food that significantly mitigate the adverse effects of reactive oxygen species and reactive nitrogen species in normal physiologic function in humans. And typical dietary antioxidants are ascorbate, tocopherols, carotenoids, and bioactive plant phenols. The potential of fruits and vegetables to promote human health, according to the literature, is due to the presence of antioxidant inclined vitamins, and the large number of phytochemicals having antioxidant properties. The most widely studied dietary antioxidants, according to Yadav et al. [30], are Vitamin C, vitamin E, ß-carotene, and other carotenoids and oxycarotenoids, e.g., lycopene and lutein. They have the potential, to reduce reactive oxygen species and reactive nitrogen species and their associated adverse effects on the body [36]. Dietary antioxidants, at times, referred to as exogenous antioxidants are derived from food eaten to complement or strengthen the activities of the endogenous antioxidants. Hence, they are either sourced from synthetic antioxidants or natural ones. Percival [37] mentioned vitamins, flavonoids, anthocyanins, and some mineral compounds as some of the naturally sourced dietary antioxidants. Table 1 highlights some of these antioxidants and their sources. Yadav et al. [30] reported that there is an increasing interest in the application of antioxidants as food preservatives, particularly dietary antioxidants intended to prevent the presumed deleterious effects of free radicals in the human body, as well as the deterioration of fats and other constituents of foodstuffs. The report of Sardarodiyan and Sani [2] indicated that vitamins C and E, carotenoids, stilbenes, phenolic acids such as benzoic and hydroxybenzoic acids, cinnamic and hydroxycinnamic acid derivatives and flavonoids—flavonols, flavones, flavanones, flavanols, flavones, and anthocyanidins (as the aglycones of anthocyanins) and others are the main dietary antioxidants. Table 1 indicates that trace elements such as selenium and zinc usually sourced from seafood, meat, and whole grains; are part of exogenous antioxidants. Although, synthetic antioxidants are not among the dietary antioxidants in Table 1, Butylated hydroxyanisole (BHA) and BHT (butylated hydroxytoluane that are frequently applied as food preservatives are examples of exogenous antioxidants (see Table 1). They are not deliberately added as food ingredients but are used as preservatives, hence are involuntarily consumed with foods and are observed to play some roles in the body system.

2.3 Synthetic Antioxidants

Synthetic antioxidants are synthesized artificially by combinations of some chemical substances in the laboratory. They are widely used as food additives to prevent rancidification, owing to their high performance and wide availability [14]. They are chemically synthesized compounds since they do not occur in nature and are added to food as preservatives to help prevent lipid oxidation [7]. The instability of the natural antioxidants occasioned their involvement as preservatives for food products. According to the literature, the predominant applications of synthetic antioxidants as food preservatives are due to their high reactivity and more efficiency and effectiveness in preserving foods. Butylated hydroxytoluene (BHT) and butylated hydroxyanisole (BHA) were originally developed to protect petroleum from oxidative gumming [38]. However, these compounds have been used as antioxidants in human foods since 1954 and are perhaps the most common antioxidants used in those foods today [39]. Though they are predominantly used, the food industry is pushing for their replacement with natural antioxidants because of the consumers' increasing preference for natural antioxidants which in addition, not only are more affordable but are eco-friendly. However, their usage is regulated by the established authorities to protect food consumers like the Nigerian Food and Drugs Administration (NAFDAC) and Standard Organization of Nigeria (SON), Food and Drug Administration (FDA) of the USA, European Food Safety Agency etc. Rashmi et al. [40] reported that the level of antioxidants permitted for use in food is usually determined by the fat content of the recipient food item, and is limited to 0.02% total antioxidants. Table 2 shows synthetic antioxidants commonly used as food preservatives and their recommended levels of usage, based on the legislations of FDA,

European Food Safety Agency (EFSA), Joint FAO/WHO Expert Committee on food additives etc (Table 2).

Table 2. Typical synthetic antioxidants used as preservatives, their legal limits in the foods.

Compound name	Limit in foods	Morphology/solubility	Food matrix
BHA (butylated hydroxyanisole)	< 200 mg/kg*	White waxy flakes, soluble in fat, insoluble in water	Cereals, chewing gum, potato chips, vegetable oils, biscuits, cakes pastries, sugar honey, meat products, spices, milk products, etc
BHT (butylated hydroxytoluane	< 100 mg/kg*	White crystalline compound /soluble in fat, insoluble in water	Vegetable oil, meat products, potato sticks, chicken soup base, chewing gum, sugar, honey, spices, milk products, etc.
PG (Propyl Gallate)	< 200 mg/kg*	White crystalline powder sparingly soluble in water	Vegetable oil, meat products, potato sticks, chicken soup base, chewing gum, sugar, honey, spices, milk products, etc.
OG (Octyl Gallate)	< 200 mg/kg*	White to creamy white crystalline solid, insoluble in water	Oils and fats, cereals, snack foods, dairy produce, sugar, honey, meat products, etc
DG (Dodecyl Gallate)	< 200 mg/kg*	White to creamy white Crystalline solid, insoluble in water	Oils and fats, cereals, snack foods, dairy produce, meat products, etc.
EDTA	75ppm**	Slowly soluble in water	Salad dressing, margarine, sandwich spreads, mayonnaise, processed fruits and vegetables, canned shellfish, soft drinks
TBHQ (tertiary butylhydroquinone)	120 mg/kg*	Beige colored powder, soluble in fats	Milk, milk products like cheese, meat and meat products, chewing gum, fish & Fish products, sea food, sugar, honey, spices, etc.
Source: Rashmi et al. [40]. ***CODEX Alimentarius**/Joint FAO/WHO Expert Committee on food additives. **US FDA.			

2.4 Action-oriented classification

Manessis et al. [27] further classified antioxidants based on the way they act in the biological system into the following; (i) primary antioxidants, (ii) oxygen scavengers, (iii) secondary antioxidants, (iv) enzymatic antioxidants, and (v) chelating agents. According to them, primary antioxidants donate electron or hydrogen to terminate free-radical chain reactions and some antioxidants in this group include phenolic compounds, tocopherols; and synthetic antioxidants such as alkyl gallates, BHA, BHQ, and TBHQ; (ii) oxygen scavengers are groups of antioxidants that remove oxygen to reduce their chance of furthering oxidative activities and they include vitamin C, ascorbyl palmitate, erythorbic acid, and its sodium salt; (iii) secondary antioxidants are group of antioxidants that breakdown lipid hydroperoxides into stable end-products such as dilauryl thiodipropionate and thiodipropionic acid; (iv) enzymatic antioxidants act by removing oxygen like glucose oxidase and removing ROS by such enzymes like superoxide dismutase, catalase etc. and (v) chelating agents act to remove metallic ions like iron and copper, known to catalyze lipid oxidation. Common chelating agents, according to, Pokorný [41] and Hudson [42] are citric acid, ethylenediaminetetraacetic acid (EDTA), and amino acids.

3. Effectiveness of Natural and Synthetic Antioxidants

Antioxidants, depending on the category they belong, differ in their delivery or operation in checkmating undesirable oxidation and their derivatives to mitigate food deterioration. Variation in effectiveness of antioxidants is connected with several factors which include the system of operation, type or the group the antioxidant belongs to, aspect of functionality etc. The literature mentioned the use of antioxidants as food preservatives, and the differences in potency observed between synthetic and natural antioxidants. For instance, Rasmi and Disha [40] reported the differences in the potency of natural and synthetic antioxidants as food preservatives and stated how both differ in performance levels. The ascertainment of performance levels of both synthetic and natural antioxidants, according to Rasmi and Disha [40], depending on the number of peroxides formed in lipids over time and what they referred to as carry-through properties i.e. the ability of the antioxidant to provide stability under different processing conditions like heat (such as frying or baking), varying solubility, etc. According to the literature, the application of antioxidants as food preservatives depend on the nature or class of food being preserved and overall price consideration. For instance, preservation of foods with high rancidity levels is better achieved by using synthetic antioxidants since these are more potent and hence, have high-performance levels, while natural antioxidants with lower potency and performance levels can suffice for hydrogenated oils with lower rancidity levels [3]. However, the effectiveness of antioxidants may be measured based on the extent to their utilization promotes food safety, in terms of promoting the health of the consumers, with respect to disabling occurrences of degenerative diseases in vivo, and in terms of preventing or minimizing lipid peroxidation which produces toxic compounds that enhance deterioration of flavor, color, texture, and nutritional values [43] which incidentally lead to overall depreciation of food quality and its consumer acceptability [44]. Also, studies carried out on the applications of natural antioxidants, according to Fernades et al. [45] showed their promotion of the palatability of food products, that is, the appetizingness of such products. The effectiveness of natural antioxidants in stabilizing food products in a manner equivalent to that of synthetic antioxidants and their contribution to longer shelf-life to meat products was, as well, reported by Jung et al. [46]. The strong correlation between diet and disease prevention instigated by applications of antioxidants in the food system which pushes industrial trend toward the development of functional food products [47] is leading to increasing adoption of many dietary and technological techniques that facilitate the use of these antioxidants in maintaining the quality of food products like meat and its derivatives [45]. The type of such technological strategy was reported by Velasco and Williams [48] to be the inclusion of plant ingredients having high bioactive potential in packaging materials or incorporation of dietary supplements in animal feeds. Hence, whereas synthetic antioxidants are considered to be more potent or effective as preservatives because of their readiness to donate electrons to food substrates, the increasing preference of consumers to natural antioxidants and the associated inclination of the food industry to satisfy the consumers' demands; is factored on the following: their good bioactive health potential, their perceived functional properties and increasing demand for healthy food products.

Also, in terms of the involvement of food in promoting the consumers' health, food processors are increasingly producing food products containing prerequisite antioxidants (natural ones) to discourage the high incidence of degenerative diseases and improve food safety. Voutilainen [49] reported the important role nutrition play in preventing many chronic diseases such as cardiovascular diseases (CVD), cancers, and degenerative brain diseases. This assertion was corroborated by Atta et al. [7] who stated that consumption of dietary antioxidants such as β-Carotene prevents muscular degeneration and cataracts. The potency of natural antioxidants in this regard overrides that of synthetic antioxidants. The overriding contributions of natural antioxidants in checkmating occurrences of lifestyle diseases aforementioned are well reported in the literature. Though the use of synthetic antioxidants in maintaining the quality of ready-to-food products has gained prominence, the increasing demand for food products that guarantee the safety of consumers has instigated the food industry to seek their replacement with natural antioxidants [50]. According to Anbudhasan et al. [14], food products containing natural antioxidants were more functional in promoting shelf-life and health of their consumers when compared with those ones whose antioxidants were removed during processing. The above reports indicated that natural antioxidants are the kernels of involvement of foods in minimizing chronic diseases, promoting health and incidentally reducing intake of drugs taken for healing, which also generate adverse side effects. Whereas synthetic antioxidants have the advantages of being readily available and affordable and are more reactive when compared with natural antioxidants, the preference of consumers for food products processed with natural antioxidants, is increasingly demeaning their applications as food preservatives.

4. The Uses of Antioxidants in Food Systems

Antioxidants are part of food additives used in the food systems primarily to infringe oxidation of lipids and proteins to elongate the keeping quality of food products, thus enhancing the shelf-life of food products. Antioxidant compounds present in food systems help to reduce the number of lifestyle diseases which may reduce the amount of drugs consumed by people who suffer from these diseases. Atta et al. [7] agreed with the above assertion with a report that antioxidants are widely used as an ingredient in a dietary supplement for promoting good health. Recent developments in food processing are indicating that food could be used as preventive and curative channels to discourage increasing occurrences of chronic diseases such as cardiovascular diseases, high blood pressure, diabetes, cancers etc. being witnessed among people. This may enliven the intention of replacing synthetic antioxidants with natural ones that have the potential to elongate food quality and enhance the health of food consumers. The use of antioxidants as a preservative and in enhancing the flavor, aroma, and color of food products is reported in the literature [8]. The addition of antioxidants to food items as preservatives can be during many different stages of food production [51], but since the antioxidants have no potential to reverse already oxidized food products; their application during the early stage of the manufacturing process may give better results. Some practical applications of antioxidants in the food system include their addition to fats and oils used in food production [14], in the preservations of vegetables and vegetable products; fruits and fruit products [52]; cereals and bakery products; milk and milk products like cheese; meat, fish and their products; spices; and other dry foods like sugar, honey, beverages, and chewing gum [53].

Also, apart from being used as preservatives, antioxidants could be utilized to enhance the benefits of food to man. Aside from the provision of nutrients, the inclusions of antioxidants or the use of antioxidant-containing ingredients in food processing improve the productivity of food processing and its products. Their use could bring about an added value to food products by giving them the potential to provide nutrients and bioactive compounds and hence, promote the tendency of food consumed to checkmate lifestyle diseases and intake of drugs occasioned by occurrences of the diseases.

The use of antioxidants in food systems is bringing up novel food products designed to take care of both the nutritional and health aspects of human life. Before now, food products were produced to provide mainly nutritional needs of the consumers with little or no attention given to using food to address the increasing occurrence of degenerative diseases and to incidentally discourage the intake of drugs because of their associated side effects.

Also, the reports of the literature have indicated that the inclusion of plants parts with a high concentration of antioxidants is more effective than the use of extracted antioxidants as food supplements, either in terms of prolongation of shelf-life or particularly of promoting health orientations of food products in the food system [54]. Anbudhasan et al. [14] corroborated the foregoing by implying that processing impacts negatively on the potency of the antioxidants. And incidentally, the literature is replete with information that most antioxidants are concentrated in the areas of plants, like the seed/seed-coat, peels, etc. as in mango fruits; usually generated as wastes/byproducts discarded as pollutants to the environment by processing operations. Kebede and Admassu [8] confirmed this assertion by stating that wastes and by-products of fruits and vegetables in the food processing industry are abundant sources of antioxidant polyphenols or phenolic compounds. This means reintegration of these components as constituents of the food system could boost the strength of food in providing required antioxidants. The exploration of antioxidants may, therefore, likely reduce the quantity of wastes/by-products generated to promote eco-friendly food processing. Some studies have already been done on by-products, which could be potential sources of antioxidants [32]. Agricultural and industrial residues are attractive sources of natural antioxidants [55]. The use of waste as a source of polyphenols and antioxidants may have considerable economic benefits to food processing industries. Therefore, a cheap, efficient, and environmentally sound utilization of these huge agro-industrial wastes is needed [56].

The use of antioxidants in the food system depends on the conditions of food processing operations applied. According to Reddy et al. [57], processing (including preparation) of food is designed to make food healthier, safer, tastier, and more shelf-stable. This is achieved by inactivating disease-causing microorganisms (pathogens) and enzymes to reduce moisture content and concentrate nutrients and bioactive compounds in processed foods, or to soften the outer tissue to separate fruit/vegetable skin [58]. This incidentally causes several changes including appearance, composition, nutrition, and sensory properties which occur during processing in terms of color, texture, and flavor. Generally, food-processing procedures are recognized as one of the major factors responsible for the destruction or changes of natural phytochemicals, which may affect the antioxidant capacity in foods [59]. Processing conditions either boost nutrients and antioxidants or reduce them depending on many factors. The conditions that are generally considered in food processing include temperature, time, the level of antioxidants in the ingredients/raw material; but for antioxidants, it is reported that genetics, environment, growing conditions (moisture, fertilization, pests, and disease burden, etc.) of the fruits and vegetables from which they are extracted, as well as processing methods and storage conditions affect the level of antioxidant activity of phytochemicals [60, 61, 62]. The understanding that over-processing or some severe processing conditions and environment could eliminate most of the antioxidants are inducing processors to explore processing techniques capable of producing food products containing assured levels of nutrients and antioxidants. One of the techniques, as shown by the recent development in food processing, is the enclosure or entrapment of antioxidants within a material or substance reported in the literature as encapsulation technology. Trifkovic et al. [63] reported on different encapsulation technologies applicable in food processing for antioxidants to include spray drying, spray chilling, spray cooling spray-drying, spray-chilling, spray-cooling, melt injection, fluidized bed coating etc. Encapsulation according to Pattnaik et al. [64], protects sensitive antioxidants from being destabilized by severe processing conditions or environment, improves their bioavailability, masks their identifiable astringent flavors, enhances their delivery in active forms to the targeted site or appropriate release in the gastrointestinal tracts.

Furthermore, apart from the use of encapsulation technology for the retention of nutrients and antioxidants in food products, the interplay of processing conditions is another way to optimize the availability of nutrients and antioxidants. Nayak et al. [58] reported that the application of kinetic models in the thermal processing of foods is important to assessing and predicting the influence of processing operations on critical quality parameters to minimize the undesirable changes and to optimize the quality of specific foods. Thus foods could be processed to provide the required nutrients and antioxidants.

5. Implications of Replacing Synthetic Antioxidants with the Natural Ones in Food Systems

Food being one of the basic needs of man must always be available as and when needed at an acceptable condition or quality. This means preserving natural or processed food products, ensuring retention of the characteristics of the foods that constitute acceptable quality to consumers. The application of antioxidants in the food system was widely reported in the literature to checkmate undesirable oxidative reactions, identified as one of the major causatives of food deterioration; in order to maintain the quality of foods. Antioxidants have become an indispensable group of food additives mainly because of their unique properties of extending the shelf-life of food products without leaving any adverse effect on their sensory or nutritional qualities [2]. Atta et al. [7] also, reported the use of antioxidants to prevent the oxidation process in foods which leads to rancidity and browning. The major segment of antioxidants, natural and synthetic antioxidants, are involved as ingredients in food systems, mainly as preservatives and then recently as the promoter of health orientation of foods. Anbudhassan et al. [14] mentioned the involvement of both aspects of antioxidants, especially the recent drastic increase in the application of natural antioxidants at the expense of synthetic ones in the food system, because of concern for the safety of food consumers. Before now, synthetic antioxidants were highly involved in the food system because they were adjudged to be more reactive and effective as food preservers than natural ones. While the use of synthetic antioxidants (such as butylated hydroxytoluene and butylated hydroxyanisole) to maintain the quality of ready-to-eat food products has become commonplace, consumer concern regarding their safety has motivated the food industry to increasingly apply more natural antioxidants [50].

Thus, the friendliness of natural antioxidants, in comparison to synthetic ones, to the body system could be the reason for increasing interest in replacing synthetic with natural antioxidants, as food preservers, in food systems. Though Anbudhassan et al. [14] reported that synthetic antioxidants are widely used as food additives to prevent rancidification, owing to their high performance and wide availability, the public opinion that natural compounds are safer and more health-beneficial per se, has motivated the meat industry, for instance, to exploit plant-derived additives in meat systems with the objective of replacing synthetic antioxidants [65]. Whereas the literature is replete with increasing replacement of natural antioxidants with synthetic ones, it is necessary to elucidate implications of this in the food system. The increasing use of natural antioxidants will promote health orientation of foods, bring up new food products, enhance food quality, promote processing of composite food product, improve safety with assured attainment of food security, improve the circular economy of nations that invest in it, ameliorate occurrences of chronic diseases and their associated reduction in drug intake and many other benefits that are discussed hereunder.

5.1 Promotion of health orientation of food

Food is an indispensable resource to a man taken to provide nutrients required for the growth of the body cells with little or no consideration for its inclination to health aspects of life. The recent development in the food systems which targets the use of food to prevent chronic diseases as afore-mentioned seem to be widening the scope of benefits uses or productivity of food to man. In recent years, considerable research has been carried out, evaluating natural substances as antioxidative additives in food products, leading to novel combinations of antioxidants and the development of novel food products. The natural antioxidants have, in addition, shown a supportive effect to the human body with documented health benefits [8]. The targets of food processors, in the recent development, are to provide food products made up of required nutrients and antioxidants to ensure that foods have added value of promoting health of the consumers [66]. Antioxidants have important preventive roles not only on undesirable changes in the flavor and nutritional quality of food but also on tissue damage in various human diseases [8]. They are potentially effective in the prevention of degenerative illnesses, such as different types of cancers, cardiovascular and neurological diseases, cataracts, and oxidative stress dysfunctions [67, 68]. Chronic diseases such as arteriosclerosis and cancer, which are the leading causes of death in the Western world, are likely to be mediated by free radical and lipid peroxidation mechanisms [69], but could be remedied with increasing consumption of dietary antioxidants processed into food products. Antioxidants have been investigated and reported to play a specific role in the prevention of these diseases/disorders [68]. In the last decades, several epidemiological studies have shown that dietary intake of foods rich in natural antioxidants was correlated with a reduced risk of coronary heart disease [70, 71]. Dietary and natural antioxidants present in foods and other biological materials have attracted considerable interest because of their presumed safety and potential nutritional and therapeutic or health effects [72, 73]. While processing food to provide required nutrients, food processors should also consider other health-related aspects of their additives and products. The quality parameters for acceptance of food should widen to include adequate availability of antioxidants in addition to those characteristics for which food quality is measured.

5.2 Production of novel food products

The attempt to include the required availability of antioxidants in food as a measure of food quality characteristics is undoubtedly throwing up a novel or new products in food systems globally. Kedebe and Admassu [8] reported changes in human lifestyle and his view of food which are occasioning shift from one nature of food to another, e.g. from convenient foods to ready to eat food products category. The deadliness of chronic diseases and the understanding that consumption of the right foods could prevent or end their occurrences may broaden the demands of consumers of inclusion of antioxidants at the required levels in food products. In a bid to meet this dynamic demands of consumers, food processors are increasingly developing new food products processed to provide nutritive and healthy values to consumers.

5.3 Enhancing food quality

The contributions of antioxidants to the enhancement of food quality are well reported in the literature. Anbudhassan et al. [14] mentioned the involvement of antioxidants, both natural and synthetic in accentuating the shelf-life and appearance of many food products to buttress the disposition of antioxidants toward promoting food quality. While the use of synthetic antioxidants (such as butylated hydroxytoluene and butylated hydroxyanisole) to maintain the quality of ready-to-eat food products has become commonplace, consumer concern regarding their safety has motivated the food industry to seek natural antioxidants [50]. The antioxidants obtained from plants are more functional toward improving the shelf life of food products and providing healthier promotion when compared to materials whose antioxidants have been removed during processing [14]. Orientating foods toward promoting the health of consumers, in addition to their nutritional roles emphasizes the widening contributions of natural antioxidants to the maintenance of food quality. Kebede and Admassu [8] alluded to the effectiveness of natural antioxidants in preventing undesirable changes in the flavor and nutritional quality of food and tissue damage that occasion incidence of various human diseases; and asserted that nutritional importance, promotion of health, and prevention against damages caused by free radicals can lead to the potential applications of antioxidants in food industries in more intensified approaches. The applications of natural antioxidants in the food system will undoubtedly improve keeping quality of foods in the food systems.

This is in agreement with reports of Arshiya et al. [50] and Singh et al. [68] on natural antioxidants such as vitamins (ascorbic acid [AA] and α-tocopherol (E306)), many herbs and spices (rosemary, thyme, oregano, sage, basil, pepper, clove, cinnamon, and nutmeg), and plant extracts (tea and grape seed) applied on meat products as preservatives. The supremacy of natural antioxidants over synthetic ones in the functionality of antioxidants as enhancers or enablers of increasing shelf-life of foods is indeed incontrovertible as attested to with the report of Kebede and Admassu [8], which stated that the antioxidants obtained from plants are more functional toward improving the shelf life of food products and providing health promotion.

5.4 Promotion of circular economy

The increasing applications of natural antioxidants will incidentally boost or signify a circular economy since most the antioxidants are derived from by-products/wastes generated during food processing or utilization. The literature is replete with the involvement of food by-products or wastes in the extractions of natural antioxidants or their recycling for their use as ingredients in the processing of some foods. Bartosz et al. [74] associated the use of food by-products/wastes as raw materials in the production and or commercialization of natural antioxidants as well as in the advancement of the circular economy. The circular economy is a regenerative system that, unlike the linear economy, involves recycling or reuse of wastes generated in the food system to boost values derivable from the food processing system. The circular economy is all about minimizing waste generation in the food system by the re-use of food, conversion of by-products, and wastes into usable products, recycling nutrients, and adopting changes in diet toward more diverse and more efficient food patterns [75]. The identification of food wastes as reservoirs of antioxidants, and increasing inclination to eco-friendly food processing culture will purvey strategies and projects required to encourage upstream waste recovery, leading to the production of downstream value-added ingredients (e.g. natural antioxidants), based on a sustainable economy, i.e. circular economy. In the concept of a circular economy, recovery and valorization of wastes allow materials to be reused and be recycled into the supply chain, allowing economic growth from environmental losses [76]. Thus increase in applications of natural antioxidants will incidentally translate to an increase in the utilization of food wastes or adoption of eco-friendliness in food processing, the purveyor of the sustainable or circular economy.

6. Conclusion

Antioxidants are substances that minimize or disable oxidative activities in food and body systems to preserve them from being damaged. Two major types of antioxidants, based on their mode of synthesis applicable in the food system for food preservation are natural and synthetic antioxidants. Though synthetic antioxidants, from the inception of food processing, are predominantly used as food preservatives to maintain the keeping quality and appearance of many foodstuffs; some reports about their carcinogenicity and mutagenicity and hence, the concern of consumers' health have motivated the food industry to seek for their replacement with natural antioxidants. The replacement is necessary because of the increasing demands of consumers for health-promoting foods globally. The use of natural antioxidants, either in the form of extracts and or parts of natural resources that contain concentrations of antioxidants; in food processing and preservation, may encourage consumption of healthy foods. Also, the discovery that natural antioxidants are mostly concentrated in the parts of raw resources usually removed as wastes during food processing and, the efforts to reintegrate the wastes into the food system, seemingly included promotion of eco-friendly food processing and a guaranteed sustainable/circular economy; as one of the implications of replacing synthetic antioxidants with natural ones in the food system. The use of plant parts, as food ingredients instead of extracted antioxidants as food supplements, heightens the effectiveness of natural antioxidants in the food system either in terms of prolonging shelf-life or promoting health orientations of food products. The replacement of synthetic antioxidants with natural ones will, thus, boost sensory, safety, and other quality parameters as well as health orientations of food products and incidentally, the values of food to man.

References

1. Abdelkarim G, Soumaya B, Naima E, Mohammed B, Abdellah H. What is a bioactive compound? A combined definition for a preliminary consensus. International Journal of Nutrition and Food Sciences. 2014;3(3):174-179. DOI: 10.11648/j.ijnfs.20140303.16
2. Sardarodiyan M, Sani MA. Natural Antioxidants: Sources, Extraction and Application in Food Systems. 2016. Available from: https://www.researchgate.net/publication/303827521 [Assessed: May 27, 2021]
3. Morton LW, Caccetta RA, Puddey IB, Croft KD. Chemistry and biological effects of dietary phenolic compounds: Relevance to cardiovascular disease. Clinical and Experimental Pharmacology and Ehysiology. 2000;27:152-159
4. Halliwell B. How to characterize a biological antioxidant. Free Radical Research Communications. 1990;9:1-32. DOI: 10.3109/10715769009148569
5. Arun R, Abdul Azeez RF. A Review on Natural Antioxidants, Traditional and Complementary Medicine. Rijeka: IntechOpen; 2019. https://www.intechopen.com/books/traditional-and-complementary-medicine/a-review-onnatueal-antioxidants [Accessed: 25 March, 2021]
6. Tuberoso CIG, Boban M, Bifulco E, Budimir D, Pirisi FM. Antioxidant capacity and vasodilatory properties of Mediterranean food: The case of Cannonau wine, myrtle berries liqueur and strawberry-tree honey. Food Chemistry. 2013;140(4):686-691
7. Atta EM, Mohamed NH, Abdelgawad AAM. Antioxidants: An overview on the natural and synthetic types. European Chemical Bulletin. 2017;6(8):365-375. DOI: 10.17628/ecb.2017.6.365-375
8. Kebede M, Admassu S. Application of antioxidants in food processing industry: Options to improve the extraction yields and market value of natural products. Advances in Food Technology and Nutritional Science Open Journal. 2019;5(2):38-49. DOI: 10.17140/AFTNSOJ-5-155
9. Becker EM, Nissen LR, Skibsted LH. Antioxidant evaluation protocols: Food quality or health effects. European Food Research Technology. 2004;219:561-571. DOI: 10.1007/s00217-004-1012-4
10. Halliwell B. Antioxidants: The basics, what they are and how to evaluate them. Advanced in Pharmacology. 1997;38:3-20. DOI: 10.1016/S1054-3589(08)60976-X 22

11. Ihekoronye AA, Ngoddy PC. Integrated Food Science and Technology for the Tropics. London: Macmillan Publishers; 1985

12. MacDonald-Wicks LK, Wood LG, Garg ML. Methodology for the determination of biological antioxidant capacity in vitro. Journal of Science and Food Agriculture. 2006;86:2046-2056. DOI: 10.1002/jsfa.2603

13. Asimi OA, Sahu NP, Pal AK. Antioxidant capacity of crude water and ethylacetate extracts of some Indian species and their antimicrobial activity against Vibrio vulnificus and Micrococcus luteus. Journal of Medicinal Plants Research. 2013;7(26):1907-1915

14. Anbudhasan P, Surendraraj A, Karkuzhali S, Sathishkumaran P. Natural antioxidants and its benefits. International Journal of Food and Nutritional Sciences. 2014;3(6):225-232

15. Bontempo P, Carafa V, Grassi R, Basile A, Tenore GC, Formisano C, et al. Antioxidant, antimicrobial and anti-proliferative activities of Solanum tuberosum L. var. Vitelotte. Food and Chemical Toxicology. 2013;55:304-312

16. Halliwell B, Gutteridge JMC. Free Radicals in Biology and Medicine. Oxford: Clarendon Press; 1999

17. Cheeseman KH, Slater TF. An introduction to free radicals' chemistry. Brithish Medical Bulletin. 1993;49:481-493

18. Shivkumar. Free radicals and antioxidants: Human and food system. Advances in Applied Science Research. 2011;2(1):129-135

19. Brewer MS. Natural antioxidants: Sources, compounds, mechanisms of action and potential applications. Comprehensive Reviews in Food Science and Food Safety. 2011;10(4):221-247. DOI: 10.1111/j.1541-4337.2011.00156

20. Nawar WF. Lipids. In: Fennema O, editor. Food Chemistry. 3rd ed. New York: Marcel Dekker, Inc.; 1996. pp. 225-320

21. Pisoschi AM, Pop A. The role of antionxidants in the chemistry of oxidative stress: A review. European Journal of Medicinal Chemistry. 2015;97:55-74. DOI: 10.1016/j.ejmech.04.040

22. Pérez JAM, Aguilar TAF. Chemistry of natural antioxidants and studies performed with different plants collected in Mexico. In: Morales-González JA, editor. Oxidative Stress and Chronic Degenerative Diseases: A Role for Antioxidants. London, UK: Intech; 2013. pp. 59-81. DOI: 10.5772/52247

23. Rosenblat M, Aviram M. Antioxidative properties of pomegranate: In vitro studies. Seeram NP, Schulman RN, Heber D, editors. Pomegranates: Ancient roots to modern medicine. New York, USA: CRC Press; 2006:31-43

24. Polumbryk M, Ivanov S, Polumbryk O. Antioxidants in food systems. Mechanism of action ukrainian Journal of Food Science. 2013;1(1):15-40

25. Anwar M, Hussain G, Mustafa I. Antioxidants from natural sources. 2018. [Accessed: December 20, 2021]. DOI: 10.5772/intechopen.75961

26. Akbarirad H, Ardabili GA, Kazemeini SM, Khaneghah MA. An overview on some of important sources of natural antioxidants. International Food Research Journal. 2016;23(3):928-933

27. Manessis G, Aphrodite IK, Lazou T, Moschova M, Bossis I, Gelasakis AI. Review: Plant-derived natural antioxidants in meat and meat products. Antioxidants. 2020;9:1215. www.mdpi.com/journal/antioxidants [Accessed on 30 March, 2021]

28. Grozea MBAI. Antioxidant (Antiradical) compounds. Journal of Bioequivalence and Bioavailability. 2012;4(6). DOI: 10.4172/jbb.10000e18

29. Da Costa L, Badawi A, El-Sohemy A. Nutrigenetics and modulation of oxidative stress. Annals of Nutritional Metabolism. 2012;60(3):37-36. DOI: 1159/000337311

30. Yadav A, Kumar R, Yadav A, et al. Antioxidants and its functions in human body: A review. Research in Environment and Life Sciences. 2016;9(11):1328-1331

31. Sies H, Stahl W, Sundquist AR. Antioxidant functions of vitamins, vitamins E and C, beta-carotene, and other carotenoids. Annals of the New York Academy of Sciences. 1992;669:7-20

32. Mohdali AAA. Evaluation of some food processing by-products as sources for natural antioxidants [master's thesis]. Chicago, IL: Technical University of Berlin; 2010

33. Balasundram N, Sundram K, Samman S. Phenolic compounds in plants and agri-industrial by-products: Antioxidant activity, occurrence, and potential uses. Food Chemistry. 2006;99:191-203. DOI: 10.1016/j.foodchem.2005.07.042

34. Rahman K. Studies on free radicals, antioxidants, and co-factors. Clinical Interventions in Aging. 2007;2(2):219-236

35. Young VR, Erdman JWJ, King JC. Dietary Reference Intakes: Proposed Definition and Plan for Review of Dietary Antioxidant and Related Compounds. Washington, DC: National Academic Press; 1998

36. Jacob RA. The integrated antioxidant system. Nutrition Research. 1995;15:755-766

37. Percival M. Antioxidants: Clinical Nutrition Insights. Advanced Nutrition Publications, Inc; 1998

38. Simec MG, Porter KM. Autoxidation in Foods and Biological Systems. New York: Plenum Press; 1980. DOI: 10.1007/978-1-4757-9351-2

39. Sherwin ER. Antioxidants for vegetable oils. Journal of the American Oil Chemists' Society. 1976;53:430-436. DOI: 10.1007/BF02605739

40. Rashmi V and Disha S. A review of the physiological implications of antioxidants in food, interactive qualifying Project Report Submitted to the Faculty of the Worcester Polytechnic Institute in partial fulfillment of the requirements for the Degree of Bachelor of Science. 2011. [Accessed on 27 February, 2021]

41. Pokorný J. Are natural antioxidants better-and safer-Than synthetic antioxidants? European Journal of Lipid Science and Technology. 2007;109:629-642

42. Hudson BJ. Food Antioxidants. Vol. Volume 41. Dordrecht, The Netherlands: Springer; 1990

43. Mohamed A, Jamilah B, Abbas K, Rahman RA. A review on lipid oxidation of meat in active and modified atmosphere packaging and usage of some stabilizers. Journal of Food, Agriculture and Environment. 2008;6:76-81

44. Shahidi F, Janitha PK, Wanasundara PD. Phenolic antioxidants. Critical Reviews in Food Science and Nutrition. 1992;32:67-103

45. Fernades RPP, Trindade MA and Melo MP. Natural antioxidants and food applications: Healthy perspectives research gate. 2018. https://www.researchgate.net/publication/324127242 [Accessed: 12 February 2022]

46. Jung S, Choe JH, Kim B, Yun H, Kruk ZA, Jo C. Effect of dietary mixture of gallic acid and linoleic acid on antioxidative potential and quality of breast meat from broilers. Meat Science. 2010;86:520-526

47. Tamjidi F, Shahedi M, Varshosaz J, Nasirpour A. Nanostructured lipid carriers (NLC): A potential delivery system for bioactive food molecules. Innovative Food Science and Emerging Technologies. 2013;19:29-43

48. Velasco V, Williams P. Improving meat quality through natural antioxidants. Chilean Journal of Agricultural Research. 2011;71:313-322

49. Voutilainen S, Nurmi T, Mursu J, Rissanen TH. American Journal of Clinical Nutrition. 2006;83:1265-1271

50. Arshiya S. The antioxidant effect of certain fruits: A eeview. Journal of Pharmaceutical Sciences and Research. 2013;5:265-268

51. Andre C, Castanheira I, Cruz JM, Paseiro P, Sanches-Silva A. Analytical strategies to evaluate antioxidants in food: A review. Trends in Food Science and Technology. 2010;21:229-246

52. Mordi RC. Mechanism of B-carotene degradation. Biochemistry Journal. 1993;292:310-312

53. Williamson G, Day AJ, Plumb GW, Couteau D. Human metabolic pathways of dietary flavonoids and cinnamates. Biochemical Society Transactions. 2000;28:16-22

54. Austistyniak A, Bartosz G, Cipak A, Duburs G, Hoakova L, Lukzaj W, et al. Natural and synthetic antioxidants: An updated review. Free Radical Research. 2010;44(10):1216-1262

55. Moure A, Cruz JM, Franco Ruiz DJ, et al. Natural antioxidants from residual sources. Food Chemistry. 2001;72:145-171. DOI: 10.1016/S0308-8146(00)00223-5

56. Immanuel G, Singh S. Extraction of antioxidants from fruit peels and its utilization in paneer. Journal of Food Processing and Technology. 2014;5:349. DOI: 10.4172/2157-7110.1000349

57. Reddy MB, Love M. The impact of food processing on nutritional quality of vitamins and minerals. Advances in Experimental Medicine and Biology. 1999;459:99-106. DOI: 10.1007/978-1-4615-4853-9_7

58. Nayak B, Liu RH, Tang J. Effect of processing on phenolic antioxidants of fruits, vegetables and grains: A review. Critical Review of Food Sciences and Nutrition. 2015;5:887-918

59. Nicoli MC, Anese M, Parpinel MT, Franceschi S, Lerici CR. Influence of processing on antioxidant property of fruits and vegetables. Trends in Food Science and Technology. 1997;10:94-100

60. Blessinton T. Effect of Cooking, Storage and Ionizing Irradiation on Carotenoids, Antioxidant Activity and Phenolic in Potato. College Station: Texas A and M; 2005

61. Dewanto V, Wu XZ, Lie RH. Processed sweet corn has higher antioxidant activity. Journal of Agricultural and Food Chemistry. 2002;50:4959-4964

62. Reyes LF, Miller JC, Cisneros-Zevallos L. Environmental influence the content and yield of anthocyanins and total phenolics in purple and red potatoes during tuber development. American Journal of Potato Research. 2004;81:187-193

63. Trifković K, Tadić G, Bugarsk B. Short overview of encapsulation technologies for delivery of bioactives to food. Journal of Engineering and Processing Management. 2016;8(1):103-111

64. Pattnaik M, Pandey P, Martin GJO, Mishra HN, Ashokkumar M. Innovative technologies for extraction and microencapsulation of bioactives from plant-based food wastes and their applications in functional food development. Foods. 2021;10:279. DOI: 10.3390/foods10020279

65. Kamala Kumari PV, Akhila S, Srinivasa Rao Y, Rama Devi B. Alternative to artificial preservatives. System Reviews in Pharmacy. 2019;10:S13-S16

66. Scott G. Antioxidants in Science, Technology, Medicine and Nutrition. Chichester: Albion Publishing; 1997. DOI: 10.1533/9780857099938

67. Mbata TI. Antioxidant nutrients: Beneficial or harmful. Internet Journal of Food Safety. 2000;7:29-33

68. Singh L, Suruchi S, Sharma SK. A review on medicinal plants having antioxidant potential. Indian Journal of Research in Pharmacy and Biotechnology. 2013;1(3):404-409

69. De Beer D, Joubert E, Gelderblom WCA, Manley M. Phenolic compounds: A review of their possible role as in vivo antioxidants of wine. South African Journal for Enology and Viticulture. 2002;23(2):48-61. DOI: 10.21548/23-2-2155

70. Tortora G and Derrickson B. Antioxidants beyond the Hype Medline Plus: Antioxidants "Introduction to the Human Body." 2010. https://www.hsph.harvard.edu/nutritionsource/antioxidants/

71. Yordi EG, Pérez EM, Matos MJ, Uriarte E. Antioxidant and pro-oxidant effects of polyphenolic compounds and structure activity relationship evidence. In: Bouayed J, editor. Nutrition, Well-Being and Health. London, UK: Intech; 2012. DOI: 10.5772/29471

72. Jongen W. Fruit and Vegetable Processing: Improving Quality. Sawston, Cambridge: Woodhead Publishing Ltd; 2000

73. Mandal S, Yadav S, Yadav S, Nema RK. Antioxidants: A review. Journal of Chemical and Pharmaceutical Research. 2009;1(1):102-104

74. Bartosz G. Nonenzymatic: Antioxidant capacity assays limitations of use in biomedicine. Free Radical Research. 2010;44:711-720

75. Jurgilevich A, Birge T, Kentala- Lehtonen J, Korhonen-Kurki K, Pietikäinen J, Saikku L, et al. Transition towards circular economy in the food system. Sustainability. 2016;8:69. www.mdpi.com/ journal/sustainability. [Accessed 20 March, 2021]

76. Ghisellini P, Cialani C, Ulgiati S. A review of circular economy: The expected transition to a balanced interplay of environmental and economic system. Journal of Cleaned Products. 2016;114:11-32. DOI: 10.1016/j.jclepro.2015.09.007

Chapter-2

Application of Green Extraction Techniques For Natural Additives Production

1. Introduction

The use of additives in the food industry has become a routine process due to the ability of this type of compound to improve the organoleptic properties (flavor, aroma, color) of foodstuff or to extend its shelf-life due to its bioactivities (mainly, antioxidant and antimicrobial) [1]. Food additives are the most useful tool to improve foodstuff quality. According to the European Food Safety Authority (EFSA), food additives are defined as "substances that are not normally consumed as food nor used as intrinsic ingredients of food, which has a technological purpose" [2]. In this context, the nature of some additives, known to be potentially hazardous if consumed in excess, has derived to an increasing consumer trend to avoid these types of products. One possible effect of these molecules is the potential allergic reactions or health risks associated with a frequent consumption. Compounds that have been subjected to this controversy are sulfites, nitrosamines or palm oil, whose presence in food have been sometimes forbidden (*e.g.*, sulfites in wines sold in the USA) [3, 4], leading to an ideological current against synthetic additives [5]. This phenomenon collides with a growing interest for developing a green and sustainable economy and reducing generated wastes. For this reason, and the food industry has focused its efforts research on new effective and feasible extraction methods of natural compounds from different raw materials (including by-products) that could be used as additives [6]. These compounds belong to different classes, being the most studied phenolic compounds (*i.e.* phenolic acids, flavonoids, tannins), pigments (*i.e.* chlorophylls, carotenoids, anthocyanins), vitamins, polysaccharides, proteins and unsaturated fatty acids [7, 8, 9, 10, 11]. They have different bioactivities, depending on the chemical structure of the molecule (Figure 1). For example, phenolic compounds, pigments or vitamins are recognized as potent antioxidants and antimicrobials, whereas pigments like β-carotene will additionally provide color to the product. In addition, there are other aromatic compounds such as terpenoid carvacrol, obtained from thyme and oregano, *d*-limonene from citrus tree or curcumin from turmeric with beneficial properties which can be also used as food additives [12].

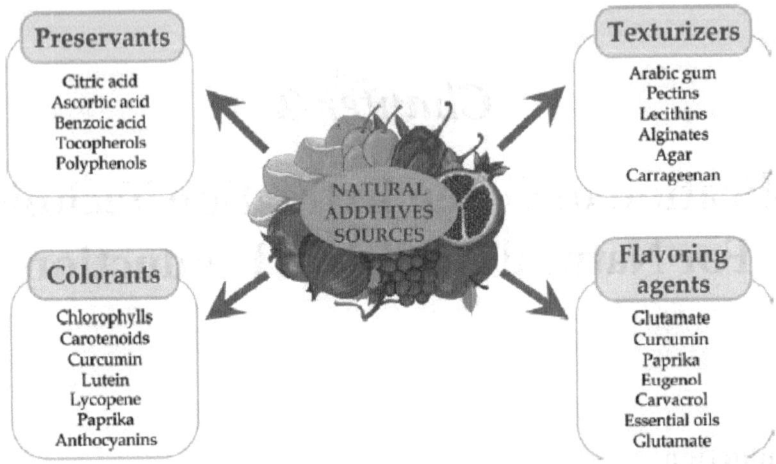

Figure 1. Properties and main applications of major additives of natural origin in foods.

Studies pointed out that natural additives may have higher bioactivity than artificial ones. For example, it has been observed that phenolic compounds (phenolic acids and flavanols) have comparable or even more potent antioxidant activity than currently used artificial additives like propyl gallate or butylhydroxytoluene in meat, fish and bakery products [13, 14, 15, 16]. Hence, there is scientific evidence to propose their use as alternative condiments. On this matter, many of these compounds and/or extracts, like essential oils, are already approved and generally recognized as safe (GRAS) flavoring food additives by the *Food* and Drug Administration (FDA) and EFSA [2, 17]. Some examples of additives directly extracted from natural sources approved by EFSA include plant and algal pigments, tocopherols, or antioxidant-rich extracts like those from rosemary. On the other hand, some natural additives, like tannic acid, are approved for their use in animal feed but are not yet approved for human consumption [18].

In parallel, the effectiveness and new potential applications of additives like food active packaging and their addition into novel food matrices are the focus of study by many research groups. To release new functional food to market, health claims and properties must be supported by scientific evidences to be legally labeled as functional food [19]. Thereupon, a great number of compounds or extracts may be obtained from natural sources, such as vegetables, fruits, algae or even by-products for their application as food and feed additives but also as cosmetic ingredients [20]. Table 1 collects a list of selected natural additives that may be directly extracted from natural sources. Although these compounds are of natural origin and may be extracted from natural sources, efficient yields, costs and steady production has led the chemical and biotechnology industry to manufacture them by biological or chemical synthesis. Representative examples are ascorbic acid and citric acid. Ascorbic acid is a potent antioxidant and an essential vitamin with many uses in agriculture and food industry. While it is present in a wide number of fruits and vegetables (*i.e.* citrus, peppers, kiwifruit), the current production approach comprises microbiological synthesis with either bacteria or fungi, chemical synthesis from *d*-sorbitol, or a combination of both [48]. Natural colors, pigments such as chlorophylls, carotenoids and anthocyanins, are more feasible to obtain from their natural sources [49]. It is of great importance to carry out optimization process according to raw material and compounds of interest. This fact leads the increase of the operation performance and the reduction of costs. Therefore, developing optimum and environmental-friendly extraction procedures and methods is of great importance to value natural sources in a cost-effective way.

Table 1. Summary of main natural additives from vegetable sources* and EU E-number reference, if they are at present approved for use in the EU.

Compound	Applications	Current & Potential Natural Sources	E-Number	Ref.
Benzoic acid	Antioxidant & antimicrobial preservatives	Cranberries, blueberries (*Vaccinium* sp.)	E-210	[21]
Ascorbic acid	Antioxidant	Peppers, kiwifruit, citrus, rosehip.	E-300	[22]
Citric acid	Antioxidant, acidifier	Citrus	E-330	[23]
Tartaric acid	Acidifier	Grapes	E-334	[24]
Chlorophylls	Colorant	Green leaves from alfalfa, nettles, spinach, green microalgae	E-140, E-141	[25]
Carotenoids (β-carotene)	Colorant, antioxidant	Pigmented vegetables (carrots, palm fruit), red microalgae	E160a	[26]
Curcumin	Colorant, flavoring, antioxidant	Turmeric (*Curcuma longa*)	E-100	[27]
Lutein	Colorant, antioxidant	Kale (*Brassica oleracea*), spinach, *Calendula officinalis, Tagetes erecta*	E-160b	[28]
Paprika extract (capsaicin, captaxanthin, capsorubin)	Colorant, flavoring	Red peppers (*Capsicum* sp.)	E-160c	[29]
Lycopene	Colorant	Tomato peels	E-160d	[30]
Anthocyanins	Colorant	Red grape skin, pomegranate, black currant	E-163	[31]
Glutamate	Flavoring	Wheat gluten, de-oiled soybeans	E-620	[32]
Limonene	Odorant, antimicrobial	Citrus peels	—	[33]
Eugenol	Flavoring, antimicrobial preservative	Clove	—	[34, 35]
Carvacrol	Flavoring, antimicrobial preservative	Oregano, thyme, rosemary	—	[35, 36]
Rosemary extract (carnosic acid, carnosol)	Preserver (antioxidant, antimicrobial), flavoring	Rosemary	E-392	[37]
Lecithins	Emulsifiers, lubricants	Soybean, sunflower kernels, rapeseed	E-322	[38]

Compound	Applications	Current & Potential Natural Sources	E-Number	Ref.
Alginates	Texturizer, binder, thickener	Brown seaweeds	E-401, E-402	[39]
Agar	Texturizer, binder, softener	Red seaweeds	E-406	[40]
Carrageenan	Texturizer, binder, softener	Red seaweeds	E-407	[41]
Pectins	Thickener, texturizer	Apple pomace, citrus peels	E-440	[42]
Arabic gum	Thickener, texturizer	*Acacia nilotica*	E-414	[43]
Tannic acid	Antioxidant, plasticizer, flavoring	Grape seeds, skins	—	[44, 45]
Phytosterols	Functional, health	Soybean	E-499	[46]
Tocopherols	Health, antioxidant	Vegetal oils, cereal germs, rapeseed, soybean	E-306, E-307, E-308, E-309	[47]

*Described natural sources are indicated and evidenced in the used references, but chemical or microbiological synthesis as well as enzymatic transformations from natural sources may be the most extended production method.

2. Conventional Extraction Technologies: Advantages and Drawbacks

Extracting compounds from complex vegetable matrixes requires the application of several techniques to extract and isolate the selected molecules. Thus, it is important to choose the best fitting extraction technique to the specific additive in need of isolation. Conventional extraction techniques, which have been applied for many years, include maceration, Soxhlet extraction, distillations, infusions, and cold-pressed extractions [50]. They are usually not eco-friendly due to the large amounts of solvents and energy required for their application [51], which also provides a safety concern for the workers and consumers involved, as well as the lack of sustainability, and green extraction protocols [52]. Besides, some of these techniques are also very time-consuming which aggravates the energy spending problem as the equipment must remain working for long periods of time, being one of the biggest expenses heating or cooling. Lastly, the yields obtained using these methodologies are usually not as rentable as the ones produced using more innovative extraction techniques, that are faster and more efficient than the previously available protocols. For this reason, conventional methodologies are started to be conjugated with new technologies resulting in Soxhlet or distillations assisted by ultrasound or microwave technologies to respond to the current needs of the industry.

Nevertheless, these techniques are still used nowadays because they allow the extraction of compounds in a cost-effectively manner with simpler equipment [52], like the distillation of essential oils [53], or they achieved a better stability of the extracted compound to be used as additive, as is the case cold pressing for oils extraction [54].

2.1 Maceration

Maceration is one of the most known and used conventional extraction techniques [55]. It is a solid–liquid extraction achieved by applying heat and agitation to a previously selected solvent, with a convenient polarity, that is in contact with the sample of interest [52, 56].

Maceration has few advantages. It can be performed using low-cost and simple equipment compared with other conventional and innovative techniques. Besides, a large range of molecules can be extracted by changing the protocols and adapting the variables like solvent [57], temperature, agitation, and time in order to optimize the extraction of the desired compounds [58]. Furthermore, this extraction technique is still used due to its easy scale-up to several applications in the industry.

In comparison, maceration also has some major drawbacks. It often requires long extraction times, large volumes of solvent (mostly organic solvents), high temperatures which translate in a big amount of energy spent, and it has to be coupled with several filtration or centrifugation steps in order to separate the extract from the biomass [52, 59, 60, 61, 62].

2.2 Soxhlet extraction

Soxhlet is a reference extraction method to evaluate the performance of other liquid–solid extraction methodologies [56]. This technique, developed in 1879, uses a particular type of condenser known as the Soxhlet apparatus [50]. The traditional Soxhlet extractor is composed by a thimble-holder where the sample is placed inside the thimble, and a distillation flask where fresh solvent is added. When the solvent reaches the boiling point, it vaporizes and enters the matrix, solubilizing compatible compounds. After that, the solvent hits the cooling tubes of the condenser and condense back into the initial flask with the extracted compounds. This operation repeats until the full extraction is completed [50, 52].

This type of extraction presents several advantages. Firstly, the constant renovation of the solvent in contact with the matrix, allows for a disequilibrium between the compounds in the sample and the lack of them in the solvent, favoring the extraction of these compounds. Secondly, the temperature of the system is maintained throughout the process. Soxhlet extraction also does not require filtration or centrifugation of the final extracts, being perfectly separated from the original biomass. And lastly, it allows for the treatment of several samples in parallel at a relatively low cost and easy operational processes, considering that the basic equipment is quite affordable and simple [52, 63].

However, Soxhlet extraction also presents some disadvantages, as a large amount of organic solvents required, long periods of extraction until the final number of cycles is completed [52], the high temperatures employed to boil the solvents that can degrade the compounds [50], and this technique cannot be accelerated by adding agitation [55, 63].

Nevertheless, the Soxhlet extraction has continued to evolve to try to compensate some of these disadvantages, by automating the process, aiming to shorten the extraction times, and even recently Soxhlet extraction has been coupled to innovative technologies like high-pressure Soxhlet extraction, supercritical fluid-Soxhlet extraction, and automated Soxhlet extraction or by applying auxiliary energies such as ultrasounds or microwaves, that results in higher efficiency than the conventional Soxhlet extraction [52, 63].

2.3 Distillation

Distillation is one of the oldest extraction techniques that has been used for over 5000 years and is still currently used. It main application is to separate liquid mixtures through the boiling point of each component of the mixture after which condensation steps take place [64]. Another ancient application is to create "distilled waters" from plant materials [65]. Distillation is still used in chemical refineries to separate and purify compounds, due to its simplicity and easy scale up [66]. Even though distillation processes are still quite common, they have many drawbacks. For example, the necessity for the consumption of large amounts of energy for long periods of time, as well as the high temperatures used which can degrade the additive of interest. In addition, the large amounts of solvent required [52] and the long extraction times [66].

Regarding the extraction of compounds from natural products, distillation is mostly used to isolate volatile molecules from mixtures of compounds of even several biological matrices, but it is only efficient in the case of thermostable molecules due to the high temperatures employed. Even so, steam distillation is still currently the most used technique in the food industry for extracting volatile compounds and essential oils to be used as food additives. Distillation can be divided into three types: water, water-steam, and steam distillations, and the extraction of essential oils is based on the latest, which the steam goes into a recipient containing the plant matrix, releasing the essential oils from the samples, then the essential oil is cooled and condensed, generating two different phases that can be separated [53].

2.4 Infusions

Infusions are very short macerations, where the plant is put in contact with boiling or sometimes cold water, for short periods of time. Therefore, infusions contain the readily soluble active chemical compounds that were present in the crude plant, in a diluted concentration. This methodology is used to obtain fresh infusions with the phytochemicals from aromatic or medicinal plants which can be further used as food additives [55]. Many infusions nowadays are prepared starting from a very concentrated infusion and diluting one volume of it to 9 volumes of water. To prepare the concentrated infusions the most common strategy is a percolation or modified maceration. In a modified maceration it is added 25% ethanol to the extraction solvent during or after the maceration process. The final solution is then diluted with water to resemble the scent and the potency of a normal fresh infusion.

One of the greatest susceptibilities of this method is that infusions are very prone to fungus and bacterial growth, due to the large amount of water they contain, so they have a very short shelf life and need to be used right away. For that reason, infusions are rarely used in industrial fields [67].

2.5 Cold pressing technique

This technique relies on pure pressure applied to the ground plant material to squeeze out what they contain. When the plant is well dried and grounded, this extraction technique is a viable, simple and eco-friendly option to consider for the extraction of some oils. During the process, cold pressers apply a large amount of pressure which translates into a mechanical rupture of the oil glands in the plant. The oil is extracted independently of its polarity and is rich in several lipophilic compounds. This process is still used in some industries because of its simplicity and easy scalability and for the fact that the obtained oils are more stable and resistant to oxidative stress than oils that have been refined or processed in any other way [54]. However, this technique can only be used to produce some specific vegetable and seed oil and it is not suitable to produce essential oils. Furthermore, the final product can many times contain other chemicals or contaminants [50].

3. Green Extraction Techniques

The application of green extraction techniques to obtain food natural additives is gaining great interest in recent years due to the growing demand of healthier and more sustainable products. Figure 2 collects the main advantages of the green extraction techniques applied to natural sources.

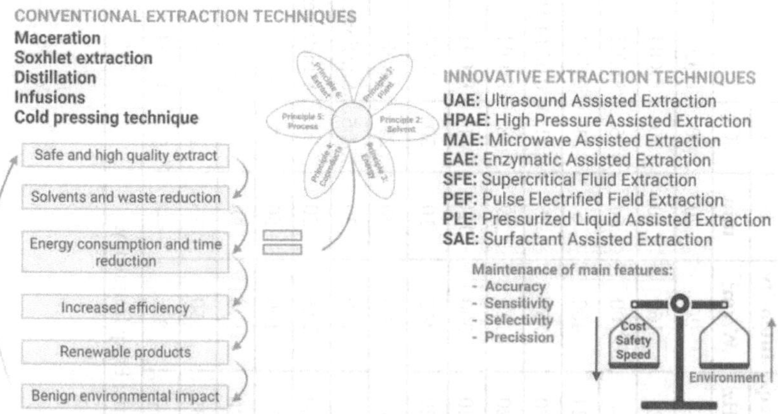

Figure 2. Conventional and green extraction techniques development.

3.1 Ultrasound assisted extraction (UAE)

Ultrasound assisted extraction (UAE) is used in different fields such as biomedicine or food technology, in which UAE is applied to obtain compounds of interest or as a pre-step in numerous technological procedures. There are several parameters to consider optimizing the method including ultrasound power intensities, frequency, wavelength and time. High frequency and low power ultrasounds are used in medical science. However, low frequency and high power ultrasounds are used in food industry [68].

Low frequency and high-power conditions produce cell disruption and the subsequent release of compounds present in the matrix [69, 70, 71]. This liberation of the compounds is based on the principle of cavitation. This physical–chemical process comprises generating bubbles that grow in such a way that they explode causing the rupture of the cell wall of plants, and the consequent release of the substances in their interior [72]. Another important characteristic of ultrasonic-assisted extraction is that the hot spots created by cavitation bubbles during the extraction process hardly generate heating and have a great capacity to cool down while the process is taking place, in this way UAE is cataloged as an extraction method suitable for extracting thermolabile compounds. However, when the extraction is longer than five minutes, at high powers it is necessary to use some refrigeration to keep the temperature constant [73].

Table 2. Different experimental conditions carried out with ultrasound assisted extraction (UAE).

Compound	Raw material	Conditions					Yield	Ref.
		Freq. kHz	Intensity W/cm^3	t min	T °C	Solvent		
Phenolics	*Nephelium lappaceum* L. fruit peel	—	20	20	50	H_2O	5.53 mg GAE/g	[75]
Phenolics	*Plinia cauliflora* (jabuticaba) peel	25	150	10	30	EtOH	92.8 mg GAE/g	[76]
Phenolics	Microalgae	40	700	60	75	EtOH	9.8 mg GAE/g	[77]
Phenolics	*Nannochloropsis* spp.	24	400	5	21	EtOH	50%	[78]
Phenolics	*Ascophyllum nodosum*	20	750	25	21	0.06 M HCl	143.12 mg GAE/g	[79]
Phenolics	*Malva sylvestris* leaves	20	110	49	48	EtOH	279.9 mg GAE/g	[80]
Phenolics	*Elaeocarpus serratus* L. leaves	40	300	120	21	EtOH	92.4 mg GAE/g	[81]
Capsaicinoids	Peppers	20	360	10	50	MeOH	448 μmol/kg	[82]
Vitamin C	*Citrus sinensis* (orange) peels	20	400	30	21	EtOH	53.78 mg AA/100 g	[83]
Carotenoids	*Punica granatum* (pomegranate) wastes	20	130	30	51.5	Sunflower oil and soy oil	93.8%	[84]
Carotenoids	*Daucus carota* (carrots)	20	22.5	20	40	Sunflower oil	334.75 mg/L	[85]
β-Carotene	*Daucus carota* (carrot) wastes	20	100	50	50	—	83.32%	[86]
Sulfated polysaccharide	*Nizamuddinia zanardinii*	20	196	58	70	EtOH	3.51%	[87]
Fucoidan	*Fucus evanescens*	35	150	15	23	H_2O	3.63%	[88]
Polysaccharides	*Silvetia compressa*	—	3.8	—	50	EtOH	23%	[89]

Abbreviations: EtOH: ethanol; MeOH: methanol; T: temperature; t: time; Freq: frequency.

Several are the parameters that have influence on the UAE extraction. They include the type of the reactor, the ultrasonic intensity and frequency, the extraction time and temperature, the solvent proportion and nature [90]. In this sense, the intensity or power is proportional to the ultrasonic amplitude, however, a greater ultrasonic amplitude is not directly related to a better efficiency of the process. On the contrary, it could be related to certain problems such as those that cause the erosion of the probe and reduce the formation of cavitation which can even promote the degradation of the extracted compounds [91]. Regarding the frequencies used, they must be selected together with the ultrasonic intensity to obtain the desired cavitation. Higher extraction yields are reported in the low frequency range (20–40 kHz). Regarding temperature, high temperatures help to interrupt the interaction of the solvent and the matrix and improve the diffusion rates of the solvent, while low temperatures improve cavitation. Extraction time is another variable to consider. Long extraction time improves extraction yields; however, it can cause changes in the extracted compound. In addition, the nature of the solvent has influence in the UAE, so that a viscous solvent reduces cavitation and a volatile solvent can be evaporated if the extraction is carried out at a higher temperature for a long period. Finally, the size of the matrix, its interaction with the solvent and the ratio of solvent to matrix are also parameters to consider when UAE is used [91].

3.2 High pressure assisted extraction (HPAE)

High pressure assisted extraction (HPAE) is a novel technique used for extraction of active ingredients from plant materials. Its mechanism of action is based on two principles: the isostatic principle and the Le Chatelier's principle. Isostatic principle establishes that the pressure is exerted uniformly on the matrix regardless of its shape or constitution. In turn, according to Le Chatelier's principle, by applying a force (pressure) that alters the equilibrium, the system acts trying to minimize said disturbance [92]. To carry out this type of extraction, it is necessary to apply pressures ranging from 100 to 800 MPa, even in some cases reaching values of 1000 MPa, which has been proved to be more effective. However, HPAE cannot be applied for the extraction of all compounds since this technique can cause some structural changes in foods, such as cellular deformation, cellular membrane damage or protein denaturation. The parameters that must be optimized to increase the extraction yield are type and amount of solvent, temperature, pressure, extraction time and number of cycles [93].

Among the advantages of HPAE are the improvement of the mass transfer rate, the enhancement of solvent permeability in cells as well as secondary metabolite diffusion. Other advantages include shorter extractions times, the process may be performed at room temperature (avoiding thermal degradation of heat labile components) and higher extraction yields. In addition, the use of solvents with different polarity allows to extract a great variety of compounds [94].

Some studies can already be observed that support the application of HPAE to obtain bioactive compounds derived from both plants and food (Table 3). However, research in the field of HPAE is in its initial stages, and more in-depth studies are still needed to determine the full potential of HPAE.

Table 3. Different experimental conditions carried out with high pressure assisted extraction (HPAE).

Compound	Raw material	Conditions				Yield	Ref.
		P	t	T	Solvent		
		MPa	min	°C			
Flavonoids	*Ficus carica* L. (fig)	600	18–29	—	40% EtOH	35%	[94]
Phenolics	*Olea europaea* (olive)	10.3	110	60	EtOH	386.42 mg GAE g/ extract	[95]
Phenolics	*Chenopodium formosanum*	600	5	25	70% EtOH	567–642 mg GAE/g	[96]
Melanoidins	*Allium sativum* (garlic)	300	5	25	DW	(595.14 ± 12.14 µg/mg melanoidins	[97]

Abbreviations: EtOH: ethanol; P: pressure; T: temperature; t: time; DW: distilled water; DPPH: 2,2-diphenyl-1-picryl-hydrazyl-hydrate free radical method; GAE: gallic acid equivalent.

3.3 Microwave assisted extraction (MAE)

Microwave assisted extraction (MAE) is a technique which combines conventional solvent extraction with microwave heating. Microwaves are composed of electric and magnetic fields with spectral frequency ranging from 300 to 300,000 MHz [98]. Many reports have been published on the extraction of secondary metabolite from plants using MAE. Some of them are collected in Table 4. The advantages of this technique include high efficiency, rapid temperature rise, short extraction time, better process monitoring, and low energy consumption and cost [104]. One of the disadvantages of MAE is the degradation of some compounds due to the heat produced by irradiation. The efficiency of the MAE depends on factors such as the extractant nature, the power of microwave irradiation, the temperature, and the extraction time as well as the characteristics of the matrices and the solvent-food relationship. The extraction efficiency usually behaves directly proportional to the microwave power due to the local heating that contributes to the rupture of the matrix. However, there is a limit in the microwave power that can cause a decrease in the extraction efficiency. In this sense, Alara and co-workers point out that a power between 400 and 500 W in the microwave, the extraction amount of phenolic compounds is greater than that achieved using a power upper 500 W [111]. This is because overexposure to radiation from the microwaves produce overheating and degradation of compounds [112]. On the other hand, the choice of solvent influences the extraction of compounds. It was reported that a mixture of organic solvents and water has a desirable impact on the extraction efficiency. On the contrary, it was observed that the presence of water in organic solvents leads to greater penetration of the extractant in the matrix molecules promoting microwave heating and causing a positive impact on the general efficiency and extraction time compared to MAE that uses only organic solvents [113].

Table 4. Different experimental conditions carried out with microwave-assisted extraction (MAE).

Compound	Raw material	Conditions of extraction				Yield	Ref.
		T	t	Power	Solvent ratios		
		°C	min	W			
Polysaccharides	*Potentilla anserina*	63.3	76.8	369	Water:raw material 14.5:1	13.33%	[98]
Polysaccharides	*Actinidia chinensis* (kiwi)	80	120	480	EtOH (80%) 1:10	2.92%	[99]
Polysaccharides	*Ascophyllum nodosum*	120	15	—	EtOH (80%) 1:10	16.08%	[100]
Polysaccharides	*Palmaria palmata*	70	10	500	Water 1:70	17.01%	[101]
Polysaccharides	*Ribes nigrum*	30	41	414	Water 1:30	10.59%	[102]
Essential oil	*O. vulgare* (oregano)	60	50	600	Water 20:1	7.1%	[103]
Crocetin	*Crocus sativus* L. (saffron)	96	30	2.45 GHz	59.59% EtOH	228 mg/g	[104]
Polyphenols	*Malus domestica* (apple) skin	150	90	60	Water 1:10	50.4 mg GAE/g	[105]
Phenolics	*Hibiscus sabdariffa (hibiscus)*	164	12.5	850	EtOH (45%)	42.4%	[106]
Phenolics	*Aristotelia chilensis*	100	2	800	MeOH (60%)	54.3 mg/g	[107]
Usnic acid	*Cladonia foliacea*	80	5	—	ACE 1:10	4.2 mg/g	[108]
Flavonolignan	*Silybum marianum*	30	12	600	EtOH (80%) 1:25	79%	[109]
Vicine	*Vicia faba* (beans)	30	0.5	1140	MeOH (50%)	—	[110]

Abbreviations: EtOH: ethanol; MeOH: methanol; ACE: acetone; T: temperature; t: time; GAE: gallic acid equivalent.

Also, the toxicity of the solvent is another important factor that must be evaluated with respect to the selection of a suitable extractant for MAE [114] since some theories highlight the efficiency and selectivity of MAE depends on the dielectric constant of the solvent mixture [115]. The polarity of organic solvents increases with the addition of water, the temperature within the sample increases due to better absorption of microwave energy, and extraction increases. The extraction time is another parameter that influences the extraction by MAE. Long times of extraction decreases the yield due to the alteration of the structural integrity of the chemically active principles. In this way, the extraction time in most of the MAE process is ranged from a few minutes to 30 min. Nevertheless, when MAE is used without solvent, longer extraction times can be employed. In this case, extraction cycles can be used to reduce the degradation of the compounds [116]. Additionally, the agitator effect influences the extraction process, reducing the negative effects of the S/F ratio on extraction recovery [116].

3.4 Enzyme assisted extraction (EAE)

Enzymes are protein molecules whose function is to catalyze chemical reactions. Due to this ability to accelerate reactions, enzymes have always been important to food technology and are widely used to transform raw materials into improved food products such as starch processing, meat processing, dairy industry, wine industry and manufacturing of predigested foods [117]. However, with the advancement of technology, new applications have been developed as well as new sources for obtaining enzymes, being microbial enzymes the preferred source due to the advantages they present, among which it is worth highlighting an easy, profitable and constant production [118]. Among the novel applications of enzymes is the extraction of compounds of interest from different raw materials to be used as additives in the food industry. However, they are typically used for a feedstock pretreatment that makes conventional solvent extraction or distillation more efficient. Enzymes promote access to the substances of interest, so its applications include, among others, the extraction of flavor and color from plant materials as a pre-treatment of the raw material before subjecting the plant material to hydro-distillation/ solvent extraction [119].

This entire process must be optimized to obtain the extract with the highest yield, that is, in a more purified and profitable way. To carry out this optimization, it is necessary to know the mechanism of action of the enzymes and, therefore, their optimal conditions of activity, being pH and temperature of vital importance. Another important factor to take into account is the majority composition of the raw material, which will determine the type of enzyme to be used (lipases, celluloses, proteases) [120]. This system has been investigated for extracting lipids, proteins, polysaccharides, phenols and oils among others (Table 5).

Table 5. Different experimental conditions carried out to obtain additives by enzyme-assisted extraction.

Compound	Raw material	Enzyme	Conditions			Yield	Ref.
			T	pH	t		
			°C	—	h		
Proteins	*Glycine max* (soy)	Protease	50	9.0	1	97%	[121]
Proteins	*Brassica napus* (rapeseed)	Pectinase, cellulase, β-glucanase	48	10.0	4	83.0%	[122]
Proteins	*Arachis hypogaea* (peanut)	Alcalase	60	9.5	1.5	88.2%	[123]
Lipids	*Rubus idaeus* (raspberry)	Proteases	60	9.0	2	38%	[124]
Polysaccharides	*Allium sativum* (garlic)	Cellulase	45	5.0	1.3	35.3%	[125]
Polysaccharides	*Medicago sativa* (alfalfa)	Cellulase, pectin, pectase	52.7	3.9	2.73	5.1%	[126]
Phenols	*Vitis vinifera* (grape) marc seeds	Pectinase	48	3.5	2.7	18–20 mg/g	[127]
Phenols	*Brassica oleracea* (cauliflower)	Viscozyme	35	4.0	0.5	0.6 mg/g	[128]
Phenols	*Ulmus pumila* (elm tree)	Cellulase, pectinase, β-glucosidase	52.6	4.6	1	16.04 mg/g	[129]
Oils	*Citrullus lanatus* (watermelon) seeds	Protex	47.1	7.9	7.8	97.9%	[130]
Oils	Yellow mustard flour	Protex	60	4.5	3	91.0%	[131]
Vitamin C	*Malpighia emarginata* (acerola)	Celluclast	50	4.5	2	23.4 g/L	[132]
Carotenoids	*Capsicum annum* (pepper)	Viscozyme L	60	4.5	1	87.0%	[133]
Carotenoids	*Capsicum annum* (pepper)	Viscozyme L	50	4.5	5	78.0%	[134]
Lycopene	*Solanum lycopersicum* (tomato)	Pectinase	60	5.0	0.4	1.1 mg/g	[135]

Abbreviations: T: temperature; t: time.

This technique has the advantages over other traditional methods of having a high selectivity and efficiency, being an environmentally friendly process with a minimum consumption of energy and chemical products, having good performance and the possibility of recycling the process. However, it also has some drawbacks such as the cost of the enzymes, the need for storage tanks that may require long-term incubation, the lack of knowledge about optimal or compatible enzyme formulations for cell disruption and the inability to fully hydrolyze the bonds in the plant cell wall [136].

3.5 Supercritical fluid extraction (SFE)

Extraction by supercritical fluids is characterized by bringing the fluid to conditions of temperature and pressure above the critical point, at which time the extracting agent behaves as a liquid and a gas simultaneously. Among the advantages of this method are the absence of toxic residues in the final product, high selectivity, short times, low consumption of solvents, high stability of the product obtained and that the remaining biomass can be treated with other techniques to continue the extraction. This technique can also be used to remove undesirable compounds such as pollutants, toxins, and pesticides [136].

The extraction process takes place in several stages. At the beginning, the plant matrix absorbs the supercritical solvent, producing a swelling of the cell structure with the consequent dilation of the intercellular channels. This results in a decrease in resistance to mass transfer. In addition, there is a simultaneous transfer of matter from the internal matrix to the surface. After that, these compounds are transported from the surface to the supercritical solvent and finally removed from the solvent [137].

To increase extraction performance, several parameters need to be optimized. These are fundamentally temperature, pressure and co-solvent type and all depend on the compound to be extracted. The most widely used solvent is CO_2 for its thermodynamic and heat transfer properties. Furthermore, it has a low critical point (31°C, 73 bar). In addition, the polarity of CO_2 can be modified by using co-solvents such as ethanol, so that the polar components are also extracted [138].

Several reviews have been published on SFE fundamentals, experimental design and specific applications on food processing, surface coating analysis, vegetable matrices, extraction of metals as complexes, functional ingredients from natural sources, constituents of fish oil and decontamination of hazardous substances [139]. Table 6 shows some examples. Despite the large number of studies using this extraction technique, the comparatively high cost of investment has kept this expertise from being broadly considered as an alternative. Nevertheless, new studies have proved that SFE is an economically viable choice [152]. Moreover, the interest in SFE is not only at the laboratory level as an analytical tool but also in industrial processing, mainly decaffeination of coffee or tea, extraction of essential oils, extraction of high added value compounds and fatty acids [153].

Table 6. Different experimental conditions carried out with supercritical-fluid extraction (SFE).

Compound	Raw material	Conditions					Yield	Ref.
		Pressure	T	Flow	t	Mass		
		bar	°C		h	g	mg/g	
Lipids	*Hypnea charoides*	241–379	40–50	1 mL/min	2	2	58 mg/g	[140]
Vitamin E	*Spirulina platensis*	361	83	50 mL/min	1.5	75	29.4 mg/g	[141]
Protein	*Spirulina platensis*	350	40	400 g/min	4	—	—	[142]
Astaxanthin	*Haematococcus pluvialis*	435	65	167 mL/min	3.5	240	87.4%	[143]
Lycopene	*Solanum lycopersicum* (tomato)	300	60	1.44 cm/min	6	1	213 mg/g	[144]
Capsaicinoids	*Capsicum chinensis* (chili pepper)	150	60	2 mL/min	1.4	2.5	0.5%	[145]
Flavonoids	*Mentha spicata* (mint)	200	60	15 g/min	1	30	60.57 mg/g	[146]
Phenols	*Vitis lambrusca*	160	45	0.2 mL/min	0.3	3	12.3%	[147]
Decaffeination	Green tea	300	80	1500 mL/min	2	10	70.2%	[148]
Decaffeination	*Ilex paraguariensis* (yerba mate)	300	60	15.83 g/min	4.25	0.65	99.97%	[149]
Aroma	Vinegar	350	50	0.42 g/min	2	200	96.6%	[150]
Sulforaphane	*Brassica oleracea* (cauliflower)	250	60	2 g/min	3	25	0.47%	[151]

Abbreviations: T: temperature; t: time.

The extraction through supercritical fluids presents greater efficiency in terms of increased yields and shorter extraction times compared to conventional methods. There are several factors that could limit its effectiveness, including solvent type, temperature, pressure, extraction time and particle size of the matrix to be studied [153]. Among them, it seems that the effectiveness of an SFE depends mainly on temperature and pressure. When temperature and pressure variations are considered, these have important repercussions on fluid hydrodynamics, solubility, and mass transfer [146]. In this sense, the low temperature is crucial for the conservation of bioactive compounds in the extracts and to achieve higher global/specific yields or greater bioactive capacities in the extracts. Numerous studies apply an extraction temperature range of 40–50°C [154, 155, 156, 157]. Regarding pressure, there is evidence that the optimum range is between 200 and 400 bars [154, 158, 159, 160, 161]. Additionally, an increase in pressure can modify the solubility of the solute, therefore, it is interesting to control the composition of the extract by pressure [162]. In addition, the optimal conditions of pressure and temperature can be influenced by the origin of the biomass and their morphology. The type of modifier and its proportion are also of extreme importance within the extraction, since they determine the solubility of the analytes, ethanol being the most applied modifier [155]. The size of the biomass particle, shape and porosity are also important factors in terms of the mass transfer rate [152]. The use of small particles would have the advantage of increasing the accessible solute, but could lead to clogging of the extraction, making the use of dispersing agents useful in such cases.

3.6 Pulse electrified field extraction (PEF)

Electric pulse is one of the newest techniques in the field of thermal food processing. Its mechanism of action is based on causing the permeability of cell membranes in a short period of time and with low energy consumption. This is achieved by applying short duration pulses (μs to ms) of moderate electric voltage (typically 0.5–20 kV/cm) to a substrate of choice placed between two electrodes, which is commonly used for preservation, enzyme and microbial inactivation purposes [163].

These characteristics have led to different studies being carried out in order to improve the extraction performance of bioactive compounds, such as polyphenols, anthocyanins and vegetable oil from plant tissues and their by-products, as well as soluble intracellular matter of microorganisms [164]. However, low to mild PEF treatment intensities are often considered an effective pretreatment method for enhancement of secondary metabolite extraction yields in cell cultures and plant systems [165].

Critical parameters of the process are electric field intensity, treatment time (number of pulses x pulse duration), pulse waveform, conductivity, pH and ionic strength of the medium [166]. Among the advantages of this method are avoidance of undesirable changes in a biological material, which are typical of other techniques such as thermal, chemical and enzymatic ones. Moreover, it is able of killing microorganism [167].

As it can be observed in Table 7, most PEF-assisted extraction studies from by-products have targeted the polyphenolic compounds; nonetheless, extraction of polysaccharides, proteins, phytosterols, alkaloids, seed and germ oil among others have also been investigated.

Table 7. Different experimental conditions carried out with pulse electrified extraction (PEF).

Compound	Raw material	Conditions						Yield	Ref.
		E	Power	Pulse t	No. pulses	Mass	Flow		
		kWh/kg	kV/cm	μs	—	g	mL/min	mg/g	
Proteins	*Saccharomyces cerevisiae*	—	40	8.3	500	0.5% w/w	166	1%	[168]
Polysaccharide	*Agaricus bisporus* (common mushroom)	—	38.4	2	136	9% w/w	336	7.9 mg/g	[169]
Polysacharide	*Zea mays* (corn) silk		30	6		50% w/w	25	7.3%	[170]
Pigments	*Beta vulgaris* (red beetroot)	7	1	10	270	—	48	90%	[171]
Betanine	*Beta vulgaris* (red beetroot)	2.5	7	2	5	—	—	90%	[171]
Anthocyanins	*Solanum tuberosum* (potato)	13.5	3.4	3	35	14 g	—	0.66 mg/g	[172]
Anthocyanins	*Citrus sinensis* (orange) peel	3.77	3.4	3	105		—	65.8 mg/g	[173]
Anthocyanins	*Vitis vinifera* (grape) skin	10	3.0	15	30	—	—	14.05 mg/g	[174]
Phenols	*Citrus sinensis* (orange)	—	7	3	20	100	—	3.1 mg/g	[173]
Phenols	*Agaricus bisporus* (common mushroom)	—	38.4	2	68	9% w/w	498	51%	[169]
Oil	*Zea mays* (corn) germ	0.62	0.6	—	120	2	1	88.4%	[175]
Copra	*Cocos nucifera* (coconut)	—	2.5	575	20	—	—	20%	[176]
Alkaloids	*Solanum tuberosum* (potato) peel	18.5	0.75	3	200	—	—	1.85 mg/g	[177]

Abbreviations: E: energy; T: temperature; t: time.

In general, according to the studies available so far, it has been demonstrated that the application of moderate electric field pulse technology either as pretreatment step or as continuous extraction system improves the extraction in the area of phytochemical extraction from by-products.

3.7 Pressurized liquid assisted extraction (PLE)

This technique is also known as Accelerated Solvent Extraction (ASE), Pressurized Hot Water Extraction (HSPE), Pressurized Fluid Extraction (PFE) or Enhanced Solvent Extraction (ESE) [178]. This technique, firstly described in 1996 as a new emerging environmentally-friendly technique, presents the following advantages [179]: reduced use of solvents, lower energy consumption and short time periods (5–10 min). It consists of an extraction procedure which uses organic liquid solvents applied at high pressures (normally up to 200 bar) and temperatures (normally up to 200°C) to extract target compounds, this is, PLE is based on applying high pressures to heat the extraction solvent above its boiling point while using low volume of organic solvents [178, 180].

Extraction efficiency and selectivity is significantly affected by the following parameters, namely: i) extraction solvent; ii) temperature; iii) pressure; iv) static extraction time and number of cycles and v) sample weight [178]. Temperature and pressure have a significant effect on PLE, therefore, they are the most important for PLE optimization. Many studies support that the use of solvents at high temperatures and pressures improve the yield in comparison to other conventional techniques [178, 180, 181].

On the other hand, two types of instrumentation can be used in PLE: static and dynamic. However, generally, the equipment is very similar between them, except for some small differences such as a more sophisticated high-pressure pump and a pressure restrictor in the case of dynamic PLE. The extraction design equipment can contain the following basic parts: i) solvent supply; ii) a pump; iii) a heater; iv) a pressure vessel where occurs the extraction and v) a collected vessel for the extract [182]. Regarding static PLE, the process occurs in one or several extraction cycles with replacement of solvent in between. When the static time finishes (normally between 5 and 10 min), part of solvent in the extraction cells is replaced with fresh solvent (new extraction cycle). After the last cycle, the nitrogen gas is used for purging the sample cell and to remove the remaining solvent. As regards dynamic PLE, the extraction solvent is continuously pumped through the extraction vessel containing the sample [183]. Currently, static PLE is the most used and there are not equipment of dynamic PLE in the market [184]. This system has been employed to recover different target compounds that can be used as food additives, including proteins, phenolic compounds or fatty acids, among others (Table 8).

Table 8. Different experimental conditions carried out with pressurized liquid extraction (PLE).

Compound	Raw material	Conditions					Yield	Ref.
		P	T	Solvent	Static cycles	t		
		MPa	°C			min		
Proteins	*Punica granatum* (pomegranate) peels	10.34	120	E 70% (v/v)	4, s.t. 3 min	12	9%	[185]
Phenolic compounds	*Hibiscus sabdariffa (hibiscus)*	10.34	200	E 100% (v/v) E	1	20	—	[186]
Fatty acids	*Anacardium occidentale* (cashew)	8	80	E-Pr	4, s.t. 5 min; 4, s.t. 10 min	60	19.2–33.1%	[187]
Cafestol and Kahweol diterpenes	*Coffea arabica* (coffee) beans	10.34	70	E (99.3%)	3, s.t. 8 min	24	9.78%	[188]
Anthocyanins	*Phaseolus vulgaris* L. (black beans) hulls	10	60	E-CA (30:70) (v/v)	1	26	3.96 mg C3GE/g DW	[189]
Phenolic, antioxidant and antimicrobial compounds	*Psidium guajava* L. (pineapple guava) peels	10	80	E	1	60	42%	[190]
				E-W			50%	
Antihyaluronidase	*Padina pavonica*	15	60	W	2, s.t. 10 min	20	0.04 mg/mL (IC50)	[191]
Polysaccharides	*Campomanesia xanthocarpa* (gabiroba) pulp	15	120	UW (pH 6.9)	1	15	5.70% of pectin	[192]

Abbreviations: P: pressure; T: temperature; t: time; C3GE: cyanidin-3-glucoside equivalents, DW: dry weight; s.t.: static time. W: water, E: ethanol, CA: citric acid, Pr: propane, UW: Ultrapure water.

Parameters such as the characteristics of the matrix, the solvent type, the extraction time, and temperature show a big influence on the extraction efficiency of PLE. It is pursued an appropriate combination of analyte and solvent to achieve high diffusion rates and mass transfer, the use of binary solvents such as ethanol: water or methanol: water is more efficient and respectful with the environment than pure reagents [193]. Regarding the characteristics of the matrix, they affect the recovery rate of the compounds, including the nature of the target compounds, the relative binding behavior of the analyte with the solvent, the particle size, and the moisture content [181]. Additionally, temperature and pressure significantly influence the selectivity and efficiency of PLE. The use of high temperatures under reduced pressure helps to break the matrix structure by overcoming the molecular bonding forces [194]. In this way the use of high temperatures reduces the activation energy and overcomes the interactive forces (cohesive and adhesive) between the matrix and the solvent molecules to cause desorption. However, the use of high pressure causes bubbling problems during extraction, causing low solubility rate [181].

3.8 Surfactant assisted extraction (SAE)

Surfactants are components able to lower the surface tension between two liquids (gas–liquid) or between liquid and solid. They are usually amphiphilic organic compounds: with a polar group in the head part (hydrophilic) and a nonpolar group in the queue part. Their distinctive characteristic is their capacity to lower surface tension of a solvent. Generally, the higher surfactant concentration in the solution, the lower surface tension. This occurs due to adsorption of the surfactant at the interface. Surfactants can be classified following two criteria: according to its structure or its hydrophilic–lipophilic balance (HLB).

Regarding its structure, some authors divide surfactants in three groups: i) ionic: within this group are anionic and cationic; ii) nonionic (without electric charge); iii) amphoteric (they contain anionic and cationic groups in the molecule) and v) no hydrocarbons. Among them, nonionic surfactants are the most used [195]. Regarding its balance, Griffin and co-workers introduced an arbitrary scale for classifying surfactants according to HLB. The HLB scale has a ranges from 0 to 20, so that a molecule is hydrophilic when HLB value is above 9, and lipophilic when HLB value is below 9 [196].

One of the main parameters used in surfactant assisted extraction (SAE) is the adsorption efficiency (pC_{20}). The higher $pC_{20,}$ the lower concentration of surfactant required to surface tension of pure solvent to decrease to 20 mM/m. Ideally, a surfactant employed to extract a compound of interest should have a high adsorption efficiency. A study verified that pC_{20} was higher when the surfactant consists of an anionic and nonionic mixture than when the surfactant was only anionic [197]. Nowadays, SAE is not used as an individual technique to extract bioactive compounds in food or nutraceutical industry. Generally, it is used as a complement in other type of extractions, mainly with MAE. A study employed surfactant-MAE method to obtain high yield bioactive compounds from fig leaves [198]; another study used a MAE - solid–liquid extraction approach using a surfactant to extract bioactive phenolic compounds from *Vitis vinifera* leaves [199]. Also, different ionic liquid based surfactants were used together with MAE to extract flavonoids from *Mangifera* sp. and *Passiflora* sp. leaves [200].

4. Conclusions

There are many natural compounds with possible potential applications as preservatives to act either antioxidants, antimicrobials or to provide specific flavor, aroma, or color to foods. A great number of compounds can be obtained from natural sources, such as vegetables, fruits, plants, macroalgae and microalgae for their application as food and feed additives but also as cosmetic ingredients.

To isolate the molecules of interest to be used as food additives, it is necessary to apply adequate extraction techniques which preserve intact the *compounds*. Among them, green extraction techniques are becoming popular due to the society growing demand in sustainable products. Green extraction methods, which comprise UAE, MAE, EAE, PEP and high-pressure methods (SFE and PLE), have some advantages over the traditional extraction methods such as lower solvent and resources consumption, as well as matching and extraction yields improvement, in some cases. Besides, these techniques are in line with the green aspects of sample preparation, they avoid the use of harmful reagents and minimize the use of organic solvents. In this context, several studies investigating the effects of the extraction parameters of these novel techniques on yield and composition pointed the importance of optimizing the extraction time, temperature and solvent type depending on the characteristics of the matrix. Also, in the case of UAE and MAE, other parameter to considered are the frequency and the ultrasonic intensity (UAE) and power of microwave irradiation (MAE). Higher extraction yields are reported in the low frequency range (20–40 kHz). Moreover, European politics on circular economy highlight the importance of food revalorization. By-products represent a rich source of valuable additives with high antimicrobial and antioxidant activities. And in many cases, these additives are being obtained through the employment of environmentally friendly extraction techniques. Thus, these techniques are being used in various types of industries due to the benefit in the decrease of the time needed to achieve sample extraction.

References

Carocho, M.; Barreiro, M.F.; Morales, P.; Ferreira, I.C.F.R. Adding molecules to food, pros and cons: A review on synthetic and natural food additives. Compr. Rev. Food Sci. Food Saf. 2014, 13, 377-399

European Commision Regulation (EC) No 1333/2008 of the European Parliament and of the Council of 16 December 2008 on food additives. Off. J. Eur. Union2008, L, 16

Song, P.; Wu, L.; Guan, W. Dietary nitrates, nitrites, and nitrosamines intake and the risk of gastric cancer: A meta-analysis. Nutrients 2015, 7, 9872-9895, doi:10.3390/nu7125505

Mancini, A.; Imperlini, E.; Nigro, E.; Montagnese, C.; Daniele, A.; Orrù, S.; Buono, P. Biological and nutritional properties of palm oil and palmitic acid: Effects on health. Molecules 2015, 20, 17339-17361, doi:10.3390/molecules200917339

Dickson-Spillmann, M.; Siegrist, M.; Keller, C. Attitudes toward chemicals are associated with preference for natural food. Food Qual. Prefer. 2011, 22, 149-156, doi:10.1016/j.foodqual.2010.09.001

Mirabella, N.; Castellani, V.; Sala, S. Current options for the valorization of food manufacturing waste: A review. J. Clean. Prod. 2014, 65, 28-41, doi:10.1016/j.jclepro.2013.10.051

Yahfoufi, N.; Alsadi, N.; Jambi, M.; Matar, C. The immunomodulatory and anti-inflammatory role of polyphenols. Nutrients 2018, 10

Rao, A.V.; Rao, L.G. Carotenoids and human health. Pharmacol. Res. 2007, 55, 207-216

Shahidi, F.; De Camargo, A.C. Tocopherols and tocotrienols in common and emerging dietary sources: Occurrence, applications, and health benefits. Int. J. Mol. Sci. 2016, 17, 1745

Al-Hijazeen, M.; Lee, E.J.; Mendonca, A.; Ahn, D.U. Effects of Tannic Acid on Lipid and Protein Oxidation, Color, and Volatiles of Raw and Cooked Chicken Breast Meat during Storage. Antioxidants 2016, 5, doi:10.3390/antiox5020019

El-Guendouz, S.; Aazza, S.; Dandlen, S.A.; Majdoub, N.; Lyoussi, B.; Raposo, S.; Antunes, M.D.; Gomes, V.; Miguel, M.G. Antioxidant activity of thyme waste extract in O/W emulsions. Antioxidants 2019, 8, 1-14, doi:10.3390/antiox8080243

Carpena, M.; Nuñez-Estevez, B.; Soria-Lopez, A.; Garcia-Oliveira, P.; Prieto, M.A. Essential oils and their application on active packaging systems: A review. Resources 2021, 10, 1-20, doi:10.3390/resources10010007

Medina, I.; Gallardo, J.M.; González, M.J.; Lois, S.; Hedges, N. Effect of Molecular Structure of Phenolic Families as Hydroxycinnamic Acids and Catechins on Their Antioxidant Effectiveness in Minced Fish Muscle. J. Agric. Food Chem. 2007, 55, 3889-3895, doi:10.1021/jf063498i

Lorenzo, J.M.; Sineiro, J.; Amado, I.R.; Franco, D. Influence of natural extracts on the shelf life of modified atmosphere-packaged pork patties. Meat Sci. 2014, 96, 526-534, doi:10.1016/j.meatsci.2013.08.007

Iglesias, J.; Pazos, M.; Andersen, M.L.; Skibsted, L.H.; Medina, I. Caffeic acid as antioxidant in fish muscle: Mechanism of synergism with endogenous ascorbic acid and α-tocopherol. J. Agric. Food Chem. 2009, 57, 675-681, doi:10.1021/jf802888w

Caleja, C.; Barros, L.; Antonio, A.L.; Oliveira, M.B.P.P.; Ferreira, I.C.F.R. A comparative study between natural and synthetic antioxidants: Evaluation of their performance after incorporation into biscuits. Food Chem. 2017, 216, 342-346, doi:10.1016/j.foodchem.2016.08.075

Food and Drug Administration Subchapter B - Food for Human Consumption. Part 182: Substances Generally Recognized as Safe. Code Fed. Regul. 2020

European Comission for Health and Consummer Protection Commission Implementing Regulation (EU) 2017/656 Of 19 December 2016. Off. J. Eur. Union 2016, L304, 20-30

Løvik, M.; Marchelli, R.; Martin, A.; Moseley, B.; Berg, H. Van Den; Loveren, H. Van; Verhagen, H. Plant stanol esters and blood cholesterol - Scientific substantiation of a health claim related to plant stanol esters and lower/reduced blood cholesterol and reduced risk of (coronary) heart disease pursuant to Article 14 of Regulation (EC) No 1924/2006 . EFSA J. 2008, 6, 1-12, doi:10.2903/j.efsa.2008.825

Carocho, M.; Morales, P.; Ferreira, I.C.F.R. Natural food additives: Quo vadis? Trends Food Sci. Technol. 2015, 45, 284-295, doi:10.1016/j.tifs.2015.06.007

del Olmo, A.; Calzada, J.; Nuñez, M. Benzoic acid and its derivatives as naturally occurring compounds in foods and as additives: Uses, exposure, and controversy. Crit. Rev. Food Sci. Nutr. 2017, 57, 3084-3103, doi:10.1080/10408398.2015.1087964

EFSA Panel on Food Additives and Nutrient Sources added to Food (ANS) Scientific Opinion on the re-evaluation of ascorbic acid (E 300), sodium ascorbate (E 301) and calcium ascorbate (E 302) as food additives. EFSA J. 2015, 13, 1-124, doi:10.2903/j.efsa.2015.4087

EFSA Panel on Food Additives and Nutrient Sources added to Food (ANS) Safety assessment of the active substances citric acid and sodium hydrogen carbonate for use in active food contact materials. EFSA J. 2016, 14, doi:10.2903/j.efsa.2016.4529

EFSA Panel on Food Additives and Flavourings Re-evaluation of l(+)-tartaric acid (E 334), sodium tartrates (E 335), potassium tartrates (E 336), potassium sodium tartrate (E 337) and calcium tartrate (E 354) as food additives. EFSA J. 2020, 18, doi:10.2903/j.efsa.2020.6030

Viera, I.; Pérez-Gálvez, A.; Roca, M. Green natural colorants. Molecules 2019, 24, doi:10.3390/molecules24010154

EFSA Panel on Food Additives and Nutrient Sources added to Food Scientific Opinion on the re-evaluation of mixed carotenes (E 160a (i)) and beta-carotene (E 160a (ii)) as a food additive. EFSA J. 2012, 10, 1-67, doi:10.2903/j.efsa.2012.2593

EFSA Panel on Food Additives and Nutrient Sources added to Food Scientific Opinion on the re-evaluation of curcumin (E 100) as a food additive. EFSA J. 2010, 8, doi:10.2903/j.efsa.2010.1679

EFSA Panel on Food Additives and Nutrient Sources added to Food (ANS) Scientific Opinion on the re-evaluation of lutein (E 161b) as a food additive. EFSA J. 2010, 8, 1-57, doi:10.2903/j.efsa.2010.1678

EFSA Panel on Food Additives and Nutrient Sources added to Food Scientific Opinion on the re-evaluation of paprika extract (E 160c) as a food additive. EFSA J. 2015, 13, 1-52, doi:10.2903/j.efsa.2015.4320

Panel on Food Additives, Flavourings, P.A. and M. in C. with F. Use of Lycopene as a food colour. EFSA J.2008, 6, 674, doi:10.2903/j.efsa.2008.674

EFSA Panel on Food Additives and Nutrient Sources added to Food Scientific Opinion on the re-evaluation of anthocyanins (E 163) as a food additive. EFSAJ.2013, 11, 3145, doi:10.2903/j.efsa.2013.3145

Sano, C. History of glutamate production. Am. J. Clin. Nutr. 2009, 90, 728-732, doi:10.3945/ajcn.2009.27462F

EFSA Panel on Food Contact Materials Enzymes Flavourings and Processing Aids Scientific Opinion on Flavouring Group Evaluation 25, Revision 2 (FGE.25Rev2): Aliphatic and aromatic hydrocarbons from chemical group 31. EFSAJ.2011, 9, 1-126, doi:10.2903/j.efsa.2011.2177

Hu, Q.; Zhou, M.; wei, S. Progress on the Antimicrobial Activity Research of Clove Oil and Eugenol in the Food Antisepsis Field. J. Food Sci.2018, 83, 1476-1483, doi:10.1111/1750-3841.14180

European Commission Regulation (EC) No 1334/2008 of the European Parliament and of the Council of 16 December 2008. Off. J. Eur. Union 2017, 10, 1-32

Ramos, M.; Jiménez, A.; Peltzer, M.; Garrigós, M.C. Characterization and antimicrobial activity studies of polypropylene films with carvacrol and thymol for active packaging. J. Food Eng. 2012, 109, 513-519, doi:10.1016/j.jfoodeng.2011.10.031

EFSA Panel on Food Additives and Nutrient Sources added to Food (ANS) Refined exposure assessment of extracts of rosemary (E 392) from its use as food additive. EFSA J. 2018, 16, doi:10.2903/j.efsa.2018.5373

Van Nieuwenhuyzen, W.; Tomás, M.C. Update on vegetable lecithin and phospholipid technologies. Eur. J. Lipid Sci. Technol. 2008, 110, 472-486, doi:10.1002/ejlt.200800041

EFSA Panel on Food Additives and Nutrient Sources added to Food Re-evaluation of alginic acid and its sodium, potassium, ammonium and calcium salts (E 400–E 404) as food additives. EFSA J. 2017, 15, doi:10.2903/j.efsa.2017.5049

EFSA Panel on Food Additives and Nutrient Sources added to Food Re-evaluation of agar (E 406) as a food additive. EFSA J. 2016, 14, doi:10.2903/j.efsa.2016.4645

EFSA Panel on Food Additives and Nutrient Sources added to Food Re-evaluation of carrageenan (E 407) and processed Eucheuma seaweed (E 407a) as food additives. EFSA J. 2018, 16, doi:10.2903/j.efsa.2018.5238

EFSA Panel on Food Additives and Nutrient Sources added to Food (ANS) Re-evaluation of pectin (E 440i) and amidated pectin (E 440ii) as food additives. EFSA J. 2017, 15, doi:10.2903/j.efsa.2017.4866

EFSA Panel on Food Additives and Nutrient Sources added to Food (ANS) Re-evaluation of acacia gum (E 414) as a food additive. EFSA J. 2017, 15, 1-51, doi:10.2903/j.efsa.2017.4741

Hager, A.S.; Vallons, K.J.R.; Arendt, E.K. Influence of gallic acid and tannic acid on the mechanical and barrier properties of wheat gluten films. J. Agric. Food Chem. 2012, 60, 6157-6163, doi:10.1021/jf300983m

EFSA Panel on Additives and Products or Substances used in Animal Feed Scientific Opinion on the safety and efficacy of tannic acid when used as feed flavouring for all animal species. EFSA J. 2014, 12, 1-18, doi:10.2903/j.efsa.2014.3828

EFSA Panel on Food additives and Nutrient Sources added to Food Scientific Opinion on the safety of stigmasterol-rich plant sterols as food additive. EFSA J. 2012, 10, 1-39, doi:10.2903/j.efsa.2012.2659

EFSA Panel on Food additives and Nutrient Sources added to Food Scientific Opinion on the re-evaluation of tocopherol-rich extract (E 306), α-tocopherol (E 307), γ-tocopherol (E 308) and δ-tocopherol (E 309) as food additives. EFSA J. 2015, 13, doi:10.2903/j.efsa.2015.4247

Pappenberger, G.; Hohmann, H.-P. Industrial Production of l-Ascorbic Acid (Vitamin C) and d-Isoascorbic Acid. In Advances in Biochemical Engineering/Biotechnology; Springer Berlin Heidelberg, 2013; Vol. 143, pp. 143-188

Qu, J.; Zhang, W.; Yu, X. A combination of elicitation and precursor feeding leads to increased anthocyanin synthesis in cell suspension cultures of Vitis vinifera. Plant Cell. Tissue Organ Cult. 2011, 107, 261-269, doi:10.1007/s11240-011-9977-8

Ashraf, R.; Ghufran, S.; Akram, S.; Mushtaq, M.; Sultana, B. Cold pressed coriander (Coriandrum sativum L.) seed oil; Elsevier Inc., 2020; ISBN 9780128181881

Thu Dao, T.A.; Webb, H.K.; Malherbe, F. Optimization of pectin extraction from fruit peels by response surface method: Conventional versus microwave-assisted heating. Food Hydrocoll. 2021, 113, 106475, doi:10.1016/j.foodhyd.2020.106475

Garcia-Vaquero, M.; Rajauria, G.; Tiwari, B. Conventional extraction techniques : Solvent extraction. In Sustainable Seaweed Technologies; Torres, M.D., Kraan, S., Dominguez, H., Eds.; Elsevier Inc., 2020; pp. 171-189 ISBN 9780128179437

Chemat, F.; Boutekedjiret, C. Extraction - Steam distillation. Ref. Modul. Chem. Mol. Sci. Chem. Eng. Elsevier Amsterdam,Netherlands2016, 1-12

Ramadan, M.F.; Kroh, L.W.; Mörsel, J.T. Radical Scavenging Activity of Black Cumin (Nigella sativa L.), Coriander (Coriandrum sativum L.), and Niger (Guizotia abyssinica Cass.) Crude Seed Oils and Oil Fractions. J. Agric. Food Chem. 2003, 51, 6961-6969, doi:10.1021/jf0346713

Manousi, N.; Sarakatsianos, I.; Samanidou, V. Extraction Techniques of Phenolic Compounds and Other Bioactive Compounds From Medicinal and Aromatic Plants; Elsevier Inc., 2019; ISBN 9780128152584

Luque de Castro, M.D.; Priego-Capote, F. Soxhlet extraction: Past and present panacea. J. Chromatogr. A 2010, 1217, 2383-2389, doi:10.1016/j.chroma.2009.11.027

Poole, C.F. Solvent selection for liquid-phase extraction; Elsevier Inc., 2019; ISBN 9780128169117

Cong, Q.; Chen, H.; Liao, W.; Xiao, F.; Wang, P.; Qin, Y.; Dong, Q.; Ding, K. Structural characterization and effect on anti-angiogenic activity of a fucoidan from Sargassum fusiforme. Carbohydr. Polym. 2016, 136, 899-907, doi:10.1016/j.carbpol.2015.09.087

Gallo, M.; Formato, A.; Ianniello, D.; Andolfi, A.; Conte, E.; Ciaravolo, M.; Varchetta, V.; Naviglio, D. Supercritical fluid extraction of pyrethrins from pyrethrum flowers (Chrysanthemum cinerariifolium) compared to traditional maceration and cyclic pressurization extraction. J. Supercrit. Fluids 2017, 119, 104-112, doi:10.1016/j.supflu.2016.09.012

Sasidharan, S.; Jothy, S.L.; Vijayarathna, S.; Kavitha, N.; Oon, C.E.; Chen, Y.; Dharmaraj, S.; Lai, N.S.; Kanwar, J.R.; others Conventional and non-conventional approach towards the extraction of bioorganic phase. In Bioorganic Phase in Natural Food: An Overview; Springer, 2018; pp. 41-57

Albuquerque, B.R.; Prieto, M.A.; Barreiro, M.F.; Rodrigues, A.E.; Curran, T.P.; Barros, L.; Ferreira, I.C.F.R. Catechin-based extract optimization obtained from Arbutus unedo L. fruits using maceration/microwave/ultrasound extraction techniques. Ind. Crops Prod. 2017, 95, 404-415, doi:10.1016/j.indcrop.2016.10.050

Ozkan, K.; Bayram, Y.; Karasu, S.; Karadag, A.; Sagdic, O. Extraction of bioactive compounds from saffron species; Elsevier Inc., 2021; ISBN 9780128212196

López-Bascón-Bascon, M.A.; Luque de Castro, M.D. Soxhlet extraction. Liq. Extr.2019, 327-354, doi:10.1016/B978-0-12-816911-7.00011-6

Górak, A.; Sorensen, E. Distillation: fundamentals and principles; Academic Press, 2014;

Kockmann, N. History of distillation. In Distillation; Elsevier, 2014; pp. 1-43

Prado, J.M.; Vardanega, R.; Debien, I.C.N.; Meireles, M.A.A.; Gerschenson, L.N.; Chemat, H.B.S.; Chemat, S. Conventional extraction Second Edition; INC, 2021; ISBN 9780128205631

Handa, S.S.; Khanuja, S.P.S.; Longo, G.; Rakesh, D.D. Extraction Technologies for Medicinal and Aromatic Plants; 2008; ISBN 978-0-85404-193-0

Chemat, F.; Rombaut, N.; Sicaire, A.G.; Meullemiestre, A.; Fabiano-Tixier, A.S.; Abert-Vian, M. Ultrasound assisted extraction of food and natural products. Mechanisms, techniques, combinations, protocols and applications. A review. Ultrason. Sonochem.2017, 34, 540-560

Dey, S.; Rathod, V.K. Ultrasound assisted extraction of β-carotene from Spirulina platensis. Ultrason. Sonochem. 2013, 20, 271-276, doi:10.1016/j.ultsonch.2012.05.010

Khan, M.K.; Abert-Vian, M.; Fabiano-Tixier, A.S.; Dangles, O.; Chemat, F. Ultrasound-assisted extraction of polyphenols (flavanone glycosides) from orange (Citrus sinensis L.) peel. Food Chem.2010, 119, 851-858, doi:10.1016/j.foodchem.2009.08.046

Luque-García, J.L.; Luque De Castro, M.D. Ultrasound-assisted Soxhlet extraction: An expeditive approach for solid sample treatment - Application to the extraction of total fat from oleaginous seeds. J. Chromatogr. A 2004, 1034, 237-242, doi:10.1016/j.chroma.2004.02.020

Chemat, F.; Zill-E-Huma; Khan, M.K. Applications of ultrasound in food technology: Processing, preservation and extraction. Ultrason. Sonochem.2011, 18, 813-835, doi:10.1016/j.ultsonch.2010.11.023

Vinatoru, M.; Mason, T.J.; Calinescu, I. Ultrasonically assisted extraction (UAE) and microwave assisted extraction (MAE) of functional compounds from plant materials. TrAC - Trends Anal. Chem. 2017, 97, 159-178, doi:10.1016/j.trac.2017.09.002

Boonkird, S.; Phisalaphong, C.; Phisalaphong, M. Ultrasound-assisted extraction of capsaicinoids from Capsicum frutescens on a lab- and pilot-plant scale. Ultrason. Sonochem. 2008, 15, 1075-1079, doi:10.1016/j.ultsonch.2008.04.010

Prakash Maran, J.; Manikandan, S.; Vigna Nivetha, C.; Dinesh, R. Ultrasound assisted extraction of bioactive compounds from Nephelium lappaceum L. fruit peel using central composite face centered response surface design. Arab. J. Chem.2017, 10, S1145–S1157, doi:10.1016/j.arabjc.2013.02.007

Rodrigues, S.; Fernandes, F.A.N.; de Brito, E.S.; Sousa, A.D.; Narain, N. Ultrasound extraction of phenolics and anthocyanins from jabuticaba peel. Ind. Crops Prod. 2015, 69, 400-407, doi:10.1016/j.indcrop.2015.02.059

Gam, D.H.; Kim, S.Y.; Kim, J.W. Optimization of ultrasound-assisted extraction condition for phenolic compounds, antioxidant activity, and epigallocatechin gallate in lipid-extracted microalgae. Molecules 2020, 25, doi:10.3390/molecules25030454

Parniakov, O.; Apicella, E.; Koubaa, M.; Barba, F.J.; Grimi, N.; Lebovka, N.; Pataro, G.; Ferrari, G.; Vorobiev, E. Ultrasound-assisted green solvent extraction of high-added value compounds from microalgae Nannochloropsis spp. Bioresour. Technol. 2015, 198, 262-267, doi:10.1016/j.biortech.2015.09.020

Kadam, S.U.; Tiwari, B.K.; Smyth, T.J.; O'Donnell, C.P. Optimization of ultrasound assisted extraction of bioactive components from brown seaweed Ascophyllum nodosum using response surface methodology. Ultrason. Sonochem. 2015, 23, 308-316, doi:10.1016/j.ultsonch.2014.10.007

Bimakr, M.; Ganjloo, A.; Zarringhalami, S.; Ansarian, E. Ultrasound-assisted extraction of bioactive compounds from Malva sylvestris leaves and its comparison with agitated bed extraction technique. Food Sci. Biotechnol. 2017, 26, 1481-1490, doi:10.1007/s10068-017-0229-5

Chen, Y.-H.; Yang, C.-Y. Ultrasound-Assisted Extraction of Bioactive Compounds and Antioxidant Capacity for the Valorization of Elaeocarpus serratus L. Leaves. Processes2020, 8, 1218, doi:10.3390/pr8101218

Barbero, G.F.; Liazid, A.; Palma, M.; Barroso, C.G. Ultrasound-assisted extraction of capsaicinoids from peppers. Talanta 2008, 75, 1332-1337, doi:10.1016/j.talanta.2008.01.046

Montero-Calderon, A.; Cortes, C.; Zulueta, A.; Frigola, A.; Esteve, M.J. Green solvents and Ultrasound-Assisted Extraction of bioactive orange (Citrus sinensis) peel compounds. Sci. Rep. 2019, 9, 1-8, doi:10.1038/s41598-019-52717-1

Goula, A.M.; Ververi, M.; Adamopoulou, A.; Kaderides, K. Green ultrasound-assisted extraction of carotenoids from pomegranate wastes using vegetable oils. Ultrason. Sonochem. 2017, 34, 821-830, doi:10.1016/j.ultsonch.2016.07.022

Li, Y.; Fabiano-Tixier, A.S.; Tomao, V.; Cravotto, G.; Chemat, F. Green ultrasound-assisted extraction of carotenoids based on the bio-refinery concept using sunflower oil as an alternative solvent. Ultrason. Sonochem. 2013, 20, 12-18, doi:10.1016/j.ultsonch.2012.07.005

Purohit, A.J.; Gogate, P.R. Ultrasound-Assisted Extraction of β-Carotene from Waste Carrot Residue: Effect of Operating Parameters and Type of Ultrasonic Irradiation. Sep. Sci. Technol. 2015, 50, 1507-1517, doi:10.1080/01496395.2014.978472

Alboofetileh, M.; Rezaei, M.; Tabarsa, M.; You, S. Ultrasound-assisted extraction of sulfated polysaccharide from Nizamuddinia zanardinii : Process optimization, structural characterization, and biological properties. J. Food Process Eng. 2019, 42, e12979, doi:10.1111/jfpe.12979

Hmelkov, A.B.; Zvyagintseva, T.N.; Shevchenko, N.M.; Rasin, A.B.; Ermakova, S.P. Ultrasound-assisted extraction of polysaccharides from brown alga Fucus evanescens. Structure and biological activity of the new fucoidan fractions. J. Appl. Phycol. 2018, 30, 2039-2046, doi:10.1007/s10811-017-1342-9

Vázquez-Rodríguez, B.; Gutiérrez-Uribe, J.A.; Antunes-Ricardo, M.; Santos-Zea, L.; Cruz-Suárez, L.E. Ultrasound-assisted extraction of phlorotannins and polysaccharides from Silvetia compressa (Phaeophyceae). J. Appl. Phycol. 2020, 32, 1441-1453, doi:10.1007/s10811-019-02013-2

Shirsath, S.R.; Sonawane, S.H.; Gogate, P.R. Intensification of extraction of natural products using ultrasonic irradiations-A review of current status. Chem. Eng. Process. Process Intensif. 2012, 53, 10-23, doi:10.1016/j.cep.2012.01.003

Tiwari, B.K. Ultrasound: A clean, green extraction technology. TrAC - Trends Anal. Chem. 2015, 71, 100-109, doi:10.1016/j.trac.2015.04.013

Moreira, S.A.; Pintado, M.; Saraiva, J.A. High hydrostatic pressure-assisted extraction: A review on its effects on bioactive profile and biological activities of extracts; Elsevier Inc., 2020; ISBN 9780128164051

Khan, S.A.; Aslam, R.; Makroo, H.A. High pressure extraction and its application in the extraction of bio-active compounds: A review. J. Food Process Eng. 2019, 42, e12896, doi:10.1111/jfpe.12896

Alexandre, E.M.C.; Araújo, P.; Duarte, M.F.; de Freitas, V.; Pintado, M.; Saraiva, J.A. High-pressure assisted extraction of bioactive compounds from industrial fermented fig by-product. J. Food Sci. Technol. 2017, 54, 2519-2531, doi:10.1007/s13197-017-2697-2

Rosa, A.D.; Junges, A.; Fernandes, I.A.; Cansian, R.L.; Corazza, M.L.; Franceschi, E.; Backes, G.T.; Valduga, E. High pressure extraction of olive leaves (Olea europaea): bioactive compounds, bioactivity and kinetic modelling. J. Food Sci. Technol. 2019, 56, 3864-3876, doi:10.1007/s13197-019-03856-w

Huang, H.W.; Cheng, M.C.; Chen, B.Y.; Wang, C.Y. Effects of high pressure extraction on the extraction yield, phenolic compounds, antioxidant and anti-tyrosinase activity of Djulis hull. J. Food Sci. Technol. 2019, 56, 4016-4024, doi:10.1007/s13197-019-03870-y

Zhao, Y.; Jiang, Y.; Ding, Y.; Wang, D.; Deng, Y. High hydrostatic pressure-assisted extraction of high-molecular-weight melanoidins from black garlic: Composition, structure, and bioactive properties. J. Food Qual. 2019, 2019, doi:10.1155/2019/1682749

Wang, J.; Zhang, J.; Zhao, B.; Wang, X.; Wu, Y.; Yao, J. A comparison study on microwave-assisted extraction of Potentilla anserina L. polysaccharides with conventional method: Molecule weight and antioxidant activities evaluation. Carbohydr. Polym.2010, 80, 84-93, doi:10.1016/j.carbpol.2009.10.073

Han, Q.-H.; Liu, W.; Li, H.-Y.; He, J.-L.; Guo, H.; Lin, S.; Zhao, L.; Chen, H.; Liu, Y.-W.; Wu, D.-T.; et al. Extraction Optimization, Physicochemical Characteristics, and Antioxidant Activities of Polysaccharides from Kiwifruit (Actinidia chinensis Planch.). Molecules 2019, 24, 461, doi:10.3390/molecules24030461

Yuan, Y.; Macquarrie, D. Microwave assisted extraction of sulfated polysaccharides (fucoidan) from Ascophyllum nodosum and its antioxidant activity. Carbohydr. Polym. 2015, 129, 101-107, doi:10.1016/j.carbpol.2015.04.057

Li, Z.R.; Wang, B.; Zhang, Q.H.; Huang, F.F.; Ma, J.H. Microwave-assisted extraction and the antioxidant activity of water-soluble polysaccharide from Palmaria palmata. In Proceedings of the 5th International Conference on Bioinformatics and Biomedical Engineering, iCBBE 2011; 2011

Yang, Y.; Lei, Z.; Zhao, M.; Wu, C.; Wang, L.; Xu, Y. Microwave-assisted extraction of an acidic polysaccharide from Ribes nigrum L.: Structural characteristics and biological activities. Ind. Crops Prod.2020, 147, 112249, doi:10.1016/j.indcrop.2020.112249

Drinić, Z.; Pljevljakušić, D.; Živković, J.; Bigović, D.; Šavikin, K. Microwave-assisted extraction of O. vulgare L. spp. hirtum essential oil: Comparison with conventional hydro-distillation. Food Bioprod. Process.2020, 120, 158-165, doi:10.1016/j.fbp.2020.01.011

Sarfarazi, M.; Jafari, S.M.; Rajabzadeh, G.; Galanakis, C.M. Evaluation of microwave-assisted extraction technology for separation of bioactive components of saffron (Crocus sativus L.). Ind. Crops Prod.2020, 145, 111978, doi:10.1016/j.indcrop.2019.111978

Casazza, A.A.; Pettinato, M.; Perego, P. Polyphenols from apple skins: A study on microwave-assisted extraction optimization and exhausted solid characterization. Sep. Purif. Technol. 2020, 240, 116640, doi:10.1016/j.seppur.2020.116640

Pimentel-Moral, S.; Borrás-Linares, I.; Lozano-Sánchez, J.; Arráez-Román, D.; Martínez-Férez, A.; Segura-Carretero, A. Microwave-assisted extraction for Hibiscus sabdariffa bioactive compounds. J. Pharm. Biomed. Anal. 2018, 156, 313-322, doi:10.1016/j.jpba.2018.04.050

Vázquez-Espinosa, M.; Espada-Bellido, E.; V. González de Peredo, A.; Ferreiro-González, M.; Carrera, C.; Palma, M.; G. Barroso, C.; F. Barbero, G. Optimization of Microwave-Assisted Extraction for the Recovery of Bioactive Compounds from the Chilean Superfruit (Aristotelia chilensis (Mol.) Stuntz). Agronomy 2018, 8, 240, doi:10.3390/agronomy8110240

Cavalloro, V.; Marrubini, G.; Stabile, R.; Rossi, D.; Linciano, P.; Gheza, G.; Assini, S.; Martino, E.; Collina, S. Microwave-Assisted Extraction and HPLC-UV-CD Determination of (S)-usnic Acid in Cladonia foliacea. Molecules 2021, 26, 455, doi:10.3390/molecules26020455

Dhobi, M.; Mandal, V.; Hemalatha, S. Optimization of microwave assisted extraction of bioactive flavonolignan-silybinin; 2009; Vol. 3;

Kumar Chaturvedi, A.; Ravidas Nagar, S.; Pradesh, U. Extraction of Neutraceuticals from Plants by Microwave Assisted Extraction. Syst. Rev. Pharm. 9, doi:10.5530/srp.2018.1.6

Alara, O.R.; Abdurahman, N.H.; Olalere, O.A. Optimization of microwave-assisted extraction of flavonoids and antioxidants from Vernonia amygdalina leaf using response surface methodology. Food Bioprod. Process. 2018, 107, 36-48, doi:10.1016/j.fbp.2017.10.007

Chan, C.H.; Yusoff, R.; Ngoh, G.C. Optimization of microwave-assisted extraction based on absorbed microwave power and energy. Chem. Eng. Sci. 2014, 111, 41-47, doi:10.1016/j.ces.2014.02.011

Alfaro, M.J.; Bélanger, J.M.R.; Padilla, F.C.; Paré, J.R.J. Influence of solvent, matrix dielectric properties, and applied power on the liquid-phase microwave-assisted processes (MAP™) extraction of ginger (Zingiber officinale). Food Res. Int. 2003, 36, 499-504, doi:10.1016/S0963-9969(02)00198-9

Casazza, A.A.; Aliakbarian, B.; Mantegna, S.; Cravotto, G.; Perego, P. Extraction of phenolics from Vitis vinifera wastes using non-conventional techniques. J. Food Eng. 2010, 100, 50-55, doi:10.1016/j.jfoodeng.2010.03.026

Moreira, M.M.; Barroso, M.F.; Boeykens, A.; Withouck, H.; Morais, S.; Delerue-Matos, C. Valorization of apple tree wood residues by polyphenols extraction: Comparison between conventional and microwave-assisted extraction. Ind. Crops Prod. 2017, 104, 210-220, doi:10.1016/j.indcrop.2017.04.038

Chan, C.H.; Yusoff, R.; Ngoh, G.C.; Kung, F.W.L. Microwave-assisted extractions of active ingredients from plants. J. Chromatogr. A 2011, 1218, 6213-6225, doi:10.1016/j.chroma.2011.07.040

Chaudhary, S.; sagar, S. The use of enzymes in food processing: A review. South Asian J. Food Technol. Environ.2015, 01, 190-210, doi:10.46370/sajfte.2015.v01i03and04.01

Raveendran, S.; Parameswaran, B.; Ummalyma, S.B.; Abraham, A.; Mathew, A.K.; Madhavan, A.; Rebello, S.; Pandey, A. Applications of microbial enzymes in food industry. Food Technol. Biotechnol. 2018, 56, 16-30

Sowbhagya, H.B.; Chitra, V.N. Enzyme-assisted extraction of flavorings and colorants from plant materials. Crit. Rev. Food Sci. Nutr. 2010, 50, 146-161, doi:10.1080/10408390802248775

Pereira, A.G.; Jimenez-Lopez, C.; Fraga, M.; Lourenço-Lopes, C.; Garcíaoliveira, P.; Lorenzo, J.M.; Perez-Lamela, C.; Prieto, M.A.; Simal-Gandara, J. Extraction, properties, and applications of bioactive compounds obtained from microalgae. Curr. Pharm. Des. 2020, 26, doi:10.2174/1381612826666200403172206

de Almeida, N.M.; de Moura Bell, J.M.L.N.; Johnson, L.A. Properties of Soy Protein Produced by Countercurrent, Two-Stage, Enzyme-Assisted Aqueous Extraction. J. Am. Oil Chem. Soc. 2014, 91, 1077-1085, doi:10.1007/s11746-014-2436-z

Zhang, S.B.; Wang, Z.; Xu, S.Y. Optimization of the Aqueous Enzymatic Extraction of Rapeseed Oil and Protein Hydrolysates. J. Am. Oil Chem. Soc. 2007, 84, 97-105, doi:10.1007/s11746-006-1004-6

Jiang, L.; Hua, D.; Wang, Z.; Xu, S. Aqueous enzymatic extraction of peanut oil and protein hydrolysates. Food Bioprod. Process. 2010, 88, 233-238, doi:10.1016/j.fbp.2009.08.002

Saad, N.; Louvet, F.; Tarrade, S.; Meudec, E.; Grenier, K.; Landolt, C.; Ouk, T.S.; Bressollier, P. Enzyme-Assisted Extraction of Bioactive Compounds from Raspberry (Rubus idaeus L.) Pomace. J. Food Sci. 2019, 84, 1371-1381, doi:10.1111/1750-3841.14625

Pan, S.; Wu, S. Cellulase-assisted extraction and antioxidant activity of the polysaccharides from garlic. Carbohydr. Polym. 2014, 111, 606-609, doi:10.1016/j.carbpol.2014.05.022

Wang, S.; Dong, X.; Tong, J. Optimization of enzyme-assisted extraction of polysaccharides from alfalfa and its antioxidant activity. Int. J. Biol. Macromol. 2013, 62, 387-396, doi:10.1016/j.ijbiomac.2013.09.029

Štambuk, P.; Tomašković, D.; Tomaz, I.; Maslov, L.; Stupić, D.; Karoglan Kontić, J. Application of pectinases for recovery of grape seeds phenolics. 3 Biotech2016, 6, 1-12, doi:10.1007/s13205-016-0537-0

Thai Huynh, N.; Smagghe, G.; Bryan Gonzales, G.; Van Camp, J.; Raes, K. Enzyme-Assisted Extraction Enhancing the Phenolic Release from Cauliflower (Brassica oleracea L. var. botrytis) Outer Leaves. 2014, doi:10.1021/jf502543c

Zhou, Z.; Shao, H.; Han, X.; Wang, K.; Gong, C.; Yang, X. The extraction efficiency enhancement of polyphenols from Ulmus pumila L. barks by trienzyme-assisted extraction. Ind. Crops Prod. 2017, 97, 401-408, doi:10.1016/j.indcrop.2016.12.060

Sui, X.; Jiang, L.; Li, Y.; Liu, S. The research on extracting oil from watermelon seeds by aqueous enzymatic extraction method. In Proceedings of the Procedia Engineering; Elsevier, 2011; Vol. 15, pp. 4673-4680

Tabtabaei, S.; Diosady, L.L. Aqueous and enzymatic extraction processes for the production of food-grade proteins and industrial oil from dehulled yellow mustard flour. Food Res. Int. 2013, 52, 547-556, doi:10.1016/j.foodres.2013.03.005

Le, H. Van; Le, V.V.M. Comparison of enzyme-assisted and ultrasound-assisted extraction of vitamin C and phenolic compounds from acerola (Malpighia emarginata DC.) fruit. Int. J. Food Sci. Technol.2012, 47, 1206-1214, doi:10.1111/j.1365-2621.2012.02960.x

Nath, P.; Kaur, C.; Rudra, S.G.; Varghese, E. Enzyme-Assisted Extraction of Carotenoid-Rich Extract from Red Capsicum (Capsicum annuum). Agric. Res. 2016, doi:10.1007/s40003-015-0201-7

Santamaría, R.I.; Reyes-Duarte, M.D.; Bárzana, E.; Fernando, D.; Gama, F.M.; Mota, M.; López-Munguía, A. Selective enzyme-mediated extraction of capsaicinoids and carotenoids from chili guajillo puya (Capsicum annuum L.) using ethanol as solvent. J. Agric. Food Chem. 2000, 48, 3063-3067, doi:10.1021/jf991242p

Choudhari, S.M.; Ananthanarayan, L. Enzyme aided extraction of lycopene from tomato tissues. Food Chem. 2007, 102, 77-81, doi:10.1016/j.foodchem.2006.04.031

Echave, J.; Pereira, A.G.; Carpena, M.; Prieto, M.A.; Simal-Gándara, J. Capsicum Seeds as a Source of Bioactive Compounds: Biological Properties, Extraction Systems, and Industrial Application. In IntechOpen; 2020; p. 37

Brunner, G. Supercritical Gases as Solvents: Phase Equilibria. In; Steinkopff, Heidelberg, 1994; pp. 59-146

Pereira, A.G.A.G.; Jimenez-Lopez, C.; Fraga, M.; Lourenço-Lopes, C.; Garcíaoliveira, P.; Lorenzo, J.M.J.M.; Perez-Lamela, C.; Prieto, M.A.M.A.; Simal-Gandara, J.; García-Oliveira, P.; et al. Extraction, Properties, and Applications of Bioactive Compounds Obtained from Microalgae. Curr. Pharm. Des. 2020, 26, 1929-1950, doi:10.2174/1381612826666200403172206

da Silva, R.P.F.F.; Rocha-Santos, T.A.P.; Duarte, A.C. Supercritical fluid extraction of bioactive compounds. TrAC - Trends Anal. Chem. 2016, 76, 40-51

Cheung, P.C.K. Temperature and pressure effects on supercritical carbon dioxide extraction of n-3 fatty acids from red seaweed. Food Chem. 1999, 65, 399-403, doi:10.1016/S0308-8146(98)00210-6

Mendiola, J.A.; García-Martínez, D.; Rupérez, F.J.; Martín-Álvarez, P.J.; Reglero, G.; Cifuentes, A.; Barbas, C.; Ibañez, E.; Señoráns, F.J. Enrichment of vitamin E from Spirulina platensis microalga by SFE. J. Supercrit. Fluids 2008, 43, 484-489, doi:10.1016/j.supflu.2007.07.021

Qiuhui, H. Supercritical carbon dioxide extraction of Spirulina platensis component and removing the stench. J. Agric. Food Chem. 1999, 47, 2705-2706, doi:10.1021/jf9812432

Wang, L.; Yang, B.; Yan, B.; Yao, X. Supercritical fluid extraction of astaxanthin from Haematococcus pluvialis and its antioxidant potential in sunflower oil. Innov. Food Sci. Emerg. Technol. 2012, 13, 120-127, doi:10.1016/j.ifset.2011.09.004

Nobre, B.P.; Gouveia, L.; Matos, P.G.S.; Cristino, A.F.; Palavra, A.F.; Mendes, R.L. Supercritical Extraction of Lycopene from Tomato Industrial Wastes with Ethane. Molecules 2012, 17, 8397-8407, doi:10.3390/molecules17078397

de Aguiar, A.C.; dos Santos, P.; Coutinho, J.P.; Barbero, G.F.; Godoy, H.T.; Martínez, J. Supercritical fluid extraction and low pressure extraction of Biquinho pepper (Capsicum chinense). LWT - Food Sci. Technol. 2014, doi:10.1016/j.lwt.2014.06.014

Bimakr, M.; Rahman, R.A.; Taip, F.S.; Ganjloo, A.; Salleh, L.M.; Selamat, J.; Hamid, A.; Zaidul, I.S.M. Comparison of different extraction methods for the extraction of major bioactive flavonoid compounds from spearmint (Mentha spicata L.) leaves. Food Bioprod. Process. 2011, 89, 67-72, doi:10.1016/j.fbp.2010.03.002

Ghafoor, K.; Park, J.; Choi, Y.H. Optimization of supercritical fluid extraction of bioactive compounds from grape (Vitis labrusca B.) peel by using response surface methodology. Innov. Food Sci. Emerg. Technol. 2010, 11, 485-490, doi:10.1016/j.ifset.2010.01.013

Sun, Q.-L.; Hua, S.; Ye, J.-H.; Lu, J.-L.; Zheng, X.-Q.; Liang, Y.-R. Decaffeination of green tea by supercritical carbon dioxide. J. Med. Plants Res. 2010, 4, 1161-1168

Santo, A.T. do E.; Siqueira, L.M.; Almeida, R.N.; Vargas, R.M.F.; Franceschini, G. do N.; Kunde, M.A.; Cappellari, A.R.; Morrone, F.B.; Cassel, E. Decaffeination of yerba mate by supercritical fluid extraction: Improvement, mathematical modelling and infusion analysis. J. Supercrit. Fluids 2021, 168, 105096, doi:10.1016/j.supflu.2020.105096

Lu, Z.-M.; Xu, W.; Yu, N.-H.; Zhou, T.; Li, G.-Q.; Shi, J.-S.; Xu, Z.-H. Recovery of aroma compounds from Zhenjiang aromatic vinegar by supercritical fluid extraction. Int. J. Food Sci. Technol. 2011, 46, 1508-1514, doi:10.1111/j.1365-2621.2011.02649.x

Dal Prá, V.; Dolwitsch, C.B.; Da Silveira, G.D.; Porte, L.; Frizzo, C.; Tres, M.V.; Mossi, V.; Mazutti, M.A.; Do Nascimento, P.C.; Bohrer, D.; et al. Supercritical CO2 extraction, chemical characterisation and antioxidant potential of Brassica oleracea var capitata against HO·, O·-2 and ROO·. Food Chem. 2013, 141, 3954-3959, doi:10.1016/j.foodchem.2013.06.098

Pereira, C.G.; Meireles, M.A.A. Supercritical fluid extraction of bioactive compounds: Fundamentals, applications and economic perspectives. Food Bioprocess Technol. 2010, 3, 340-372

Uwineza, P.A.; Waśkiewicz, A. Recent advances in supercritical fluid extraction of natural bioactive compounds from natural plant materials. Molecules 2020, 25

Kviecinski, M.R.; Benelli, P.; Felipe, K.B.; Correia, J.F.G.; Pich, C.T.; Ferreira, S.R.S.; Pedrosa, R.C. SFE from Bidens pilosa Linné to obtain extracts rich in cytotoxic polyacetylenes with antitumor activity. J. Supercrit. Fluids 2011, 56, 243-248, doi:10.1016/j.supflu.2010.12.011

Roseiro, L.B.; Duarte, L.C.; Oliveira, D.L.; Roque, R.; Bernardo-Gil, M.G.; Martins, A.I.; Sepúlveda, C.; Almeida, J.; Meireles, M.; Gírio, F.M.; et al. Supercritical, ultrasound and conventional extracts from carob (Ceratonia siliqua L.) biomass: Effect on the phenolic profile and antiproliferative activity. Ind. Crops Prod. 2013, 47, 132-138, doi:10.1016/j.indcrop.2013.02.026

Costa, P.; Grosso, C.; Gonçalves, S.; Andrade, P.B.; Valentão, P.; Gabriela Bernardo-Gil, M.; Romano, A. Supercritical fluid extraction and hydrodistillation for the recovery of bioactive compounds from Lavandula viridis L'Hér. Food Chem. 2012, 135, 112-121, doi:10.1016/j.foodchem.2012.04.108

Bitencourt, R.G.; Queiroga, C.L.; Duarte, G.H.B.; Eberlin, M.N.; Kohn, L.K.; Arns, C.W.; Cabral, F.A. Sequential extraction of bioactive compounds from Melia azedarach L. in fixed bed extractor using CO2, ethanol and water. J. Supercrit. Fluids 2014, 95, 355-363, doi:10.1016/j.supflu.2014.09.027

Liau, B.C.; Shen, C.T.; Liang, F.P.; Hong, S.E.; Hsu, S.L.; Jong, T.T.; Chang, C.M.J. Supercritical fluids extraction and anti-solvent purification of carotenoids from microalgae and associated bioactivity. J. Supercrit. Fluids 2010, 55, 169-175, doi:10.1016/j.supflu.2010.07.002

Pasquel Reátegui, J.L.; Machado, A.P.D.F.; Barbero, G.F.; Rezende, C.A.; Martínez, J. Extraction of antioxidant compounds from blackberry (Rubus sp.) bagasse using supercritical CO2 assisted by ultrasound. J. Supercrit. Fluids 2014, 94, 223-233, doi:10.1016/j.supflu.2014.07.019

Scodro, R.B.D.L.; Espelho, S.C.; Agostinho Pires, C.T.; Garcia, V.A.D.S.; Cardozo-Filho, L.; Cortez, L.E.R.; Pilau, E.J.; Ferracioli, K.R.C.; Siqueira, V.L.D.; Cardoso, R.F.; et al. A new benzoic acid derivative from Piper diospyrifolium and its anti-Mycobacterium tuberculosis activity. Phytochem. Lett. 2015, 11, 18-23, doi:10.1016/j.phytol.2014.10.015

Espinosa-Pardo, F.A.; Martinez, J.; Martinez-Correa, H.A. Extraction of bioactive compounds from peach palm pulp (Bactris gasipaes) using supercritical CO2. J. Supercrit. Fluids 2014, 93, 2-6, doi:10.1016/j.supflu.2014.05.010

Gomes, P.B.; Mata, V.G.; Rodrigues, A.E. Production of rose geranium oil using supercritical fluid extraction. J. Supercrit. Fluids 2007, 41, 50-60, doi:10.1016/j.supflu.2006.08.018

Kumari, B.; Tiwari, B.K.; Hossain, M.B.; Brunton, N.P.; Rai, D.K. Recent Advances on Application of Ultrasound and Pulsed Electric Field Technologies in the Extraction of Bioactives from Agro-Industrial By-products. Food Bioprocess Technol. 2018, 11, 223-241

Zeng, X.; Zhang, Z. Pulsed Electric Field Assisted Extraction of Bioactive Compounds. In Advances in Food Processing Technology; Springer Singapore, 2019; pp. 125-135

Balasa, A.; Janositz, A.; Knorr, D. Electric field stress on plant systems. In Encyclopedia of Biotechnology in Agriculture and Food; 2011; pp. 208-211

Bobinaitė, R.; Pataro, G.; Lamanauskas, N.; Šatkauskas, S.; Viškelis, P.; Ferrari, G. Application of pulsed electric field in the production of juice and extraction of bioactive compounds from blueberry fruits and their by-products. J. Food Sci. Technol. 2015, 52, 5898-5905, doi:10.1007/s13197-014-1668-0

Vorobiev, E.; Lebovka, N. Pulse electric field-assisted extraction. In Enhancing Extraction Processes in the Food Industry; 2012; pp. 25-80

Liu, D.; Lebovka, N.I.; Vorobiev, E. Impact of Electric Pulse Treatment on Selective Extraction of Intracellular Compounds from Saccharomyces cerevisiae Yeasts. Food Bioprocess Technol. 2013, 6, 576-584, doi:10.1007/s11947-011-0703-7

Xue, D.; Farid, M.M. Pulsed electric field extraction of valuable compounds from white button mushroom (Agaricus bisporus). Innov. Food Sci. Emerg. Technol. 2015, 29, 178-186, doi:10.1016/j.ifset.2015.03.012

Zhao, W.; Yu, Z.; Liu, J.; Yu, Y.; Yin, Y.; Lin, S.; Chen, F. Optimized extraction of polysaccharides from corn silk by pulsed electric field and response surface quadratic design. J. Sci. Food Agric. 2011, 91, 2201-2209, doi:10.1002/jsfa.4440

Fincan, M.; DeVito, F.; Dejmek, P. Pulsed electric field treatment for solid-liquid extraction of red beetroot pigment. J. Food Eng. 2004, 64, 381-388, doi:10.1016/j.jfoodeng.2003.11.006

Puértolas, E.; Cregenzán, O.; Luengo, E.; Álvarez, I.; Raso, J. Pulsed-electric-field-assisted extraction of anthocyanins from purple-fleshed potato. Food Chem. 2013, 136, 1330-1336, doi:10.1016/j.foodchem.2012.09.080

Luengo, E.; Álvarez, I.; Raso, J. Improving the pressing extraction of polyphenols of orange peel by pulsed electric fields. Innov. Food Sci. Emerg. Technol. 2013, 17, 79-84, doi:10.1016/j.ifset.2012.10.005

Corrales, M.; Toepfl, S.; Butz, P.; Knorr, D.; Tauscher, B. Extraction of anthocyanins from grape by-products assisted by ultrasonics, high hydrostatic pressure or pulsed electric fields: A comparison. Innov. Food Sci. Emerg.Technol.2008

Guderjan, M.; Töpfl, S.; Angersbach, A.; Knorr, D. Impact of pulsed electric field treatment on the recovery and quality of plant oils. J. Food Eng. 2005, 67, 281-287, doi:10.1016/j.jfoodeng.2004.04.029

Ade-Omowaye, B.I.O.; Angersbach, A.; Eshtiaghi, N.M.; Knorr, D. Impact of high intensity electric field pulses on cell permeabilisation and as pre-processing step in coconut processing. Innov. Food Sci. Emerg. Technol. 2000, 1, 203-209, doi:10.1016/S1466-8564(00)00014-X

Hossain, M.B.; Aguiló-Aguayo, I.; Lyng, J.G.; Brunton, N.P.; Rai, D.K. Effect of pulsed electric field and pulsed light pre-treatment on the extraction of steroidal alkaloids from potato peels. Innov. Food Sci. Emerg. Technol. 2015, 29, 9-14, doi:10.1016/j.ifset.2014.10.014

Nieto, A.; Borrull, F.; Pocurull, E.; Marcé, R.M. Pressurized liquid extraction: A useful technique to extract pharmaceuticals and personal-care products from sewage sludge. TrAC - Trends Anal. Chem. 2010, 29, 752-764, doi:10.1016/j.trac.2010.03.014

Santoyo, S.; Rodríguez-Meizoso, I.; Cifuentes, A.; Jaime, L.; García-Blairsy Reina, G.; Señorans, F.J.; Ibáñez, E. Green processes based on the extraction with pressurized fluids to obtain potent antimicrobials from Haematococcus pluvialis microalgae. LWT - Food Sci. Technol. 2009, 42, 1213-1218, doi:10.1016/j.lwt.2009.01.012

Richter, B.E.; Jones, B.A.; Ezzell, J.L.; Porter, N.L.; Avdalovic, N.; Pohl, C. Accelerated solvent extraction: A technique for sample preparation. Anal. Chem. 1996, 68, 1033-1039, doi:10.1021/ac9508199

Mustafa, A.; Turner, C. Pressurized liquid extraction as a green approach in food and herbal plants extraction: A review. Anal. Chim. Acta 2011, 703, 8-18, doi:10.1016/j.aca.2011.07.018

Pronyk, C.; Mazza, G. Design and scale-up of pressurized fluid extractors for food and bioproducts. J. Food Eng. 2009, 95, 215-226, doi:10.1016/j.jfoodeng.2009.06.002

Turner, C.; Waldebäck, M. Principles of pressurized fluid extraction and environmental, food and agricultural applications; Woodhead Publishing Limited, 2013;

Bautz, H.; Polzer, J.; Stieglitz, L. Comparison of pressurised liquid extraction with Soxhlet extraction for the analysis of polychlorinated dibenzo-p-dioxins and dibenzofurans from fly ash and environmental matrices. J. Chromatogr. A 1998, 815, 231-241, doi:10.1016/S0021-9673(98)00453-1

Hernández-Corroto, E.; Plaza, M.; Marina, M.L.; García, M.C. Sustainable extraction of proteins and bioactive substances from pomegranate peel (Punica granatum L.) using pressurized liquids and deep eutectic solvents. Innov. Food Sci. Emerg. Technol.2020, 60, 102314, doi:10.1016/j.ifset.2020.102314

Pimentel-Moral, S.; Borrás-Linares, I.; Lozano-Sánchez, J.; Alañón, M.E.; Arráez-Román, D.; Segura-Carretero, A. Pressurized GRAS solvents for the green extraction of phenolic compounds from hibiscus sabdariffa calyces. Food Res. Int. 2020, 137, 109466, doi:10.1016/j.foodres.2020.109466

Zanqui, A.B.; da Silva, C.M.; Ressutte, J.B.; de Morais, D.R.; Santos, J.M.; Eberlin, M.N.; Cardozo-Filho, L.; da Silva, E.A.; Gomes, S.T.M.; Matsushita, M. Extraction and assessment of oil and bioactive compounds from cashew nut (Anacardium occidentale) using pressurized n-propane and ethanol as cosolvent. J. Supercrit. Fluids 2020, 157, 104686, doi:10.1016/j.supflu.2019.104686

Albertina, N.; Oliveira, D.; Cornelio-santiago, H.P.; Fukumasu, H.; Lopes, A.; Oliveira, D. Green coffee extracts rich in diterpenes e Process optimization of pressurized liquid extraction using ethanol as solvent. J. Food Eng. 2018, 224, 148-155, doi:10.1016/j.jfoodeng.2017.12.021

Fialho, R.; Benvenutti, L.; Maria, V.; Margarida, T.; Zielinski, F.; Regina, S.; Ferreira, S. An eco-friendly pressure liquid extraction method to recover anthocyanins from broken black bean hulls. 2021, 67, doi:10.1016/j.ifset.2020.102587

Henrique, P.; Helena, D.; Ribeiro, B.; Amadeu, G.; Vitali, L.; Hense, H. The Journal of Supercritical Fluids Extraction of bioactive compounds from feijoa (Acca sellowiana (O . Berg) Burret) peel by low and high-pressure techniques. J. Supercrit. Fluids 2019, 145, 219-227, doi:10.1016/j.supflu.2018.12.016

Fayad, S.; Nehmé, R.; Tannoury, M.; Lesellier, E.; Pichon, C.; Morin, P. Macroalga Padina pavonica water extracts obtained by pressurized liquid extraction and microwave-assisted extraction inhibit hyaluronidase activity as shown by capillary electrophoresis. J. Chromatogr. A 2017, 1497, 19-27, doi:10.1016/j.chroma.2017.03.033

Pereira, I.; Fernanda, S.; Estuardo, D.; Fetzer, L.; Lúcio, M.; Léa, J.; Silveira, M. International Journal of Biological Macromolecules Effects of pressurized hot water extraction on the yield and chemical characterization of pectins from Campomanesia xanthocarpa Berg fruits. Int. J. Biol. Macromol. 2020, 146, 431-443, doi:10.1016/j.ijbiomac.2019.12.261

Plaza, M.; Abrahamsson, V.; Turner, C. Extraction and neoformation of antioxidant compounds by pressurized hot water extraction from apple byproducts. J. Agric. Food Chem. 2013, 61, 5500-5510, doi:10.1021/jf400584f

Wang, L.; Weller, C.L. Recent advances in extraction of nutraceuticals from plants. Trends Food Sci. Technol. 2006, 17, 300-312, doi:10.1016/j.tifs.2005.12.004

Guy Broze; Uri Zoller Handbook of detergents; Marcel Denkker, Ed.; New York, 1999;

W.C. Griffin Calculation of HLB values of nonionic surfactants. J. Soc. Cosmet. Chem.1954, 5, 249-256

Kesarwani, H.; Saxena, A.; Mandal, A.; Sharma, S. Anionic / Nonionic Surfactant Mixture for Enhanced Oil Recovery through the Investigation of Adsorption, Interfacial, Rheological, and Rock Wetting Characteristics. 2021, doi:10.1021/acs.energyfuels.0c03767

Yu, L.; Meng, Y.; Wang, Z.; Cao, L.; Liu, C.; Gao, M. Sustainable and efficient surfactant-based microwave-assisted extraction of target polyphenols and furanocoumarins from fig (Ficus carica L.) leaves. J. Mol. Liq.2020, 318, 114196, doi:10.1016/j.molliq.2020.114196

Surfactants, I.L.; Mastellone, G.; Pacheco-fern, I.; Rubiolo, P.; Cagliero, C. Sustainable Micro-Scale Extraction of Bioactive Phenolic Compounds from Vitis vinifera Leaves with. Molecules 2020, 25, 3072

Kristýna Moučková, Idaira Pacheco-Fernández, Juan H. Ayala, Petra Bajerová, V.P. Evaluation of Structurally Different Ionic Liquid-Based Surfactants in a Green. Molecules 2020, 25, 1-17

Chapter-3

Carotenoids as Natural Colorful Additives For The Food Industry

1. Introduction

1.1 Carotenoids: natural pigments for coloring in food industry

Carotenoids are a class of natural pigments broadly distributed in nature and synthesized by plants, certain bacteria, fungi and algae. These molecules are classified in two main groups: carotenes, which are pure hydrocarbons (α–/β-carotenes and lycopene), and xanthophylls, which represent the oxygenated derivatives (lutein, zeaxanthin, astaxanthin, fucoxanthin and cryptoxanthin) [1]. These hydrocarbons are formed by eight five-carbon isoprenoid units with conjugated double bonds, responsible of multiple geometrical isomers (cis/trans), although carotenoids are mainly found in the most stable configuration, the all-trans one [2, 3]. These double bonds act as chromophores and are responsible for light absorption in the visual range of the spectrum [4], providing yellow, orange and red coloration [5]. Among the main biological properties described for carotenoids, they stand out for their antioxidant capacity and ability to quench singlet oxygen species [6]. Carotenoids have also been described to wield anti-inflammatory, antimicrobial and anti-hyperglycemic activities, to prevent cardiovascular and/or neurodegenerative diseases and to stimulate the immune system [7, 8]. These beneficial properties made them emerge as a promising alternative to synthetic additives, which have been related with negative side-effects. Besides, these pigments improve the nutritional value of foodstuff and can be used for food coloring. These reasons have boosted carotenoids' market size, which is expected to reach $300 billion by 2024, due to the interest shown by food, animal feeding, pharmaceutical, nutraceutical and cosmetic industries [9].

1.2 Carotenes

1.2.1 Alpha and beta-carotene

Found primarily in microalgae species such as *Dunaliella sp.* or *Arthrospira sp.* and vegetables like carrots and pumpkins, β-carotene is an isomer form of α-carotene (Figure 1). The latter compound is also found in these vegetables and in cereals like corn and fruits like peaches and apples [10]. Since β-carotene bioconversion efficiencies surpass those of α-carotene, it is more abundantly distributed among the vegetal kingdom [11]. Carotenes have been approved as Group II food additives by the European Commission under the

E160 number [12], being used as an orange-red pigment on non-alcoholic beverages, cheese, pastry and ice cream [13]. Moreover, both pigments are known for being vitamin A (retinal, retinol and retinoic acid) precursors [5]. Vitamin A intake has proven to prevent the development of ocular diseases associated with its deficiency [11, 14] and systemic affections involving an increase of the oxidative status as in immunological diseases or cancers [5]. Thus, their versatility and multiple benefits have made them emerge as high-valued food additives with large economic significance, catching the attention of food industry.

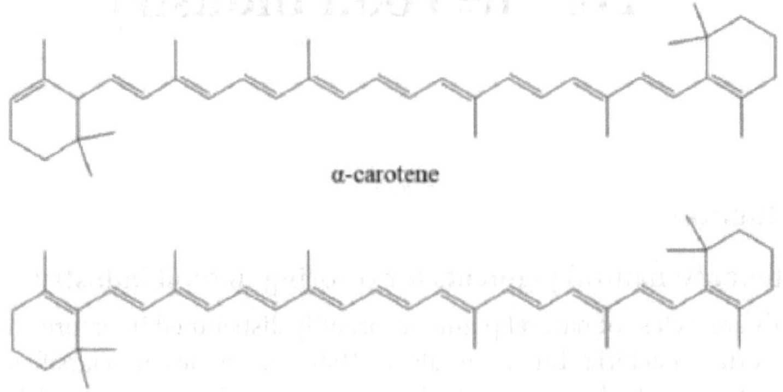

α-carotene

β-carotene

Figure 1. Chemical structure of α-carotene and β-carotene.

1.2.2 Lycopene

Lycopene can be found in fruits and vegetables, especially in tomatoes (Figure 2), being the carotenoid with the highest antioxidant capacity. It has been seen that this pigment is involved in modulating many anti-inflammatory processes, and some authors have linked it with the prevention of bone diseases, such as osteoporosis [15]. Furthermore, lycopene also shown anticancer effects against several tumoral and normal cell lines, particularly prostate cancer cell lines (PrEC and PC-3), in *in vitro* and *in vivo* studies [16, 17]. For all these health benefits and for being easily obtainable, lycopene is widely used by the food industry as a colorant, being applied into many foodstuffs like cheese, sausages or dairy drinks, among many others [10]. However, the major drawback of lycopene is its bioavailability, which depends on several factors, including the food source, people's metabolism and even the interaction with other food [18].

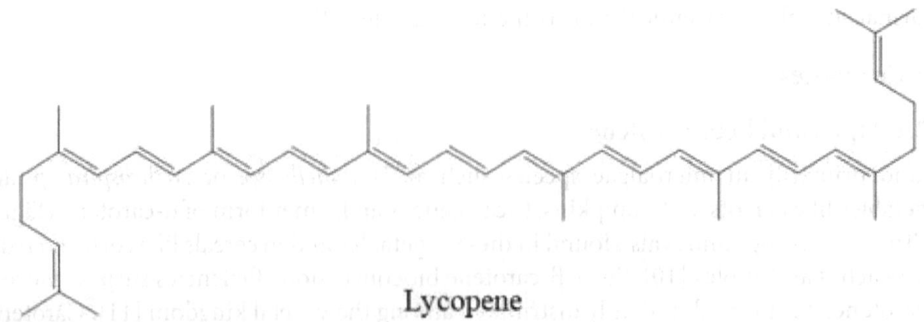

Lycopene

Figure 2. Chemical structure of lycopene.

1.3 Xanthophylls

Xanthophylls comprise oxidized derivatives of carotenoids, being broadly available in nature. These pigments are characterized for having yellow, orange or red coloration. Some of the most common xanthophylls present in nature include lutein, zeaxanthin, astaxanthin, β-cryptoxanthin and fucoxanthin (Figure 3) [6]. These compounds are polar molecules and, unlike non-polar carotenes, they get accumulated, contributing to skin pigmentation [1]. Antioxidant, neuroprotective, antiplasmodial or anticancer are some of the biological activities that pointed xanthophylls as a promising nutraceutical. These beneficial bioactivities may have preventive effects in an extensive variety of diseases such as oral, allergic, neurologic, ophthalmologic and immune affections [6]. Moreover, beneficial properties may be transferred to food. Hence, these characteristics have prompted the incorporation of xanthophylls as natural additives to obtain products with a better appearance according to the consumers' standards [1, 19].

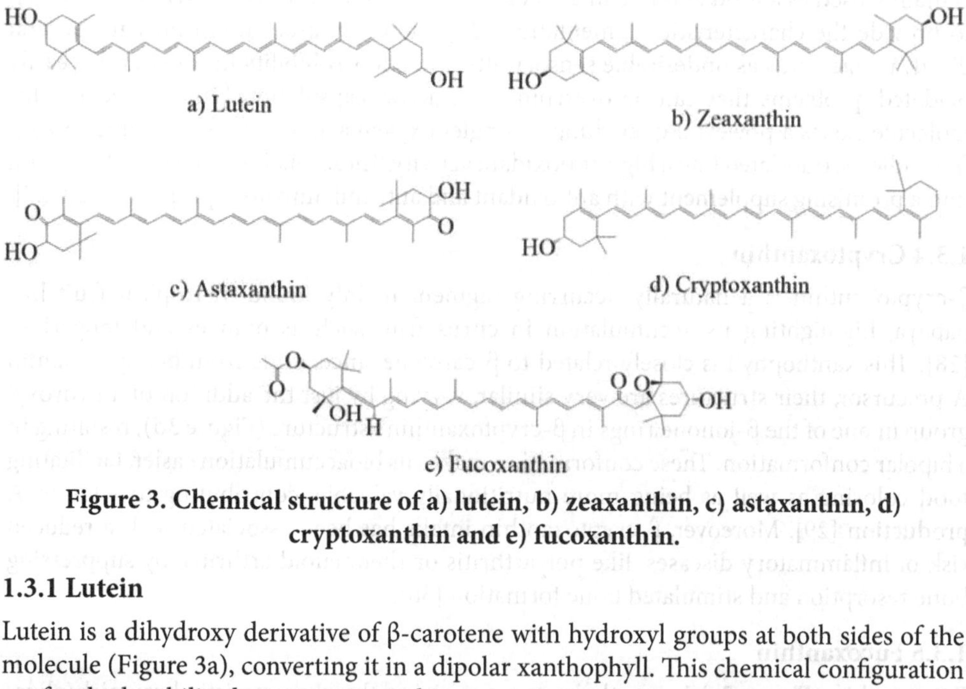

Figure 3. Chemical structure of a) lutein, b) zeaxanthin, c) astaxanthin, d) cryptoxanthin and e) fucoxanthin.

1.3.1 Lutein

Lutein is a dihydroxy derivative of β-carotene with hydroxyl groups at both sides of the molecule (Figure 3a), converting it in a dipolar xanthophyll. This chemical configuration confers hydrophilic characteristics and improves its capacity to scavenge free radicals [6, 20]. The most common chemical configuration of lutein is acylated with different fatty acids [1], such as lauric (C12:0) or palmitic acid (C16:0), becoming mono- and diacylated derivatives [21]. Leafy vegetables and plants, flower petals and yellow and orange fruits are the most important sources of lutein. Its extraction is mainly carried out with organic solvents from flower petals that have been previously fermented and/or dried [1].

1.3.2 Zeaxanthin

Zeaxanthin (Figure 3b) is a structural isomer of lutein with a darker yellow tone, closer to orange [20]. It is naturally found in leaves of green vegetables, flower petals, in some yellow and orange fruits, corn and even in microbial *Flavobacterium* sp. Also, it can be transferred into animal products such as egg yolks [6, 20, 22]. Zeaxanthin's poor stability in presence of oxygen and light and its lipophilic nature limit its applications in food industry. Nanoencapsulation of the molecule seems to be a promising strategy to improve zeaxanthin stability in the final product [22]. Currently, in the food industry, zeaxanthin is used as colorant and feed additive in birds to color skin and egg's yolk, and in swine and fish for skin pigmentation [20].

1.3.3 Astaxanthin

Astaxanthin (Figure 3c) is a lipophilic carotenoid with a reddish-orange color [23]. This pigment is found in high concentrations in microalgae like *Haematococcus pluvialis*, and *Chlorella zofingiensis*. Furthermore, it is also encountered in red yeast like *Xanthophyllomyces dendrorhous* and some bacteria like *Agrobacterium aurantiacum* [24]. Ascending in the food chain, astaxanthin gets accumulated in crustaceans like krill, shrimps, lobsters or crabs, and fish flesh like in salmon [20, 25]. This xanthophyll is mainly used as a food additive in aquaculture for animal feeding, as well as in poultry, to provide the characteristic pigmentation [24]. However, astaxanthin presents several disadvantages such as undesirable sensory attributes, low solubility in water and is easily oxidated, problems that can be overcome by microencapsulation [26]. Moreover, this molecule exerts a powerful quenching of singlet oxygen activity and scavenging oxygen free radicals, translated into a high antioxidant activity. These qualities convert astaxanthin into a promising supplement with antioxidant and anti-inflammatory properties [24, 27].

1.3.4 Cryptoxanthin

β-cryptoxanthin is a naturally occurring pigment mainly found in tropical fruit like papaya, highlighting its accumulation in citrus fruit such as oranges and tangerines [28]. This xanthophyll is closely related to β-carotene since, aside from being a vitamin A precursor, their structures are very similar, varying by just the addition of a hydroxyl group in one of the β-ionone rings in β-cryptoxanthin's structure (Figure 3d), resulting in a bipolar conformation. These conformation makes its bioaccumulation easier, facilitating food coloring as well as being more nutritionally valuable, contributing to vitamin A production [29]. Moreover, β-cryptoxanthin intake has been associated with a reduced risk of inflammatory diseases, like polyarthritis or rheumatoid arthritis, by suppressing bone resorption and stimulated bone formation [30].

1.3.5 Fucoxanthin

Fucoxanthin (Figure 3e) is mostly known for giving the characteristic brownish/olive-green color to brown algae (Phaeophyceae), as in species belonging to the genus *Undaria*, *Sargassum* and *Laminaria*, although some microalgae, mainly diatoms and *Chrysophyta*, can accumulate higher concentrations [31, 32]. This pigment has been related with antioxidant, anticancer, antihypertensive, anti-inflammatory, anti-diabetic, anti-obesity, neuroprotective, anti-angiogenic and photoprotective bioactivities [33], being considered as a non-toxic and safe bioactive ingredient for coloring and food supplementation purposes [34]. However, some limitations arise due to its low water-solubility, reduced bioavailability, and sensitivity to temperature, light and oxygen [35]. In addition, its synthesis is complex and expensive and extraction procedures from algae have not been standardized yet [36, 37]. Despite these drawbacks, some studies reported great fucoxanthin stability after encapsulation [38, 39].

2. Natural Sources of Carotenoids

Generally, natural sources of carotenoids are divided into: i) fruits, vegetables and flowers; ii) microorganisms (microalgae, cyanobacteria, fungi, bacterial and yeasts); and iii) by-products (peels, seeds and skin).

2.1 Fruits, vegetables and flowers

There is a wide variety of fruits and vegetables recognized as natural sources of carotenoids in human diet. Besides, flowers, recently introduced in gastronomy, appeared to be a suitable source of carotenoids (Table 1). In general terms, the most relevant carotenoid found among these groups is β-carotene, although lutein, β-cryptoxanthin, lycopene and zeaxanthin are also highlighted as major carotenoids. Reviewed literature showed very different ranges of carotenoid concentrations depending on the analyzed tissue, variety, ripening stage, geographical origin, etc. [46, 47]. Nevertheless, Table 1 points to fruit as the most relevant source of carotenoids.

Table 1. Quantitative and qualitative analysis of carotenoids content in different species of fruits, vegetables, and flowers.

Source		Main Carotenoids	Carotenoid Content (mg/g)	Ref.
Fruits				
Apricot	*Prunus armeniaca*	β-car, β-crypt, Lut, Zea	0.07–0.08 (DW)	[40]
'Gac' oil	*Momordica cochinchinensis*	α–/β-Car, Lyc	1.8–11 (FW)	[41, 42]
Goji	*Lycium barbarum*	β-car, β-crypt, Zea	0.04–0.51 (FW)	[43]
Kaki	*Diospyros kaki*	β-car, β-crypt, Lut, Zea	0.03–0.07 (DW)	[40]
Banana and plantain	*Musa* sp.	α–/β-Car, Lut	0.01–0.04 (DW)	[44, 45]
Mandarin juice	*Citrus reticulata*	ζ–/ β-car, β-crypt	0.01 (DW)	[46]
Mango	*Mangifera indica*	α–/β-Car, β-crypt, Lut, Zea	3–129 (FW)	[47]
Orange	*Citrus sinensis*	α–/ β–/ζ-car, β-crypt, Lut, Zea	0.01–0.030.01–0.02 (DW)	[46, 48]
Papaya	*Carica papaya*	β–/ζ-car, β-crypt, Lyc, Vio, Zea	0.14–4.13 (FW)	[49, 50]
Peach	*Prunus persica*	β-car, β-crypt, Lut, Zea	0.04–0.09 (DW)	[40]
Vegetables and cereals				
Broccoli	*Brassica oleracea* var. *italiaca*	Lut, Neo	8.5–11.6 (DW)	[51]
Carrot	*Daucus carota*	α–/β-car, Lut, Lyc	0.01–0.8 (DW)(TC's)	[52, 53]
Kale	*B. oleracea var. sabellica*	Zea	1.6–2.5 (DW)	[54]
		β-car	0.10 (DW)	
		Neo	0.12 (DW)	
Lettuce	*Lactuca sativa*	Lut	0.1–0.13 (DW)	[55]
		β-car	2.2–2.9 (DW)	

Source		Main Carotenoids	Carotenoid Content (mg/g)	Ref.
Pea	*Pisum sativum*	Lut	0.01–0.02 (DW)	[56]
		β-car	0.01–0.02 (DW)	
Pepper	*Capsicum annuum*	β-car	0.39–0.71 (DW)	[57]
		Zea	0.31–0.73 (DW)	
Spinach	*Spinacia oleracea*	Neo	0.1–0.2 (FW)	[58]
		Lut	0.34–0.53 (FW)	
		β-car	0.2–0.32 (FW)	
Sweet corn	*Zea mays*	Lut	0.02 (DW)	[59]
		β-car	0.01 (DW)	
Tomato	*Lycopersicon esculentum*	β-car	0.01 (FW)	[60]
		Lyc	0.05–0.08 (FW)	
		TCs	0.04–0.2 (FW)	
Flowers				
Blue centaurea	*Centaurea cyanus*	β-car, Lut	0.06 (DW)	[61]
Blue borage	*Borago officinalis*	β-car, Lut	1.8 (DW)	
Camelia	*Camelia japonica*	β-car, Lut	0.2 (DW)	
Carnation	*Dianthus caryophyllus*	Xanthophylls	0.001–0.003 (P) (DW) 0.04–0.07 (L) (DW)	[62]
Marigold	*Tagetes* sp.	Lut	0.0002–0.006 (DW)	[63]
Nasturtium	*Tropaeolum majus*	Lut	0.4–1.2 (DW)	[64]
Pansies	*Viola x wittrockiana*	β-car, Lut, Zea	0.2–1.1 (DW)	[61, 65]
Snapdragon	*Antirrhinum majus*	β-car, Lut, Zea	0.03 (DW)	[65]

Abbreviations: DW: dry weight, FW: fresh weight, L: leaves, P: petals. Carotenoids: α–/β–/γ–/ζ-car: α–/β–/γ–/ζ-carotene, β-crypto: β-cryptoxanthin, lyc: lycopene, lut: lutein, neo: neoxanthin, TCs: total carotenoids content, vio: violaxanthin, zea: zeaxanthin.

Carotenoids extracted from fruits, vegetables and flowers become too expensive due to high production costs associated with large production areas required. Besides, the supply of carotenoids extracted from plants becomes unstable, since it is dependent on unpredictable climatologic conditions [66]. Therefore, more sustainable and green approaches have been explored for a more efficient carotenoids' collection, including the use of microorganism or the reutilization of agricultural by-products

2.2 Microorganisms

Nowadays, the interest on microbial carotenoids has increased because of their low production area requirements when compared to plants. Besides, microbial cultures are nearly independent of climatic conditions, seasonality and soil composition. Current technological advances permit a tight control of culturing conditions, which improves the efficiency of microbial carotenoid production and reduces costs. Examples of efficient production of carotenoids using microalgae, bacteria, yeasts or fungi are displayed in Table 2, that demonstrates the huge variability of microorganisms capable of producing specific types of carotenoids being the most relevant β-carotene, lutein, astaxanthin, canthaxanthin and torulene (Table 2).

Table 2. Quantitative and qualitative analysis of carotenoids content in different species microorganisms such as microalgae, bacteria, yeasts, filamentous fungi and cyanobacteria.

Species		Most abundant Carotenoids	Content (mg/g)	Ref.
Microalgae				
Dunaliella tertiolecta		β-car	0.001–0.0045 (DW)	[67]
Haematococcus pluvialis		Ast	2–20 (DW)	[68]
Haematococcus alpinus		Ast	6–19 (DW)	[69]
Nostoc commune		Canthaxanthin	N.D	[70]
Scenedesmus almeriensis β-car		Lut	0.01 (DW)	[71]
		1.50 (DW)		
Bacteria				
Arthrobacter sp. P40		Decaprenoxanthin and derivatives mono–/diglucosides; Lyc	0.3–0.4 (DW)	[72]
Corynebacterium glutamicum		β-Car	0.01–3.1 (CDW)	[73]
Zea		0.01–0.9 (CDW)		
Cryobacterium sp. P19		Carotenoids, glucoside derivatives	0.4–0.5 (DW)	[72]
Chryseobacterium sp. P36		Zea; β-Crypto; β-Car; β-Zeacarotene	0.5–0.6 (DW)	
Flavobacterium sp.P33		Zea; β-Crypto; β-Car; β-Zeacarotene	0.7–0.8 (DW)	
Planococcus sp. 48		Carotenoids and glucoside derivatives	0.7 (DW)	[72]
Salinibacterium sp. P15		Carotenoids and glucoside derivatives	0.5 (DW)	[72]
Yeasts and filamentous fungi				
Blakeslea trispora	Fungi	β-car	30 (DW)	[68]
		Lyc	>900 (DW)	
Mucor circinelloides		β-car	0.275–0.698 (DW)	[74]
Phycomyces blakesleeanus		β-car	0.05–10 (DM)	[68]
Phaffia rhodozyma	Yeasts	Ast	0.000725–0.007642 (DW)	[75, 76]
Rhodotorula minuta		β-car	0.0172 (DW)	[77]
Rhodotorula glutinis		Torulene	5–14 (DW)	[78, 79]
		Torularhodin	32.2 (DW)	
Rhodotorula graminis		Torulene	18.2 (DW)	[79, 80]
		Torularhodin	9.3 (DW)	
Sporobolomyces sp.		Torulene	0.0001 (DW)	[79]
		Torularhodin	0.00001 (DW)	
Xanthophyllomyces dendrorhous		Ast	0.0026–0.001 (DW)	[81]

Abbreviations: DW: dry weight; CDW: cold-water-dispersible; N.D: not determined. Carotenoids: ast: astaxanthin, β-car: β-carotene, β-crypto: β-cryptoxanthin, lut: lutein, lyc: lycopene, TCs: total carotenoids content; zea: zeaxanthin.

2.3 By-products

Food waste has been increased in the last years driven by an increasing population, expected to reach 10 billion people by 2050, and inefficient and unsustainable production systems [82]. These factors boosted waste production, which is usually composted or burnt, emitting high amounts of CO_2 to the atmosphere. To counteract this situation, multiple strategies have been explored in the last decades, such as the revalorization of wastes as source of biomolecules. In fact, peels, seeds, husks, pomace or pulp are recognized as alternative sources of compounds with diverse biological properties [83].

Table 3 collects information about potential agricultural and food by-products as sustainable sources of carotenoids.

Table 3. Quantitative and qualitative analysis of carotenoids content in different by-products derived from agricultural and food industries.

By-product	Most abundant Carotenoids	Content (μg/g)	Ref.
Tucumã peels	β-car	68–88 (FW)	[84]
Peach palm peel		71–75 (FW)	
Mandarin epicarp	β-car	1397–1417 (DW)	[85]
Melon peels	β-car	67–915 (DW)	[86]
	β-crypto	3–49 (DW)	
Atlantic shrimp cooked shell	Ast	57.3–284.5 (DW)	[87]
Grape canes	Lut; β-car	0.3–2.4 (DW)	[88]
Peels and pulp of persimmon	β-crypto	6500–167,000 (DW)	[89]
	β-car	6900–45,000 (DW)	
Pressed palm fibers	α-car	142–305 (DW)	[90]
	β-car	317–713 (DW)	
Mango peel	α–/β-car; crypto	5600 (β-car) (DW)	[91]
Skin and seeds of tomatoes	Lyc	3.8–166.4 (DW)	[92]
	β-car	0.6–26.4 (DW)	
	Lut	0.8–10.8 (DW)	
Carrot by-products	β-car	230 (FW)	[93]
Carrot juice processing waste	β-car	240 (DW)	[94]
Tomato peels and seeds	Lyc	410 (P);28 (S) (FW)	[95]
	β-car	31 (P); 5.2 (S) (FW)	
Abbreviations: P: peel; S: seeds. Carotenoids: ast: astaxanthin, α–/β-car: α–/β-carotene, β-crypto: β-cryptoxanthin, lut: lutein, lyc: lycopene. FW – Fresh weight, DW – Dry weight.			

2.4 Macroalgae

In the last decades, macroalgae have been pointed out as a promising source of carotenoids. These photosynthetic organisms contain high amounts of pigments involved in light absorption for nourishment. However, they also have a secondary role related with damage protection from UV exposition. The main xanthophylls found in macroalgae include fucoxanthin, lutein, or zeaxanthin, being fucoxanthin the most abundant one, while β-carotene stands out from carotenes (Table 4). The main advantage of using macroalgae, is that invasive species can be used as an alternative source of carotenoids.

Table 4. Quantitative and qualitative analysis of carotenoids content in different macroalgae species.

Species	Most abundant Carotenoids	Content (mg/g)	Ref.
Cystoseira sp.	Fuco	2.0–3.5 (DW)	[96]
Dictyota sp.	Fuco	0.4–6.4 (DW)	[97]
Eisenia bicyclis	Fuco	0.42 (DW)	[98]
Fucus serratus	Fuco, Lut	5.2 (DW) 0.3 (DW)	[99]
Laminaria digitata		1.4 (DW) 0.1 (DW)	
Himanthalia elongata	Fuco	18.6 (DW)	[100]
Hypnea musciformis	β-car, lut, zea	0.0029 (TCs, FW)	[101]
Monostroma nitidum	Lut	0.3 (FW)	[102]
Sargassum muticum	Fuco	0.0084 (TCs, DW)	[103]
Abbreviations: DW: dry weight, FW: fresh weight, β-car: β-carotene, fuco: fucoxanthin lut: lutein, TCs: total carotenoids content, zea: zeaxanthin.			

3. Extraction and Production Techniques for Carotenoids Recovery

3.1 Conventional extraction

In the last century, pigment extraction has been performed using solid–liquid extraction with different organic solvents. Extracts were later purified via semipreparative high-performance liquid chromatography (HPLC) [104] or clean up and separation columns using organic solvents such as hexane or dichloromethane [105]. The use of non-polar solvents for carotenoid extraction like petroleum ether or hexane has been linked with toxicity, having a negative impact in the environment in the long term. In addition, in the current legislation regarding the use of these solvents for the production of food ingredients is not allowed. For this reason, in the latest years, novel "greener" extraction processes have been developed for pigment recovery, including supercritical fluid extraction (SFE), pressurized liquid extraction (PLE), ultrasonic assisted extraction (UAE) and microwave assisted extraction (MAE) (Table 5). Implementing these techniques improved, among other things, extraction times, yields and solvent usage [114].

Table 5. Novel extraction techniques to efficiently recover carotenoids from natural sources.

Source	Carotenoids	Conditions	Recovery (μg/g)	Ref.
SFE				
Dunaliella salina	β-car	CO_2, 60 °C, 300 bar	15,000 (DW)	[106]
Hachiyakaki sp. (Pe)	α–/β-car, β-crypto, lyc, lut, zea	CO_2 + EtOH, 30 MPa	392 (TCs)	[107]
Rosmarinus officinalis (L)	Carotenoids	CO_2 + EtOH, 25°C, 20 min, 20 MPa	47,000–53,000	[108]

Scenedesmus almeriensis	Lut	CO_2, 65°C, 55 MPa	3000 (DW)	[109]
Scenedesmus sp.	Ast, β-car, lut, neo, zea	CO_2 + 10% EtO, 25°C, 20 min, 20 Mpa	73; 60, 436, 671, 90	[110]
Tomato, apricot, peach, pumpkin (Fl, Pe), pepper (Fl, wastes)	β-car, lut, lyc	CO_2/EtOH, 59°C, 30 min, 350 bar	88–100% β-car	[111]
PLE				
Carrot by-products	β-car	EtOH 99%, 60–180°C, 5 min, 50 bar, 1–5 cycles of 2 min	120–230 (FW) (soft soggy carrots) 80–190 (FW) (orange carrots)	[93]
Diospyros kaki, P. armeniaca	β-crypto, β-car, lut, zea	MeOH: THF 2:8 (v:v), 40 °C, 5 min, 103 bar	Kaki: β-crypto ≤29, lut ≤ 13, zea ≤ 18 Apricot: ≤48 (β-car)	[40]
Eisenia bicyclis	Fucoxanthin	90% EtOH, 110°C, 5 min, 1500 psi	420	[98]
Porphyridium cruentum	β-car, zea	125°C, 20 min, 10.5 MPa	Zea ≤14,000 β-car ≤8000	[112]
UAE				
Dark red tomato	Lyc	EtAc: tomato paste 8:1 (V/W), 86°C, 29 min	89,000	[113]
MAE				
Carrot juice wastes	β-car	Flaxseed oil: wastes 8:1 g/g, 165 W, 9.4 min	775,000	[94]

Abbreviations: DW: dry weight, EtOH: ethanol, EtAc: ethyl acetate, Fl: flesh, FW: fresh weight, L: leaves, MeOH: methanol, Pe: peels, THF: tetrahydrofuran. Carotenoids: ast: astaxanthin, α–/β-car: β-carotene, β-crypto: β-cryptoxanthin, lyc: lycopene, lut: lutein, neo: neoxanthin, TCs: total carotenoids, zea: zeaxanthin.

3.2 Novel techniques

Supercritical fluid extraction (SFE) emerged in 80s decade as a promising alternative to conventional organic extractions [115]. This is a process where a compound is separated from its matrix making use of the unique properties of supercritical fluids as solvents, being CO_2 the most commonly used. Supercritical fluid technology applies pressures and temperatures above the critical point of the extracting solvent, leading to a balanced state between liquid and gas phases. This balance confers low viscosity, high diffusivity, enhanced solubility and no surface tension, facilitating mass transfer [116]. However, this process involves a high cost, due to high temperatures and pressures requirements. Moreover, CO_2 only dissolves non-polar molecules, although using a co solvent overcomes this issue, being ethanol the most employed [107, 108].

As well as in SFE, pressurized liquid extraction (PLE), also called accelerated solvent extraction (ASE), makes use of high temperature and pressure, although along with a liquid solvent to accelerate the extraction of specific analytes from solid matrices. In this system, pressure is high enough to keep the solvent liquid without hampering extraction performance. However, extraction time, temperature, solvent type and volume have influence on extraction performance, especially temperature and solvent type. Temperature range is mostly comprised from 40 to 180°C and it has been seen that the use of *Generally Recognized as Safe* (GRAS) solvents, like water, ethanol or its mixtures enhance extraction efficiencies [117]. Nevertheless, other more non-polar solvent mixtures like methanol: tetrahydrofuran can also be used [40].

Ultrasonic assisted extraction (UAE) also emerged as a novel technique, which employs using ultrasonic waves that propagate causing the implosion of bubbles, phenomenon known as cavitation. This perturbation leads to a diffusion of the solute from the porous matrix to the solvent. Nowadays, UAE is used for extracting various compounds including carotenoids from a wide diversity of matrices, such as macroalgae, microalgae and plants. This technique is environment-friendly, simple, cheap and efficient, reporting high yields when compared to conventional techniques, although the reproducibility of the samples is jeopardized by equipment's aging [118].

Microwave assisted extraction (MAE) is a relatively new extraction technique that combines microwave and traditional solvent extraction. Since the late 1980s, MAE has become one of the most popular and cost-effective extraction methods [81]. This technique is based on the application of microwaves for heating both solvents and matrices, increasing the kinetic of the extraction. Compared to conventional and novel (SFE and PLE) techniques, MAE reduced extraction time and solvent usage, leading to higher extraction rates and reduced costs [94].

4. Carotenoids' Incorporation Into Food: Reported and Future Applications

Color is an important sensory attribute associated with safety and nutritional values of food, reason why, in the last years, consumer awareness regarding the use of synthetic food coloring has been increased. In order to develop a more natural food industry, natural pigment demand has raised, as is the case for carotenoids. These pigments are used for their coloring properties as well as for their antioxidant potential and biological functions. Carotenoids can be either applied directly into food matrices like beverages or pasta, among others [119, 120], or indirectly, into animal feeding to improve pigmentation of final products as in eggs or fish flesh [1, 29, 121]. Moreover, carotenoids have pointed out as promising ingredients in active packaging films. Their inclusion in protective films can improve the storing properties of the package, extending the shelf-life of the product, as well as transferring carotenoids' nutritional values [122, 123]. Several applications of carotenoids in the food industry have been collected in Table 6.

Table 6. Carotenoids applications in food industry.

Carotenoid	Origin	Application	Properties	Ref.
β-Carotene	Fruits & vegetables	Free and encapsulated	Yellow colorant and antioxidant	[120]
		Incorporated into polymer materials	Antioxidant, O_2 and light barrier	[123]
α-Carotene		Free and encapsulated	Yellow colorant, anti-carcinogenic and antioxidant	[120, 124]
α-Tocopherol	High fat vegetables	Incorporated into polymer materials	Antioxidant	[123]
Astaxanthin	Marigold flower	Incorporated into packaging material	Fish feeding for antioxidation and flesh coloring	[125]
Bixin	Annatto seeds	Incorporated into polymer materials	Antioxidant, O_2 and light barrier	[123]
Canthaxanthin	Mushrooms	Alginate-pectin microencapsulation	Red colorant and antioxidant	[126]
Cryptoxanthin	Mandarin, papaya, orange	Free and encapsulated	Orange colorant and antioxidant	[120]
Lutein	Green vegetables	Encapsulation in food	Eye protection against AMD development or cataracts. Anticancer	[124]
Lycopene	Tomato, watermelon, pink grapefruit, guava	Free and encapsulated	Red colorant, eye UV-protection, antioxidant	[120, 123, 124]
		Incorporated into polymer materials	Antioxidant, O_2 and light barrier	
Zeaxanthin	Mandarin, papaya, orange	Orange colorant	Eye protection against macular degeneration and cataracts	[120, 124]

Since the late 1980s, carotenoids implementation into food has significantly increased. Among all, β-carotene is the most applied one, being used for coloring oils and butters, providing a yellowish color. In addition, it has been also used to fortify different food matrices for its provitamin A activity [127]. Apart from β-carotene, other carotenoids have been incorporated as free molecules into food matrices (Table 6). However, the direct application of these natural pigments is limited by their low stability, so micro- and nanoencapsulation technologies have been applied. Multiple encapsulation technologies including spray or freeze drying, emulsion, spray chilling, extrusion coating, liposome entrapment, coacervation and ionic gelation [128] have been applied to improve solubility, chemical stability and bioavailability of pigments, as well as for masking unpleasant organoleptic properties [129]. Most of these technologies have been applied to encapsulate carotenoids, generally on a nanometric scale (≤100 nm). The type of encapsulation materials used for food applications have to be food-graded biopolymers such carbohydrates or gums (Persian gum), proteins (gelatin or whey), and animal or vegetal lipids [22, 26, 130]. Emulsion is also a prominent encapsulation processes, which results in an improved bio accessibility and bioavailability [131]. Lutein emulsions, β-carotene oil-in-water emulsions or microcapsules containing lycopene are just some examples of microencapsulation found in the food industry to improve the stability, bio accessibility and bioavailability of these pigments [129, 132, 133]. Similarly, multiple studies support carotenoids' nanoencapsulation [134, 135].

A different way to incorporate natural pigments in human food is through animal feeding. By doing so, pigments get incorporated in foodstuff such as in fish flesh or eggs, giving a characteristic pigmentation and an increased nutritional value that will be further transferred to humans or animals [121, 136]. One of the main industries where carotenoids have been implemented is aquaculture. Fish factories have been making use of pigments such as β-carotene as an important source of provitamin A, which has been shown to improve the antioxidant capacity and immune system of fish, enhancing growth and preventing lipid peroxidation [137]. In fact, in aquaculture, different biological sources of xanthophylls such as green microalgae, yeast, krill, or crab waste have been utilized as feeding supplements. This complementary pigmentation enhances the nutritional value of fish products by providing strong antioxidant activity and higher amounts of provitamin A [1]. Other industry where pigments are widely use is poultry. EU approved egg yolk and poultry tissues pigmentations with yellow and red carotenoids, including lutein, zeaxanthin, β-cryptoxanthin, violaxanthin and capsanthin [136].

Natural pigments can also be incorporated into packaging materials to improve food preservation. Carotenoids such as lycopene or β-carotene prevent color alterations due to oxidation processes and UV-induced damage, providing stability to packaging polymers [123]. Besides, pigment migration from active packaging into food matrices has been reported, transferring the beneficial properties. As could be seen in Table 6, several carotenoids such have been included in active packaging, achieving promising results.

5. Conclusions

Synthetic pigments have been frequently used as food additives to improve food appearance since colorful products have been associated with healthy and high-quality properties by consumers. However, tendency has slowly shifted towards a stronger presence of natural ingredients due to a raising concern about the negative side effects associated with synthetic molecules. In this context, carotenoids have come up as an attractive replacement of synthetic pigments, being found in multiple sources, like plants, algae, fungi, microorganisms and by-products. Moreover, carotenoids have been linked with diverse beneficial properties, such as antioxidant, prevention of degenerative diseases, cancer and stimulation of the immune system. For all these reasons, carotenoids have caught the attention of many industries, including food, nutraceutical and cosmetic industries.

In order to extract these pigments, novel technologies emerged to improve the extraction rates of traditional techniques, mostly based on maceration. Among these new strategies, SFE and PLE highlight. Equipment may result into an initial economic expense, but they offer satisfactory extraction rates while minimizing solvent usage and experimental times.

Regarding food industry, carotenoids have been widely used for their application into food matrices or as part of packaging materials. Their inclusion as food additives or feed supplements for animals is the most extended and explored application, improving the organoleptic properties and nutritional values, aiming for a higher commercial acceptance. Besides, carotenoids have also been used as ingredients for active packaging films to extend products' shelf-life. Regardless the matrix of inclusion, natural carotenoids have been incorporated as free molecules or encapsulated. This last strategy prolongs the stability and bio accessibility of carotenoids, protecting core ingredients from chemical degradation.

Furthermore, due to their extensive bioactivities, carotenoids are very useful to formulate new cosmetic ingredients. Besides, its antioxidant properties that can benefit the skin and promote skin regeneration and healthy aging, carotenoids also mitigate the harmful effects of UV radiation, which makes them excellent candidates for their application in cosmetic formulations as preservatives with photoprotective, antioxidant and anti-aging properties.

References

1. Breithaupt DE. Modern application of xanthophylls in animal feeding - a review. Trends in Food Science and Technology. 2007;18:501-506
2. Domonkos I, Kis M, Gombos Z, Ughy B. Carotenoids, versatile components of oxygenic photosynthesis. Progress in Lipid Research. 2013;52:539-561
3. Rodriguez-Concepcion M, Avalos J, Bonet ML, Boronat A, Gomez-Gomez L, Hornero-Mendez D, et al. A global perspective on carotenoids: Metabolism, biotechnology, and benefits for nutrition and health. Progress in Lipid Research. 2018;70:62-93
4. Rodriguez DA. Guide to Carotenoid Analysis in Foods. Washington, D.C: ILSI Press; 2001 ISBN 1578810728
5. Nagarajan J, Ramanan RN, Raghunandan ME, Galanakis CM, Krishnamurthy NP. Carotenoids. In: Galanakis CM, editor. Nutraceutical and Functional Food Components: Effects of Innovative Processing Techniques. Cambridge, Massachusetts: Academic Press; 2017. pp. 259-296 ISBN 9780128052570
6. Aziz E, Batool R, Akhtar W, Rehman S, Shahzad T, Malik A, et al. Xanthophyll: Health benefits and therapeutic insights. Life Sciences. 2020;240:117104
7. Bhatt T, Patel K. Carotenoids: Potent to prevent diseases review. Natural Products and Bioprospecting. 2020;10:109-117
8. Maoka T. Carotenoids as natural functional pigments. Journal of Natural Medicines. 2020;74:1-16
9. Data, Reports and. 2019. "Carotenoids Market." Reports and Data. 2016-2027
10. Jeevanandam J, Choudhary V, Selvam JD, Danquah MK. The Bioeconomy of Production of Microalgal Pigments. In: Jacob-Lopes E, Queiroz M, Zepka L, editors. Pigments from Microalgae Handbook. Cham: Springer; 2020. pp. 325-362 ISBN 9783030509705
11. Chiu HF, Shen YC, Venkatakrishnan K, Wang CK. Food for eye health: Carotenoids and omega-3 fatty acids. Encyclopedia of Food Chemistry. Elsevier. 2018:313-322. DOI: 10.1016/B978-0-08-100596-5.21740-X

12. European Parliament and Council. Directive 94/36/EC of 30 June 1994 on Colours for Use in Foodstuffs. European Union: European Parliament, Council of the European Union; 1994. pp. 13-29

13. Mezzomo N, Ferreira SRS. Carotenoids functionality, sources, and processing by supercritical technology: A review. Journal of Chemistry. 2016;2016:3164312

14. Von Lintig J. Provitamin A metabolism and functions in mammalian biology. The American Journal of Clinical Nutrition. 2012;96:1234-1244

15. Domínguez R, Gullón P, Pateiro M, Munekata PES, Zhang W, Lorenzo JM. Tomato as potential source of natural additives for meat industry. A review. Antioxidants. 2020;9:73

16. Rafi MM, Kanakasabai S, Reyes MD, Bright JJ. Lycopene modulates growth and survival associated genes in prostate cancer. The Journal of Nutritional Biochemistry. 2013;24:1724-1734

17. Mirahmadi M, Azimi-Hashemi S, Saburi E, Kamali H, Pishbin M, Hadizadeh F. Potential inhibitory effect of lycopene on prostate cancer. Biomedicine & Pharmacotherapy. 2020;129:110459

18. Costa-Rodrigues J, Pinho O, Monteiro PRR. Can lycopene be considered an effective protection against cardiovascular disease? Food Chemistry. 2018;245:1148-1153

19. Baker R, Günther C. The role of carotenoids in consumer choice and the likely benefits from their inclusion into products for human consumption. Trends in Food Science and Technology. 2004;15:484-488

20. Mohd Hatta, F.A.; Othman, R. Carotenoids as potential biocolorants: A case study of astaxanthin recovered from shrimp waste. In Carotenoids: Properties, Processing and Applications; Galanakis, C.M., Ed.; Academic Press: Cambridge, Massachusetts, 2020; pp. 289-325 ISBN 9780128170670

21. Breithaupt DE, Wirt U, Bamedi A. Differentiation between lutein monoester regioisomers and detection of lutein diesters from marigold flowers (Tagetes erecta L.) and Several fruits by liquid chromatography-mass spectrometry. Journal of Agricultural and Food Chemistry. 2002;50:66-70

22. Zhang J, Jia G, Wanbin Z, Minghao J, Wei Y, Hao J, et al. Nanoencapsulation of zeaxanthin extracted from Lycium barbarum L. by complex coacervation with gelatin and CMC. Food Hydrocolloids. 2021;112:106280

23. Ambati RR, Moi PS, Ravi S, Aswathanarayana RG. Astaxanthin: Sources, extraction, stability, biological activities and its commercial applications - A review. Marine Drugs. 2014;12:128-152

24. Solymosi K, Latruffe N, Morant-Manceau A, Schoefs B. Food colour additives of natural origin. In: Scotter MJ, editor. Colour Additives for Foods and Beverages. Sawston, Cambridge: Woodhead Publishing; 2015. pp. 4-34 ISBN 9781782420200

25. Khanafari A, Saberi A, Azar M, Vosooghi G, Jamili S, Sabbaghzadeh B. Extraction of astaxanthin esters from shrimp waste by chemical and microbial methods. Journal of Environmental Health Science & Engineering. 2007;4:93-98

26. Bassijeh A, Ansari S, Hosseini SMH. Astaxanthin encapsulation in multilayer emulsions stabilized by complex coacervates of whey protein isolate and Persian gum and its use as a natural colorant in a model beverage. Food Research International. 2020;137:109689
27. Lorenz RT, Cysewski GR. Commercial potential for Haematococcus microalgae as a natural source of astaxanthin. Trends in Biotechnology. 2000;18:160-167
28. Tanumihardjo SA. Carotenoids: Health effects. In: Encyclopedia of Human Nutrition. Madison, WI, USA: Elsevier Inc.; 2012; Vol. 1-4. pp. 292-297 ISBN 9780123848857
29. Liu YQ, Davis CR, Schmaelzle ST, Rocheford T, Cook ME, Tanumihardjo SA. β-cryptoxanthin biofortified maize (Zea mays) increases β-cryptoxanthin concentration and enhances the color of chicken egg yolk. Poultry Science. 2012;91:432-438
30. Ozaki K, Okamoto M, Fukasawa K, Iezaki T, Onishi Y, Yoneda Y, et al. Daily intake of β-cryptoxanthin prevents bone loss by preferential disturbance of osteoclastic activation in ovariectomized mice. Journal of Pharmacological Sciences. 2015;129:72-77
31. Peng J, Yuan J-PP, Wu C-FF, Wang J-HH. Fucoxanthin, a marine carotenoid present in brown seaweeds and diatoms: Metabolism and bioactivities relevant to human health. Marine Drugs. 2011;9:1806-1828
32. Wang L-J, Fan Y, Parsons RL, Hu G-R, Zhang P-Y, Li F-L. A rapid method for the determination of fucoxanthin in diatom. Marine Drugs. 2018;16:33
33. Lourenço-Lopes C, Garcia-Oliveira P, Carpena M, Fraga-Corral M, Jimenez-Lopez C, Pereira AG, et al. Scientific approaches on extraction , purification and stability for the commercialization of fucoxanthin recovered from brown algae. Food. 2020;9:1113
34. Beppu F, Niwano Y, Tsukui T, Hosokawa M, Miyashita K. Single and repeated oral dose toxicity study of fucoxanthin (FX), a marine carotenoid, in mice. The Journal of Toxicological Sciences. 2009;34:501-510
35. Ma Z, Zhao Y, Khalid N, Shu G, Neves MA, Kobayashi I, et al. Comparative study of oil-in-water emulsions encapsulating fucoxanthin formulated by microchannel emulsification and high-pressure homogenization. Food Hydrocolloids. 2020;108:105977
36. Yamamoto K, Ishikawa C, Katano H, Yasumoto T, Mori N. Fucoxanthin and its deacetylated product, fucoxanthinol, induce apoptosis of primary effusion lymphomas. Cancer Letters. 2011;300:225-234
37. Kanda H, Kamo Y, Machmudah S, Wahyudiono; Goto, M. Extraction of fucoxanthin from raw macroalgae excluding drying and cell wall disruption by liquefied dimethyl ether. Marine Drugs. 2014;12:2383-2396
38. Oliyaei N, Moosavi-Nasab M, Tamaddon AM, Fazaeli M. Encapsulation of fucoxanthin in binary matrices of porous starch and halloysite. Food Hydrocolloids. 2020;100:105458
39. Wang C, Chen X, Nakamura Y, Yu C, Qi H. Fucoxanthin activities motivate its nano/micro-encapsulation for food or nutraceutical application: A review. Food & Function. 2020;11:9338-9358
40. Zaghdoudi K, Pontvianne S, Framboisier X, Achard M, Kudaibergenova R, Ayadi-Trabelsi M, et al. Accelerated solvent extraction of carotenoids from: Tunisian Kaki (Diospyros kaki L.), peach (Prunus persica L.) and apricot (Prunus armeniaca L.). Food Chemistry. 2015;184:131-139

41. Wimalasiri D, Brkljača R, Piva TJ, Urban S, Huynh T. Comparative analysis of carotenoid content in Momordica cochinchinensis (Cucurbitaceae) collected from Australia, Thailand and Vietnam. Journal of Food Science and Technology. 2017;54:2814-2824

42. Müller-Maatsch J, Sprenger J, Hempel J, Kreiser F, Carle R, Schweiggert RM. Carotenoids from gac fruit aril (Momordica cochinchinensis [Lour.] Spreng.) are more bioaccessible than those from carrot root and tomato fruit. Food Research International. 2017;99:928-935

43. Liu Y, Zeng S, Sun W, Wu M, Hu W, Shen X, et al. Comparative analysis of carotenoid accumulation in two goji (Lycium barbarum L. and L. ruthenicum Murr.) fruits. BMC Plant Biology. 2014;14:269

44. Amah D, Alamu E, Adesokan M, van Biljon A, Maziya-dixon B, Swennen R, et al. Variability of carotenoids in a Musa germplasm collection and implications for provitamin A biofortification. Food Chemistry X. 2019;2:100024

45. Mbabazi R, Harding R, Khanna H, Namanya P, Arinaitwe G, Tushemereirwe W, et al. Pro-vitamin A carotenoids in East African highland banana and other Musa cultivars grown in Uganda. Food Science & Nutrition. 2020;8:311-321

46. Giuffrida D, Cacciola F, Mapelli-Brahm P, Stinco CM, Dugo P, Oteri M, et al. Free carotenoids and carotenoids esters composition in Spanish orange and mandarin juices from diverse varieties. Food Chemistry. 2019;300:125139

47. Maldonado-Celis ME, Yahia EM, Bedoya R, Landázuri P, Loango N, Aguillón J, et al. Chemical Composition of Mango (Mangifera indica L.) Fruit: Nutritional and Phytochemical Compounds. Frontiers in Plant Science. 2019;10:1073

48. Brasili E, Chaves DFS, Xavier AAO, Mercadante AZ, Hassimotto NMA, Lajolo FM. Effect of pasteurization on flavonoids and carotenoids in citrus sinensis (L.) Osbeck cv. Cara Cara' and Bahia' Juices. Journal of Agricultural and Food Chemistry. 2017;65:1371-1377

49. Chan-León AC, Estrella-Maldonado H, Dubé P, Fuentes Ortiz G, Espadas-Gil F, Talavera May C, et al. The high content of β-carotene present in orange-pulp fruits of Carica papaya L. is not correlated with a high expression of the CpLCY-β2 gene. Food Research International. 2017;100:45-56

50. Farina V, Tinebra I, Perrone A, Sortino G, Palazzolo E, Mannino G, et al. Physicochemical, nutraceutical and sensory traits of six papaya (Carica papaya L.) cultivars grown in greenhouse conditions in the mediterranean climate. Agronomy. 2020;10:501

51. Moreira-Rodríguez M, Nair V, Benavides J, Cisneros-Zevallos L, Jacobo-Velázquez DA. UVA, UVB light, and methyl jasmonate, alone or combined, redirect the biosynthesis of glucosinolates, phenolics, carotenoids, and chlorophylls in broccoli sprouts. International Journal of Molecular Sciences. 2017;18:2330

52. Behsnilian D, Mayer-Miebach E. Impact of blanching, freezing and frozen storage on the carotenoid profile of carrot slices (Daucus carota L. cv. Nutri Red). Food Control. 2017;73:761-767

53. Fuentes P, Pizarro L, Moreno JC, Handford M, Rodriguez-Concepcion M, Stange C. Light-dependent changes in plastid differentiation influence carotenoid gene expression and accumulation in carrot roots. Plant Molecular Biology. 2012;79:47-59

54. Mageney V, Baldermann S, Albach DC. Intraspecific Variation in Carotenoids of Brassica oleracea var. sabellica. Journal of Agricultural and Food Chemistry. 2016;64:3251-3257

55. Kim HJ, Fonseca JM, Choi JH, Kubota C. Effect of methyl jasmonate on phenolic compounds and carotenoids of romaine lettuce (Lactuca sativa L.). Journal of Agricultural and Food Chemistry. 2007;55:10366-10372

56. Edelenbos M, Christensen LP, Grevsen K. HPLC determination of chlorophyll and carotenoid pigments in processed green pea cultivars (Pisum sativum L.). Journal of Agricultural and Food Chemistry. 2001;49:4768-4774

57. Mínguez-Mosquera MI, Hornero-Méndez D. Comparative Study of the Effect of Paprika Processing on the Carotenoids in Peppers (Capsicum annuum) of the Bola and Agridulce Varieties. Journal of Agricultural and Food Chemistry. 1994;42:1555-1560

58. Bunea A, Andjelkovic M, Socaciu C, Bobis O, Neacsu M, Verhé R, et al. Van Total and individual carotenoids and phenolic acids content in fresh, refrigerated and processed spinach (Spinacia oleracea L.). Food Chemistry. 2008;108:649-656

59. Žilić S, Serpen A, Akilliog⊠lu G, Gökmen V, Vančetović J. Phenolic compounds, carotenoids, anthocyanins, and antioxidant capacity of colored maize (Zea mays L.) kernels. Journal of Agricultural and Food Chemistry. 2012;60:1224-1231

60. Abushita AA, Daood HG, Biacs PA. Change in Carotenoids and Antioxidant Vitamins in Tomato as a Function of Varietal and Technological Factors. Journal of Agricultural and Food Chemistry. 2000;48:2075-2081

61. Fernandes L, Ramalhosa E, Pereira JA, Saraiva JA, Casal S. Borage, camellia, centaurea and pansies: Nutritional, fatty acids, free sugars, vitamin E, carotenoids and organic acids characterization. Food Research International. 2020;132:109070

62. Iijima L, Kishimoto S, Ohmiya A, Yagi M, Okamoto E, Miyahara T, et al. Esterified carotenoids are synthesized in petals of carnation (Dianthus caryophyllus) and accumulate in differentiated chromoplasts. Scientific Reports. 2020;10:1-12

63. Abdel-Aal ESM, Rabalski I. Composition of lutein ester regioisomers in marigold flower, dietary supplement, and herbal tea. Journal of Agricultural and Food Chemistry. 2015;63:9740-9746

64. Niizu PY, Rodriguez-Amaya DB. Flowers and leaves of Tropaeolum majus L. as rich sources of lutein. Journal of Food Science. 2005;70:605-609

65. González-Barrio R, Periago MJ, Luna-Recio C, Garcia-Alonso FJ, Navarro-González I. Chemical composition of the edible flowers, pansy (Viola wittrockiana) and snapdragon (Antirrhinum majus) as new sources of bioactive compounds. Food Chemistry. 2018;252:373-380

66. Foong LC, Loh CWL, Ng HS, Lan JCW. Recent development in the production strategies of microbial carotenoids. World Journal of Microbiology and Biotechnology. 2021;37:1-11

67. Fazeli MR, Tofighi H, Samadi N, Jamalifar H. Effects of salinity on β-carotene production by Dunaliella tertiolecta DCCBC26 isolated from the Urmia salt lake, north of Iran. Bioresource Technology. 2006;97:2453-2456

68. Britton G, Liaaen-Jensen S, Pfander H. Carotenoids. Volume 5: Nutirition and Health. Basel: Birkhäuser; 2009 ISBN 9783764375003

69. Mazumdar N, Gopalakrishnan KK, Visnovsky G, Novis PM. A novel alpine species of Haematococcus (Chlamydomonadales: Chlorophyta) from New Zealand. New Zealand Journal of Botany. 2018;56:216-226

70. Valduga E, Tatsch PO, Tiggemann L, Treichel H, Toniazzo G, Zeni J, et al. Carotenoids production: Microorganisms as source of natural dyes. Quimica Nova. 2009;32:2429-2436

71. Macías-Sánchez MD, Fernandez-Sevilla JM, Fernández FGA, García MCC, Grima EM. Supercritical fluid extraction of carotenoids from Scenedesmus almeriensis. Food Chemistry. 2010;123:928-935

72. Vila E, Hornero-Méndez D, Azziz G, Lareo C, Saravia V. Carotenoids from heterotrophic bacteria isolated from Fildes Peninsula, King George Island, Antarctica. Biotechnology Reports. 2019;21:e00306

73. Heider SAE, Peters-Wendisch P, Netzer R, Stafnes M, Brautaset T, Wendisch VF. Production and glucosylation of C50 and C40 carotenoids by metabolically engineered Corynebacterium glutamicum. Applied Microbiology and Biotechnology. 2014;98:1223-1235

74. Naz T, Nosheen S, Li S, Nazir Y, Mustafa K, Liu Q, et al. Comparative analysis of β-carotene production by mucor circinelloides strains CBS 277.49 and WJ11 under light and dark conditions. Metabolites. 2020;10:38

75. Ramírez J, Gutierrez H, Gschaedler A. Optimization of astaxanthin production by Phaffia rhodozyma through factorial design and response surface methodology. Journal of Biotechnology. 2001;88:259-268

76. Liu YS, Wu JY, Ho KP. Characterization of oxygen transfer conditions and their effects on Phaffia rhodozyma growth and carotenoid production in shake-flask cultures. Biochemical Engineering Journal. 2006;27:331-335

77. da Silva SRS, Stamford TCM, Albuquerque WWC, Vidal EE, Stamford TLM. Reutilization of residual glycerin for the produce β-carotene by Rhodotorula minuta. Biotechnology Letters. 2020;42:437-443

78. Buzzini P, Martini A. Production of carotenoids by strains of Rhodotorula glutinis cultured in raw materials of agro-industrial origin. Bioresource Technology. 2000;71:41-44

79. Kot AM, Błazejak S, Gientka I, Kieliszek M, Bryś J. Torulene and torularhodin: "New" fungal carotenoids for industry? Microbial Cell Factories. 2018;17:1-14

80. Buzzini P, Martini A, Gaetani M, Turchetti B, Pagnoni UM, Davoli P. Optimization of carotenoid production by Rhodotorula graminis DBVPG 7021 as a function of trace element concentration by means of response surface analysis. Enzyme and Microbial Technology. 2005;36:687-692

81. Delazar A, Nahar L, Hamedeyazdan S, Sarker SD. Microwave-assisted extraction in natural products isolation. Methods in Molecular Biology. 2012;864:89-115

82. García-Oliveira P, Fraga-Corral M, Pereira AG, Prieto MA, Simal-Gandara J. Solutions for the sustainability of the food production and consumption system. Critical Reviews in Food Science and Nutrition. Bellwether Publishing, Ltd., 2020:1-17. DOI: 10.1080/10408398.2020.1847028

83. Jimenez-Lopez C, Fraga-Corral M, Carpena M, García-Oliveira P, Echave J, Pereira AG, et al. Agriculture waste valorisation as a source of antioxidant phenolic compounds within a circular and sustainable bioeconomy. Food & Function. 2020;11:48-53

84. Noronha Matos KA, Praia Lima D, Pereira Barbosa AP, Zerlotti Mercadante A, Campos Chisté R. Peels of tucumã (Astrocaryum vulgare) and peach palm (Bactris gasipaes) are by-products classified as very high carotenoid sources. Food Chemistry. 2019;272:216-221

85. Ordóñez-Santos LE, Esparza-Estrada J, Vanegas-Mahecha P. Ultrasound-assisted extraction of total carotenoids from mandarin epicarp and application as natural colorant in bakery products. LWT. 2021;139:110598

86. Gómez-García R, Campos DA, Oliveira A, Aguilar CN, Madureira AR, Pintado M. A chemical valorisation of melon peels towards functional food ingredients: Bioactives profile and antioxidant properties. Food Chemistry. 2021;335:127579

87. Dave D, Liu Y, Pohling J, Trenholm S, Murphy W. Astaxanthin recovery from Atlantic shrimp (Pandalus borealis) processing materials. Bioresource Technology Reports. 2020;11:100535

88. Ferreyra SG, Antoniolli A, Bottini R, Fontana A. Bioactive compounds and total antioxidant capacity of cane residues from different grape varieties. Journal of the Science of Food and Agriculture. 2020;100:376-383

89. Gea-Botella S, Agulló L, Martí N, Martínez-Madrid MC, Lizama V, Martín-Bermudo F, et al. Carotenoids from persimmon juice processing. Food Research International. 2020;141:109882

90. Cardenas-Toro FP, Alcázar-Alay SC, Coutinho JP, Godoy HT, Forster-ized liquid extraction and low-pressure solvent extraction of carotenoids from pressed palm fiber: Experimental and economical evaluation. Food and Bioproducts Processing. 2015;94:90-100

91. Garcia-Mendoza MP, Paula JT, Paviani LC, Cabral FA, Martinez-Correa HA. Extracts from mango peel by-product obtained by supercritical CO2 and pressurized solvent processes. LWT- Food Science and Technology. 2015;62:131-137

92. Strati IF, Oreopoulou V. Recovery and isomerization of carotenoids from tomato processing by-products. Waste and Biomass Valorization. 2016;7:843-850

93. Mustafa A, Trevino LM, Turner C. Pressurized hot ethanol extraction of carotenoids from carrot by-products. Molecules. 2012;17:1809-1818

94. Elik A, Yanık DK, Göğüş F. Microwave-assisted extraction of carotenoids from carrot juice processing waste using flaxseed oil as a solvent. LWT. 2020;123:109100

95. Kehili M, Schmidt LM, Reynolds W, Zammel A, Zetzl C, Smirnova I, et al. Biorefinery cascade processing for creating added value on tomato industrial by-products from Tunisia. Biotechnology for Biofuels. 2016;9:1-12

96. Susanto E, Fahmi AS, Abe M, Hosokawa M, Miyashita K. Lipids, fatty acids, and fucoxanthin content from temperate and tropical brown seaweeds. Aquatic Procedia. 2016;7:66-75

97. Agatonovic-Kustrin S, Morton DW. High-performance thin-layer chromatography HPTLC-direct bioautography as a method of choice for alpha-amylase and antioxidant activity evaluation in marine algae. Journal of Chromatography. A. 2017;1530:197-203

98. Shang YF, Kim SM, Lee WJ, Um BH. Pressurized liquid method for fucoxanthin extraction from Eisenia bicyclis (Kjellman) Setchell. Journal of Bioscience and Bioengineering. 2011;111:237-241

99. Heffernan N, Smyth TJ, FitzGerald RJ, Vila-Soler A, Mendiola J, Ibáñez E, et al. Comparison of extraction methods for selected carotenoids from macroalgae and the assessment of their seasonal/spatial variation. Innovative Food Science and Emerging Technologies. 2016;37:221-228

100. Rajauria G, Foley B, Abu-Ghannam N. Characterization of dietary fucoxanthin from Himanthalia elongata brown seaweed. Food Research International. 2017;99:995-1001

101. Schmidt ÉC, Pereira B, dos Santos RW, Gouveia C, Costa GB, Faria GSM, et al. Responses of the macroalgae Hypnea musciformis after in vitro exposure to UV-B. Aquatic Botany. 2012;100:8-17

102. Kanda H, Wahyudiono, Machmudah S, Goto M. Direct extraction of lutein from wet macroalgae by liquefied dimethyl ether without any pretreatment. ACS Omega. 2020;5:24005-24010

103. Vieira FA, Guilherme RJR, Neves MC, Rego A, Abreu MH, Coutinho JAP, et al. Recovery of carotenoids from brown seaweeds using aqueous solutions of surface-active ionic liquids and anionic surfactants. Separation and Purification Technology. 2018;196:300-308

104. Rauscher R, Edenharder R, Platt KL. In vitro antimutagenic and in vivo anticlastogenic effects of carotenoids and solvent extracts from fruits and vegetables rich in carotenoids. Mutation Research, Genetic Toxicology and Environmental Mutagenesis. 1998;413:129-142

105. Yokoyama A, Miki W, Izumida H, Shizuri Y. New trihydroxy-keto-carotenoids isolated from an astaxanthin-producing marine bacterium. Bioscience, Biotechnology, and Biochemistry. 1996;60:200-203

106. Macías-Sánchez MD, Mantell C, Rodríguez M, Martínez de la Ossa E, Lubián LM, Montero O. Comparison of supercritical fluid and ultrasound-assisted extraction of carotenoids and chlorophyll a from Dunaliella salina. Talanta. 2009;77:948-952

107. Takahashi M, Watanabe H, Kikkawa J, Masaki OTA, Watanabe M, Sato Y, et al. Carotenoids extraction from Japanese persimmon (Hachiyakaki) peels by supercritical CO_2 with ethanol. Analytical Sciences. 2006;22:1441-1447

108. Lefebvre T, Destandau E, Lesellier E. Sequential extraction of carnosic acid, rosmarinic acid and pigments (carotenoids and chlorophylls) from Rosemary by online supercritical fluid extraction-supercritical fluid chromatography. Journal of Chromatography. A. 2020;1639:461709

109. Mehariya S, Iovine A, Di Sanzo G, Larocca V, Martino M, Leone GP, et al. Supercritical fluid extraction of lutein from Scenedesmus almeriensis. Molecules. 2019;24:1-15

110. Abrahamsson V, Rodriguez-Meizoso I, Turner C. Determination of carotenoids in microalgae using supercritical fluid extraction and chromatography. Journal of Chromatography. A. 2012;1250:63-68

111. de Andrade Lima M, Kestekoglou I, Charalampopoulos D, Chatzifragkou A. Supercritical fluid extraction of carotenoids from vegetable waste matrices. Molecules. 2019;24:446

112. Gallego R, Martínez M, Cifuentes A, Ibáñez E, Herrero M. Development of a green downstream process for the valorization of Porphyridium cruentum biomass. Molecules. 2019;24:15642

113. Lianfu Z, Zelong L. Optimization and comparison of ultrasound/microwave assisted extraction (UMAE) and ultrasonic assisted extraction (UAE) of lycopene from tomatoes. Ultrasonics Sonochemistry. 2008;15:731-737

114. Otero P, Quintana SE, Reglero G, Fornari T, García-Risco MR. Pressurized Liquid Extraction (PLE) as an innovative green technology for the effective enrichment of galician algae extracts with high quality fatty acids and antimicrobial and antioxidant properties. Marine Drugs. 2018;16:156

115. Abbas KA, Mohamed A, Abdulamir AS, Abas HA. A review on supercritical fluid extraction as new analytical method. American Journal of Biochemistry and Biotechnology. 2008;4:345-353

116. Jodhner JCF, Meireles MA d A. Construction of a supercritical fluid extraction (SFE) equipment: Validation using annatto and fennel and extract analysis by thin layer chromatography coupled to image. Food Science and Technology. 2016;36:210-247

117. Rudke AR, Mazzutti S, Andrade KS, Vitali L, Ferreira SRS. Optimization of green PLE method applied for the recovery of antioxidant compounds from buriti (Mauritia flexuosa L.) shell. Food Chemistry. 2019;298:125061

118. Roohinejad S, Nikmaram N, Brahim M, Koubaa M, Khelfa A, Greiner R. Potential of novel technologies for aqueous extraction of plant bioactives. In: Domínguez González H, González Muñoz MJ, editors. Water Extraction of Bioactive Compounds: From Plants to Drug Development. Amsterdam: Elsevier; 2017. pp. 399-419 ISBN 9780128096154

119. Giuffrida D, Zoccali M, Mondello L. Recent developments in the carotenoid and carotenoid derivatives chromatography-mass spectrometry analysis in food matrices. TrAC, Trends in Analytical Chemistry. 2020;132:116047

120. Cheng S-H, Khoo HE, Kong KW, Prasad KN, Galanakis CM. Extraction of carotenoids and applications. In: Galanakis CM, editor. Carotenoids: Properties, Processing and

Applications. Cambridge, Massachusetts: Academic Press; 2020. pp. 259-288 ISBN 9780128170670

121. Moreno JA, Díaz-Gómez J, Fuentes-Font L, Angulo E, Gosálvez LF, Sandmann G, et al. Poultry diets containing (keto)carotenoid-enriched maize improve egg yolk color and maintain quality. Animal Feed Science and Technology. 2020;260:114334

122. Xu J, Wei R, Jia Z, Song R. Characteristics and bioactive functions of chitosan/gelatin-based film incorporated with ε-polylysine and astaxanthin extracts derived from by-products of shrimp (Litopenaeus vannamei). Food Hydrocolloids. 2020;100:105436

123. Stoll L, Rech R, Flôres SH, Nachtigall SMB, de Oliveira Rios A. Poly(acid lactic) films with carotenoids extracts: Release study and effect on sunflower oil preservation. Food Chemistry. 2019;281:213-221

124. Jaswir I, Noviendri D, Hasrini RF, Octavianti F. Carotenoids: Sources, medicinal properties and their application in food and nutraceutical industry. Journal of Medicinal Plant Research: Planta Medica. 2011;5:7119-7131

125. Cardoso LAC, Karp SG, Vendruscolo F, Kanno KYF, Zoz LIC, Carvalho JC. Biotechnological production of carotenoids and their applications in food and pharmaceutical products. In: Carotenoids. Rijeka: InTech; 2017

126. Arab M, Razavi SH, Hosseini SM, Nayebzadeh K, Meybodi NM, Khanniri E, et al. Production and characterization of functional flavored milk and flavored fermented milk using microencapsulated canthaxanthin. LWT. 2019;114:108373

127. Van Wayenbergh E, Struyf N, Rezaei MN, Sagalowicz L, Bel-Rhlid R, Moccand C, et al. Cereal bran protects vitamin A from degradation during simmering and storage. Food Chemistry. 2020;331:127292

128. Mart Ó, Calvo MM. Recent advances in astaxanthin micro / nanoencapsulation to improve its stability and functionality as a food ingredient. Marine Drugs. 2020;18:1-25

129. Liu W, Wang J, McClements DJ, Zou L. Encapsulation of β-carotene-loaded oil droplets in caseinate/alginate microparticles: Enhancement of carotenoid stability and bioaccessibility. Journal of Functional Foods. 2018;40:527-535

130. Gomez-Estaca J, Comunian TA, Montero P, Ferro-Furtado R, Favaro-Trindade CS. Encapsulation of an astaxanthin-containing lipid extract from shrimp waste by complex coacervation using a novel gelatin-cashew gum complex. Food Hydrocolloids. 2016;61:155-162

131. Zhang Z, Zhang R, McClements DJ. Encapsulation of β-carotene in alginate-based hydrogel beads: Impact on physicochemical stability and bioaccessibility. Food Hydrocolloids. 2016;61:1-10

132. Xavier AAO, Mercadante AZ. The bioaccessibility of carotenoids impacts the design of functional foods. Current Opinion in Food Science. 2019;26:1-8

133. Pal S, Bhattacharjee P. Spray dried powder of lutein-rich supercritical carbon dioxide extract of gamma-irradiated marigold flowers: Process optimization, characterization and food application. Powder Technology. 2018;327:512-523

134. Steiner BM, Shukla V, McClements DJ, Li YO, Sancho-Madriz M, Davidov-Pardo G. Encapsulation of Lutein in Nanoemulsions Stabilized by Resveratrol and Maillard Conjugates. Journal of Food Science. 2019;84:2421-2431

135. Medeiros AK de OC, Gomes C. de C, Amaral MLQ de A, Medeiros LDG de, Medeiros I, Porto DL, et al. Nanoencapsulation improved water solubility and color stability of carotenoids extracted from Cantaloupe melon (Cucumis melo L.). Food Chemistry. 2019;270: 562-572

136. Grashorn M. Feed additives for influencing chicken meat and egg yolk color. In: Carle R, Schweiggert RM, editors. Handbook on Natural Pigments in Food and Beverages: Industrial Applications for Improving Food Color. Amsterdam: Elsevier Inc.; 2016. pp. 283-302 ISBN 9780081003923

137. Hassaan MS, Mohammady EY, Soaudy MR, Sabae SA, Mahmoud AMA, El-Haroun ER. Comparative study on the effect of dietary β-carotene and phycocyanin extracted from Spirulina platensis on immune-oxidative stress biomarkers, genes expression and intestinal enzymes, serum biochemical in Nile tilapia, Oreochromis niloticus. Fish & Shellfish Immunology. 2021;108:63-72

Chapter-4

Flavonoids: A Group of Potential Food Additives with Beneficial Health Effects

1. Introduction

The growing interest in a healthy lifestyle has led the food industry to establish an alliance with the scientific community to create a viable and effective alternative for the consumer, carrying out several studies about the bioactive potential of various compounds present in natural matrices [1].

The recently discovered properties of phenolic compounds have been exploited, and the food industry has launched numerous new functional products whose health functionality is closely connected with their polyphenols content [2].

The scientific community have been developing several studies to determine the presence of phenolic compounds natural matrices, namely the presence of flavonoids. For example, in cereals, several kinds of flavonoids (principally glycosylated flavones) are distributed in these grass crops; in legumes, the presence of a total of 690 isoflavonoids have been reported; and in medicinal plants these molecules are a major constituent in lists of metabolites responsible for the bioactivities [3].

Flavonoids are a subdivision of polyphenols that are abundant in the human diet and can be found in several matrices; specifically, they are commonly found in fruits, vegetables, nuts, teas, dark chocolate, red wine and legumes [3].

These compounds are divided into principal subclasses of flavanols, including flavanol monomers (flavan-3-ols) and flavanol polymers (called proanthocyanidins), flavonols, flavanones, flavones, isoflavones, and anthocyanins (depending on the substitution at the heterocyclic ring (C-ring)) [4]. Regarding their physiological potential, flavonoids have a vast range of bioactivities, namely antioxidant, anti-inflammatory, vasorelaxant, anticoagulant, cardio-protective, anti-obesity and anti-diabetic, chemoprotective, neuroprotective, and antidepressant properties that are progressively being clarified [5]. These beneficial properties are strongly dependent on the polyphenols chemical structure [2].

1.1 Flavonols

Flavonols are the most important subgroup of flavonoids. Chemically, these compounds (as other flavonoids) have a characteristic 15-carbon skeleton (C6-C3-C6), two benzene rings constitute its structure (catechol B ring and resorcinol A ring) joined together by a 4-pyrone heterocyclic ring C (Figure 1) [6].

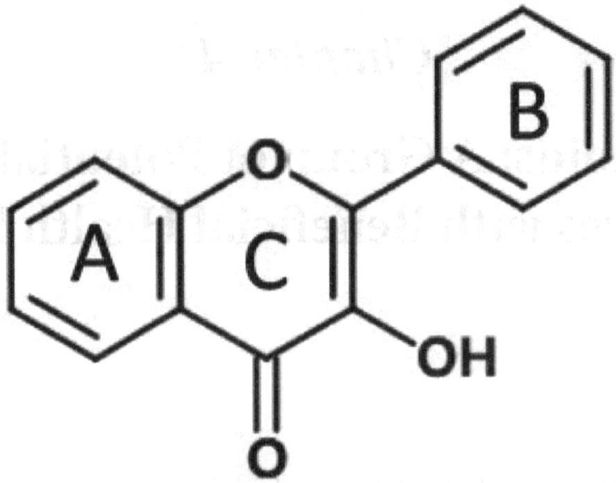

Figure 1. Chemical structure of flavonols. Designed with eMolecules (https://www.emolecules.com/).

Some compounds of this subclass include quercetin, myricetin, kaempferol, galangin, and fisetin [7, 8, 9]. These molecules represent the most ubiquitous and abundant flavonoids in the plant kingdom (dicotyledonous plants, especially flowers and leaves of woody) [10, 11] and occur abundantly in fruits (*e.g.*, apples, bananas, several berries, pomegranate), vegetables (broccoli, red and white onion, tomato, spinach), cocoa and chocolate, and beverages (such as tea and red wine) [7, 12].

The scientific community has widely studied the positive effects of flavonols on human health. These molecules have been reported as important antioxidants due to their abilities to suppress free radical formation, scavenge free radicals, and upregulate or protect antioxidant systems. They also inhibit the enzymes associated with free radical production, reduce lipid peroxidation, and chelate metal ions in reducing free radical generation [10, 11]. In addition to the antioxidant potential, these molecules have shown other target biological activities such as antimicrobial, anti-viral (interruption of virus's entry and replication cycle) hepatoprotective, nephroprotective (effective for the treatment of chronic kidney disease), anti-inflammatory, vasodilatation effects, and cardiovascular protective effects (preventative role in coronary diseases). They also have been considered as potential anticancer agents [8, 13, 14].

However, despite all the flavonols have a broad spectrum of biological activities, kaempferol, myricetin, and quercetin are the main representatives and have been widely studied due to their health-promoting functions. Both kaempferol and quercetin have unique biological properties as anticarcinogenic, antimicrobial, antidiabetic, anti-viral, anti-allergic, antioxidant, and anti-inflammatory [7, 8, 11].

1.2 Flavanols (flavan-3-ols)

Flavanols or flavan-3-ols are another flavonoid subclass with a hydroxyl group at position 3 and a fully saturated carbon ring structure (Figure 2) [9].

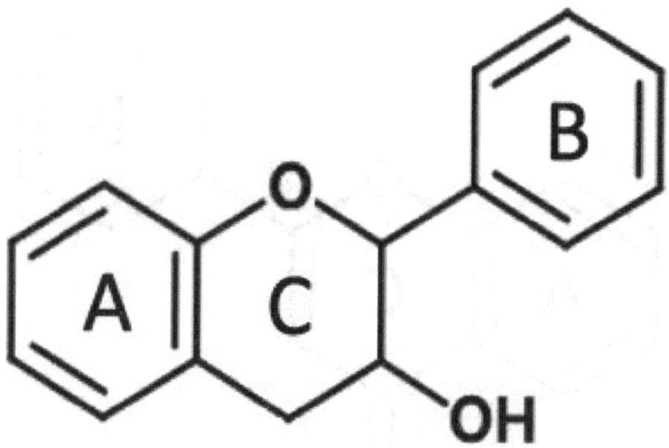

Figure 2. Chemical structure of flavanol. Designed with eMolecules (https://www.emolecules.com/).

The most common flavan-3-ol monomers are catechin, epicatechin, catechin gallate, epicatechin gallate, gallocatechin, epigallocatechin, gallocatechin gallate and epigallocatechin gallate [2]. These compounds are widely spread in nature and can be found in a wide range of natural matrices as apples, peaches, cocoa powder, nuts, dark chocolate, grapes, berries and beverages (such as red wine, tea, and cider) [9]. Furthermore, they can also be found in certain food plants, such as *Vitis vinifera*, *Camellia sinensis* and *Theobroma cacao* [15]. In these foods, flavanols can exist as monomers, such as epicatechin, or in oligomeric forms referred to as procyanidins or, more broadly, proanthocyanidins [16]. The presence of flavanols in food affects food quality parameters, principally the astringency, bitterness, sourness, sweetness, salivary viscosity, aroma, and color formation [17].

Over time, the interest in flavanols has grown, and different studies have reported these compounds' health benefits. These compounds present several beneficial effects in consumers' health, acting as antioxidant (scavenging of free radicals, chelation of transition metals, as well as the mediation and inhibition of enzymes), anticarcinogen, cardio-preventive (modulation of vascular homeostasis), anti-microbial, anti-viral, and neuro-protective agents [4, 18]. Besides, dietary intervention studies demonstrated that consuming certain flavanol-containing foods results in improved arterial function, a decrease in blood pressure, positive modulation of hemostasis, and improved insulin sensitivity [15]. In this sense, diets enriched in flavan-3-ol containing foodstuffs may provide beneficial health effects [17].

1.3 Flavones

Flavones are also a subgroup of the flavonoid class based on the backbone of 2-phenylchromen-4-one (2-phenyl-benzopyran-4-one). The molecular formula of the flavone molecule is $C_{15}H_{10}O_2$. It has a three-ring skeleton, C6-C3-C6, and the rings are referred to as A-, C-, and B-rings, respectively (Figure 3). These compounds are also characterized by the presence of three functional groups, including hydroxy, carbonyl, and a conjugated double bond. Consequently, they exhibit characteristic reactions of all three functional groups [19].

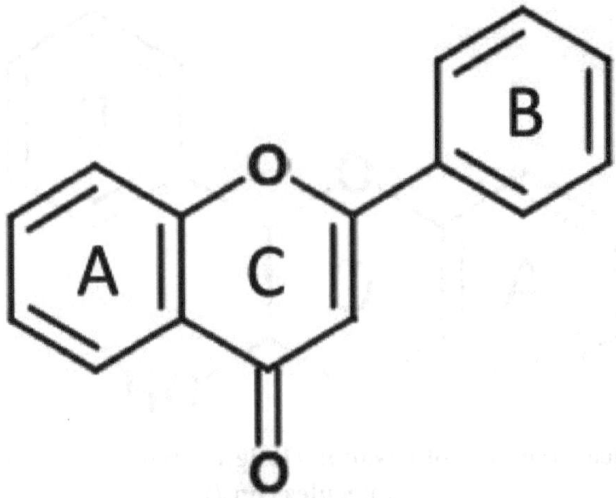

Figure 3. Chemical structure of flavones. Designed with eMolecules (https://www.emolecules.com/).

The most abundant types of flavones are luteolin, apigenin and chrysin [20]. These compounds are commonly found in edible vegetables, fruits, nuts, seeds and plant-derived beverages and cereals, which are ingested inadvertently in our daily diet and positively impact consumers' health without significant side effects [3, 20].

The scientific community has carried out several studies to determine the biological potential of flavones. These molecules have received broad interest for their antioxidant potential [21] and their ability to modulate several enzyme systems involved in many diseases [22]. Also, these compounds have demonstrated to have other biological properties beneficial to health, namely anti-inflammatory activities [23], antibacterial [24], antifungal [25], antiviral [26] and anti-carcinogenic [27]. Furthermore, they also have immunomodulatory effects [28], and they intervene in the reduction of total cholesterol [29]. Recent studies in numerous disease areas (osteoporosis, prostate hyperplasia, endocrinology, and others) have shown that many disorders, specifically in the metabolic area, are multi-factorial and are better treated with combinations of drugs and natural products [19]. However, all these therapeutic actions depend and differ according to the different compounds belonging to the subclass of flavones [30].

1.4 Flavanones

Flavones are another subgroup of flavonoids and have a C6-C3-C6 skeleton composed of 3 rings, A-, C-, and B-, respectively, and a chiral carbon at the C-3 position (Figure 4) [31].

Formerly, flavanones were considered minor flavonoids, like chalcones, dihydrochalcones, dihydroflavonols and aurones; nevertheless, in the past 15 years, the total number of known flavanones has increased, and they are now considered a major flavonoid class like flavones, isoflavones, flavanols, flavonols and anthocyanidins. Nowadays, in nature, up to 350 flavanone aglycones and 100 flavanone glycosides have been identified [32].

Flavanones are mainly divided into naringenin, hesperetin and eriodicthiol [20]. They are characteristic compounds of citrus fruit, principally lemon, lime, mandarin (tangerine), sweet orange, grapefruit, sour (bitter) orange and tomato [33, 34, 35]. These compounds are also widely distributed in around 42 plant families (*Compositae, Leguminosae* and *Rutaceae*). They can be found in all plant parts, above and below ground, from vegetative part to generative organs: stem, branches, bark, flowers, leaves, roots, rhizomes, seeds, fruits, peels, and others [32].

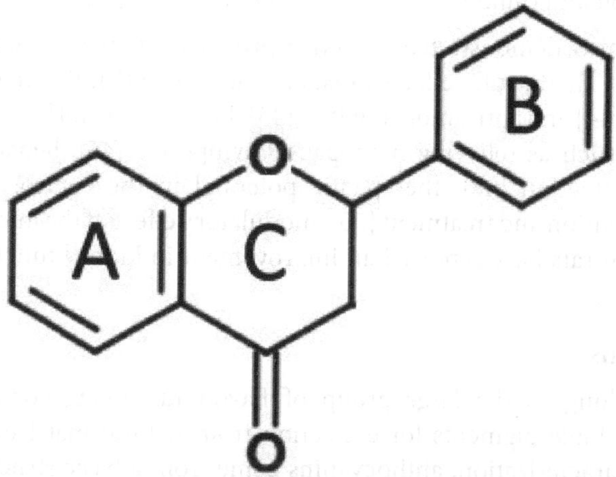

Figure 4. Chemical structure of flavanones. Designed with eMolecules (https://www.emolecules.com/).

As in the other subgroups of flavonoids, flavanones also exhibit biological properties, which positively affect consumers' health. Properties associated with flavanone intake include antioxidant [34], anti-inflammatory [36], antitumor, antiviral [37] and antimicrobial activities [38]. Furthermore, flavanones are related to some beneficial effects, such as improved gastrointestinal function [39], decreased blood cholesterol level [38], cardioprotective effect [40] and reduction of inflammatory responses caused by SARS-CoV-2 infection [35].

1.5 Isoflavones

Some structural variation from flavones are presented in isoflavones, which differs from flavones in the location of the phenyl group's location at C3 rather than C2 position (Figure 5) [9].

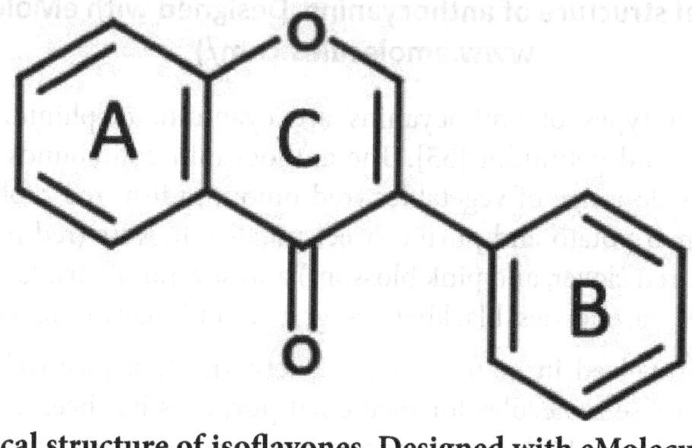

Figure 5. Chemical structure of isoflavones. Designed with eMolecules (https://www.emolecules.com/).

Isoflavones are divided into genistein, daidzein and glycitein [20]. These compounds are naturally-occurring plant compounds and are usually found in legumes from the *Fabaceae* family (chickpeas, beans, lupine and soybean) and red clover (*Trifolium pratense*), as well as small amounts of isoflavones are also contained in other plant products, fruits, vegetables (broccoli and cauliflower), barley and nuts [20, 41, 42].

Isoflavones have also demonstrated bioactive properties that intervene beneficially in human health. The therapeutic effects of isoflavones are anti-inflammatory, antioxidant [43], anti-obesity [44] and antitumor activities [45]. Besides, several benefits are associated with isoflavones, such as relieving menopausal symptoms [46], hepatoprotective [47], cardiovascular protection [48], therapeutic potential in the control of diabetes [49], osteoporosis prevention and treatment [50], modulatory effect of the intestinal microbiota [51] and studies in rats have reported an improvement in kidney function in obese rats [52].

1.6 Anthocyanins

Anthocyanins belong to the large group of flavonoids, being considered the most revealing water-soluble pigments for extraction from natural matrices [53]. Regarding their chemical characterization, anthocyanins come from a basic structure of 3 rings of the C6-C3-C6 shape, defined as an aglycone portion, called the flavylic cation (Figure 6). When associated with chemical groups in the R positions, it is called anthocyanidin [53, 54].

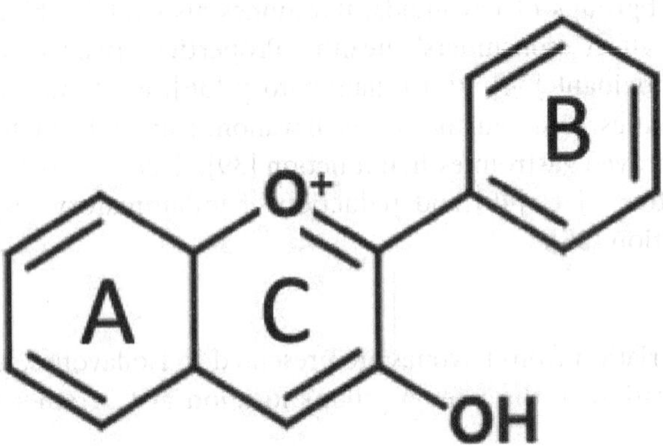

Figure 6. Chemical structure of anthocyanins. Designed with eMolecules (https://www.emolecules.com/).

The most common types of anthocyanins are cyanidin, delphinidin, pelargonidin, peonidin, malvidin and petunidin [55]. The anthocyanin compounds are present in a composition of a wide range of vegetables (red onion, radish, red cabbage, red lettuce, eggplant, red-skinned potato and purple sweet potato), flowers (red hibiscus, red rose, red pineapple sage, red clover, and pink blossom) and several red fruits, such as: cherries, plums, strawberries, raspberries, blackberries, grapes, and many others [56, 57].

Anthocyanins are involved in many biological activities that positively impact human health. The use of these molecules for medicinal purposes has been long supported by epidemiological evidence. Still, just in recent years, some of the specific, measurable pharmacological properties of isolated anthocyanin pigments have been proven by controlled *in vitro*, *in vivo* or clinical research studies [57]. According to several authors, the health benefits are associated with the increase of sight acuteness, anti-carcinogenic activity, antioxidant capacity, antiulcer activity and the maintenance of normal vascular permeability (vitamin C_2), as well as acting in the prevention of various diseases, such as coronary and degenerative diseases, diabetes, inflammation or reduction of the risk of obesity, among others [58, 59, 60].

2. Biological Properties of Flavonoids

The bioactive properties of flavonoids are directly linked to the functions they exert. Flavonoids, present in higher plants' cells, have a protective role against parasites and other pathogens (participating in allelopathy processes), herbivores, and ultraviolet (UV) radiation [61]. They have a regulatory function like most lipid-soluble vitamins and act as pollinating agents. The varied colors they can have attract pollinators, thus contributing to plant seeds' dispersion [62]. The following sections display a set of functions attributed to these compounds, seeking to relate their bioactivity to their chemical features and/or possible mechanism of action.

2.1 Antioxidant activity

The antioxidant properties of flavonoids have been recognized over the years. Given the wide presence of flavonoids in various fruits, vegetables, legumes, grains and nuts, these compounds represent approximately two-thirds of the phenols consumed in the diet, being the class predominantly described [63, 64]. The mechanisms underlying the antioxidant properties of flavonols include eliminating free radicals and the chelating activity of transition metal ions, being the preventive action and chain-breaking mechanisms responsible for the high bioactivity of flavonoids [65, 66]. In fact, flavonoids can eliminate free radicals and reduce their formation and/or their effects. As expected, the chemical structure plays a key role in the antioxidant activity of flavonoids. That is, due to the reducing capacity of the phenolic hydroxyl groups (presence of hydrogen-/electron-donating substituents), flavonoids can donate hydrogen; thanks to the ability to delocalize the unpaired electron leading to the formation of a stable phenoxyl radical, flavonoids can protect against damage caused by reacting oxygen species (ROS), and flavonoids can chelate transition metals capable of promoting the formation of hydroxyl radicals in reduced forms through the Fenton reaction under abnormal conditions. This property is strongly dependent on the arrangement of hydroxyls and carbonyl group around the molecule [66]. Considering these characteristics that underlie the antioxidant potential of flavonoids, studies carried out over the years have shown that the flavonoids with greater antioxidant activity have the following structure features: (i) a certain hydroxylation pattern, particularly in the ring B, namely 3`,4`-dihydroxyl group (*e.g.*, 3`, 4`-catechol group); (ii) the 3 – OH moiety in the C ring; (iii) the C2 = C3 double bond in the C ring conjugated with a C4-carbonyl group in the ring, causing electron delocalization from the B ring; and (iv) both 3 – OH group in C ring and 5 – OH group in A ring combined with a 4-carbonyl group and C2 = C3 double bond [65, 66, 67, 68]. The influence of the chemical structure on the bioactivity of flavonoids was also demonstrated by Novaes et al. [69]. The authors tested the antioxidant potential of 11 flavonols extracted from the leaves of *Annona coriacea* Mart. using four approaches, and attributed the differences obtained in the assays to the B-ring substitution pattern of flavonols. Recent studies also suggest that the C – H bonds may contribute to the antioxidant activity of flavonols since they evaluated the antioxidant properties of 13 flavonoids with the hydroperoxyl radical (HOO) and concluded that the C – H bonds (C3 – H of the flavonoid backbone structures) play a fundamental role in the antioxidant properties of flavonoids containing 4-carbonyl and/or 3-hydroxyl groups. These groups release the single electron on the C3 radical (the C3 – H bond) into the O – C3 – C4 – O system and form intermolecular bonds to stabilize the radicals, yielding reduced bond dissociation energy (C3 – H) and increased the antioxidant activity of the flavonoids. This study also showed that the hydrogen atom transfer (HAT) mechanism is the main pathway for flavonoids' antioxidant activity [68].

If the abovementioned features favor the antioxidant potential of flavonoids, on the other hand, the presence of saccharide groups seems to reduce the antioxidant properties of these molecules. Still, some studies showed that *C*-glycosyl flavonoids have greater potential than *O*-glycosides [70, 71]. It should be highlighted that although the antioxidant activity of glycosides is weaker than the corresponding aglycone, the bioavailability is reasonably increased *in vivo* due to the cleavage of glycosidic bonds that frequently occur. Therefore, under these conditions, the antioxidant activity is increased [66].

However, the abundant consumption of flavonoids, as polyphenols in general, through the daily diet does not always correspond to obtaining the effects observed *in vitro* in the natural matrix of origin. Nevertheless, there are already some studies carried out in this direction. Some works proved that even given the sensitivity of these biomolecules (to food processes and storage), it is possible to obtain positive results when inserted in foods, namely as ingredients/additives with preservative capacity, and bring potential health benefits, as described in the following sections.

2.2 Antimicrobial activity

The antimicrobial properties of natural products rich in flavonoids have been reported and recognized since antiquity. Of the best-known products, propolis can be highlighted, whose healing properties have been mentioned for thousands of years and used to treat wounds and ulcers. In fact, propolis's antimicrobial properties have been attributed to its high content of flavonoids, particularly galangin and pinocembrin [72]. As previously mentioned, flavonoids' bioactivity is related to their function in nature, namely protecting plants against pathogens. In this way, plant-derived flavonoids have different antibacterial mechanisms of action than conventional drugs, and generally, their bioactivity does not confer resistance. In fact, to the best of our knowledge, no report claims to have observed bacteria developing resistance to plant-based antimicrobials. In this way, antibacterial agents based on natural extracts rich in flavonoids represent an important alternative in developing new antibacterial formulations, both from a clinical perspective, as in any other application such as the food sector [73]. The possible mechanisms of antimicrobial action of flavonoids are briefly: (i) cell envelop synthesis inhibition (*e.g.*, quercetin, myricetin, luteolin); (ii) inhibition of nucleic acid synthesis (*e.g.*, quercetin, kaempferol, apigenin); (iii) bacterial motility inhibition (*e.g.*, sinensetin, luteolin, epigallocatechin gallate); (iv) inhibition of ATP synthesis on electron transport chain (*e.g.*, baicalein, silibinin, silymarin); (v) bacterial toxins inhibition (*e.g.*, naringenin, kaempferol, quercetin 6-hydroxyflavone); (vi) biofilm formation inhibition (*e.g.*, genistein, apigenin, naringenin); (vii) inhibition of bacterial enzyme-dependent virulence (*e.g.*, amoradicin, kaempferol-3-rutinoside, baicalin); (viii) membrane disruption (*e.g.*, apigenin, catechin, quercetin); (ix) inhibition of bacterial efflux pumps (*e.g.*, luteolin, morin, rutin); and (x) inhibition of bacterial quorum sensing (*e.g.*, naringin, taxifolin, chrysin) [74]. Therefore, when relating the chemical structure of flavonoids with their antimicrobial activity through various mechanisms, quercetin, apigenin, kaempferol, fisetin, myricetin, luteolin, taxifolin, or naringenin features can be highlighted, among others.

2.3 Prebiotic activity

Polyphenols' prebiotic effects have also been explored, with available reports from pre-clinical and clinical studies. Flavonoids have been the most investigated phenolic compounds in terms of their effects on the composition of the intestinal microbiota and the health benefits of the host [75]. It must be highlighted that the current definition of prebiotic recognizes that, in addition to the stimulation of *Bifidobacterium* and *Lactobacillus* bacteria, prebiotic targets include other microorganisms, such as *Roseburia*, *Eubacterium* and *Faecalibacterium* spp. However, they are not limited to these genera [76]. The health benefits associated with prebiotics are well known and comprise immunomodulation, increased mineral absorption, improved intestinal function and positive effects on glucose homeostasis, inflammation, blood lipid profile, satiety and defense against pathogens [77]. The main studied flavonoids that revealed potential prebiotic effects proven in pre-clinical studies include anthocyanins (bilberry extract, grape pomace extract, and Arctic berry extracts), proanthocyanidins (Arctic berry extracts, and cranberry extract), proanthocyanidin A (cinnamon bark extract), catechins and caffeine (green and black tea extracts, decaffeinated green and black tea polyphenol extracts, oolong tea water extracts, aqueous, raw and ripe Pu-erh tea extract), isoflavones (soy extract), naringenin (*S. chinensis* pollen extract), polymeric and oligomeric procyanidin (apple), genistein and hydroxysafflor yellow A [75]. In addition to the beneficial effects on the intestinal microbiota, authors also found other beneficial health effects such as the decrease in: (i) digestive enzymes activity; (ii) fat mass gain; (iii) liver steatosis; (iv) adiposity; (v) body weight gain and metabolic endotoxemia; (vi) serum lipid profile, glucose and insulin; (vii) serum triglycerides; (viii) oxidative damage and inflammation; and (ix) fasting blood glucose. On the other hand, they found an increase in insulin sensitivity, expression of hepatic lipid metabolism genes, glucose tolerance, and glycolysis.

In clinical trials, anthocyanins consumed in a wild blueberry drink have been studied, in a dose of 25 g/250 mL of water. The study included 20 healthy male individuals, with a 6-week consumption of the drink. After this period, an increase of *Lactobacillus acidophilus* and *Bifidobacterium* spp. in the human gut was observed [78]. In another study, they inserted punicalagins and ellagic acid (from pomegranate extract) in capsules with 0.45 g or 1.8 g of extract. These capsules were administered to 49 overweight-obese individuals (mild hyperlipidemia) for 9 weeks. After this time, the authors verified an increase of *Faecalibacterium* and decreased lipopolysaccharide-binding protein in the gut microbiota [79].

2.4 Colorant activity

Color is the most important sensory perception that defines consumer expectations about foods' organoleptic properties [80]. Thus, adding or improving food color has been one of the food industry's commitments to make products more appealing. Although each coloring agent used by the food industry in the European Union is subjected to a rigorous safety assessment, some problems of intolerance and/or allergies or hyperactivity have been related to its consumption [81], which may justify the consumer's preference for natural additives to the detriment of the artificial counterparts. In this way, the scientific community has been looking for natural alternatives to the artificial colors widely used in the industry. The major natural pigments obtained from nature include chlorophylls, carotenoids, betalains and flavonoids. Among the main classes of flavonoids used as coloring agents, the flavonols and anthocyanins stand out, obtaining a range of colors between cream, yellow, pink, red, blue and black [82]. There is already a natural coloring agent, based on flavonoids, approved by the regulatory authorities for its use in the food sector, namely anthocyanins (E163). The approved anthocyanin extract is obtained from the natural strains of vegetables and edible fruits, including blackcurrant pomace and grape skin [83]. The pH strongly influences the color of anthocyanins. In acidic conditions, anthocyanins are red, while in basic pH, they appear blue, being purple in solutions with neutral pH. Hence, grapes are one of the best sources of this natural red pigment, since their anthocyanins are largely methylated, leading to an increase in color intensity and higher stability [84, 85]. After consulting the literature, we found that most studies on natural food colors based on flavonoids are strongly focused on anthocyanins. Other classes of flavonoids have been studied as copigments. Given the sensitivity of anthocyanins to various factors, not only pH but also temperature, light, oxygen, among others. Copigments or other co-solutes can be added, even when colorless, since they trigger a hyperchromic effect [82]. Copigmentation may occur by forming (in the presence or absence of metal ions) noncovalent complexes involving an anthocyanin or anthocyanin-derived pigment (*e.g.*, a pyranoanthocyanin or anthocyanin-flavanol adduct) on the one hand and a copigment on the other; and by subsequent changes in optical properties of the pigment. Among these interactions, intramolecular and intermolecular copigmentation are the most important mechanisms of copigmentation [82, 86]. Also, the effects of flavonoid *C*-glycoside extracts from pigeon pea leaves and its main components vitexin (apigenin-8-*C*-glucoside) and orientin (luteolin-8-*C*-glucoside) on the color and anthocyanins stability of blueberry juice were evaluated. Authors verified that the addition of flavonoid *C*-glycoside markedly extended the half-life of anthocyanin and enhanced the juice quality. The need to explore other target flavonoids to obtain additional coloring agents to be applied in food formulations seems apparent [87].

2.5 Other activities

Besides the properties previously described, some other bioactivities have been assigned to flavonoids, such as anti-diabetic, anti-inflammatory or anticancer activities. For instance, a new approach to treat diabetes with enhanced antidiabetic activity from a flavonoid nanoparticulate system has been proposed. This system with new biodegradable releasers would increase the solubility of flavonoids and consequently their bioavailability, preventing flavonoid from first-pass metabolism and intestinal absorption in the form of a flavonoid nanoparticulate system. Flavonoids exert their antidiabetic properties by enhancing insulin secretion via regeneration of pancreatic β-cells, enhancing insulin-mediated glucose uptake by target cells, inhibiting aldose reductase and increasing Ca^{2+} uptake [88]. Antidiabetic activity of flavonoids depends on the chemical criterion

(C-2-C-3 double bond and ketonic group at C-4 position on ring B) which is fundamental for the bioactivity of polyphenols [89]. The antioxidant and anti-inflammatory activity of the flavonol fisetin (7, 3⊠, 4⊠-flavon-3-ol) has been evaluated, showing that it could improve the plasma insulin and antioxidant levels in diabetic rats and significantly decrease the levels of blood glucose. Therefore, the authors suggested that fisetin could be considered as an adjunct for the treatment of diabetes [90]. Regarding anthocyanins, in addition to their coloring capacity, other studies suggest that these flavonoids also have bioactivity with a potential impact on human health, namely antioxidant activity, chemopreventive potential, anti-inflammatory and immunomodulatory properties [91]. Some of the bioactivities attributed to flavonoids have been specifically pointed to flavonoid glycosides. For example, the *C*-glycosidation improved the intracellular antioxidation performance of apigenin [92]. Moreover, many *in vitro* assays reported the positive role of flavonoid glycosides on the immune response (*e.g.*, quercetin 3-*O*-xyloside). Other flavonoid glycosides have shown anticancer properties (quercetin 6-*C*-glucoside), anti-inflammatory and analgesic effects (luteolin 6-*O*-rhamnoside) and anti-parasitic activity (acyl flavonoid glycosides) [92]. Furthermore, it has been suggested that some isoflavonoids may exert antiophidic potential against *Bothrops jararacussu* snake venom [93].

3. Extraction and Production Techniques for Flavonoids Recovery

According to the current and continuously increasing demand for new healthy products for a better lifestyle of the population, an increase in the number of techniques used to extract bioactive compounds has occurred. Choose the best extraction technique for each sample is essential in terms of the quality and quantity of the target molecules obtained, that is, flavonoids. Nowadays, there is many extraction techniques that can be employed. These techniques are selected depending on the characteristics of the raw material, which are, in general, plants, food or liquid samples such as wine, tea or olive oil [94]. Independently of the source, the samples must be homogenized. Hence, the most used methods are grinding, milling, filtration, pulverizing and mechanical stirring. Depending on the raw material, before or after homogenization, a pretreatment could be used to facilitate or improve the homogenization and the extraction process. The pretreatments usually used are freezing (in a freezer or by liquid nitrogen), different drying process and freeze-drying [95]. However, the use of these pretreatments can affect the extract characteristics, limiting the optimization of the extraction process [96, 97, 98]. There is no standard for every source of raw material, so selected pretreatments must be chosen depending on the physical and chemical characteristics of the samples. Moreover, depending on the pretreatment and homogenization method, the sample must be stored in the appropriate conditions or perform the extraction immediately before the homogenization and the pretreatment [99, 100]. The extraction techniques can be divided into two groups, conventional and novel approaches.

3.1 Conventional extraction techniques

Conventional extraction techniques are characterized using conventional solvents, with or without heat and usually under agitation. In a standard conventional extraction, the sample is homogenized and submerged in a solvent or a mix of solvents. Using this extraction methodology, flavonoids are obtained through the diffusion and mass-transfer phenomena [101, 102]. The most used methods are maceration and Soxhlet. Nevertheless, other techniques like percolation, hydro-distillation, boiling, reflux and soaking can be used [103, 104]. The advantages of using these techniques are their simplicity and low cost, so they are preferred by companies [102]. On the other hand, these methods have several disadvantages: high volumes of solvent, low extraction yields and long times. Furthermore, due to the sensitivity of flavonoids to high temperature, in extraction assisted by heat, the compounds' biological properties could be affected [105, 106].

For its simplicity and low cost, extraction by Soxhlet is the most used [106]. The main advantages of this method are three. The first one is that through repeated cycles, the sample is in contact with fresh solvent almost all the time, helping the displacement of the mass transfer equilibrium. In the second place, after the extraction, it is not necessary to filter the sample. Lastly, the amount of sample extracted can be improved easily by simultaneous parallel extraction, which needs very little investment. However, Soxhlet extraction has some disadvantages concerning other conventional extractions. The main disadvantages of this technique are its duration (*i.e.,* between 6 and 12 h or higher) and the large amount of solvent used. In addition, after extraction, the high amount of solvent in the sample is usually needed to evaporate. Moreover, the high operating temperature could produce the thermal decomposition of some compounds. Despite the weaknesses, nowadays, Soxhlet extraction could be combined with other novel techniques to improve extraction efficiency [107].

Significant differences in the extraction yields between different conventional extractions can be observed. There are also differences between the number of compounds obtained and their bioactivity within the same method. These variations are caused by the parameters directly implicated in the extraction process like temperature, time, number of extractions (cycles), the ratio of solvent to raw material or type of solvent [108]. Table 1 shows diverse examples of extractions carried out under different conditions and different methods to compare the conventional extraction methods.

Table 1. Summary of studies of extraction of flavonoids from different sources.

Type (cycles)	Substrate	Solvent (%)	Temperature (°C)	Time (min)	Yields	Ref.
Batch	*A. marmelos* (fruit)	Ethanol	50	50	2.04 mg/g dw	[109]
HRE	*G. affine*	Eth : W (80:20)	—	150	1.16% (Flavonoid extraction yield)	[110]
Mac	*P. oleracea* L.	Eth : W (70:20)	25	2880	5.6 mg/g dw	[111]
Reflux	*P. oleracea* L.	Eth : W (70:20)	—	150	6.8 mg/g dw	[111]
SAE	*Moringa oleifera* L. (leaves)	Eth : W (52:48)	30	30	12.77 ± 0.65 mg/g dw	[112]
SBE	*A. marmelos* (fruit)	Ethanol	50	450	48.15 mg RE/g	[109]
Soxhlet	*A. marmelos* (fruit)	Ethanol	60	460	67.85 mg RE/g	[109]
Soxhlet	*G. affine*	Eth : W (80:20)	—	300	1.48% (Flavonoid extraction yield)	[110]

Type (cycles)	Substrate	Solvent (%)	Temperature (°C)	Time (min)	Yields	Ref.
Soxhlet	*M. oleifera* L. (leaves)	Ethanol	90	180	10.67 ± 0.27 mg/g dw	[112]
Soxhlet	*P. oleracea* L.	Eth : W (70:30)	—	300	7.0 mg/g dw	[111]
Soxhlet	*Pleurotus florida*	Methanol	—	300	0.40 ± 0.03 mg QE/g dw	[113]
Soxhlet	*Humulus lupulus* (hops)	Ethyl acetate	77	480	11.4 ± 0.6 (wt% extract)	[114]
Soxhlet	*H. lupulus* (hops)	Metanhol	65	480	25.8 ± 0.7 (wt% extract)	[114]
Soxhlet	*H. lupulus* (hops)	N-hexane	69	480	6.7 ± 0.2 (wt% extract)	[114]
Soxhlet	*H. lupulus* (hops)	Ethanol	78	480	25.8 ± 0.9 (wt% extract)	[114]
Soxhlet	*Mentha spicata* L. (leaves)	Methanol	40	360	267.33 ± 3.12 mg/g dw	[115]
Soxhlet	*M. spicata* L. (leaves)	Ethanol		360	218 ± 4.24 mg/ dw	[115]
Soxhlet	*M. spicata* L. (leaves)	Petroleum ether		360	30.47 ± 2.34 mg/g dw	[115]
Soxhlet	*M. spicata* L. (leaves)	Eth : W (70:30)		360	257 ± 3.47 mg/g dw	[115]
Soxhlet	*Populus temula*	Methanol	—	2880	11.5 mg/g dw	[116]
ASE	*Rheum palmatum* L	Met : W (80:20)	80	10	3.9 mg/g dw	[117]
ASE (3)	*Passiflora* species (leaves)	Eth : W (40:60)	40	30	49.22 mg GAE/g dw	[118]
ASE (3)	*Passiflora* species (leaves)	Eth : W (40:60)	80	30	41.46 mg GAE/g dw	[118]
ASE (3)	*Passiflora* species (leaves)	Ethanol	40	30	110.89 mg GAE/g dw	[118]
ASE (3)	*Passiflora* species (leaves)	Ethanol	80	30	101.61 mg GAE/g dw	[118]
ASE (1)	*Passiflora* species (leaves)	Eth : W (70:30)	40	10	72.60 mg GAE/g dw	[118]
ASE (1)	*Passiflora* species (leaves)	Eth : W (70:30)	80	10	79.43 mg GAE/g dw	[118]
ASE (5)	*Passiflora* species (leaves)	Eth : W (70:30)	40	50	65.78 mg GAE/g dw	[118]
ASE (5)	*Passiflora* species (leaves)	Eth : W (70:30)	80	50	84.53 mg GAE/g dw	[118]
ASE (1)	*Passiflora* species (leaves)	Eth : W (40:60)	60	10	58.54 mg GAE/g dw	[118]

Type (cycles)	Substrate	Solvent (%)	Temperature (°C)	Time (min)	Yields	Ref.
ASE (1)	*Passiflora* species (leaves)	Ethanol	60	10	85.83 mg GAE/g dw	[118]
ASE (5)	*Passiflora* species (leaves)	E t h : W (40:60)	60	50	60.21 mg GAE/g dw	[118]
ASE (5)	*Passiflora* species (leaves)	Ethanol	60	50	80.31 mg GAE/g dw	[118]
ASE (3)	*Passiflora* species (leaves)	E t h : W (70:30)	60	30	60.57 mg GAE/g dw	[118]
ASE (3)	*Passiflora* species (leaves)	E t h : W (70:30)	60	30	63.70 mg GAE/g dw	[118]
ASE (3)	*Passiflora* species (leaves)	E t h : W (70:30)	60	30	60.89 mg GAE/g dw	[118]
ASE	*Durio zibethinus* M. (leaves)	N-hexane	200	13	839.2 ± 232.3 QE/100 mg dw	[119]
ASE (2)	*Morus. atropurpurea* Roxb.	Water	100	20	4.83 ± 1.52 mg/ GAE g dw	[120]
ASE (2)	*Morus. atropurpurea* Roxb.	Methanol	100	20	15.3 ± 0.6 mg/GAE g dw	[120]
ASE (2)	*Morus. Atropurpurea* Roxb.	M e t : W (50:50)	100	20	4.28 ± 0.24 mg/ GAE g dw	[120]
ASE (2)	*Morus. atropurpurea* Roxb.	M e t : A A (99.5:0.5)	100	20	11.58 ± 0.5 mg/ GAE g dw	[120]
ASE (2)	*Morus. atropurpurea* Roxb.	Acetone	100	20	13.8 ± 0.6 mg/GAE g dw	[120]
ASE (2)	*Morus. atropurpurea* Roxb.	A:W (50:50)	100	20	13.2 ± 0.3 mg/GAE g dw	[120]
ASE (2)	*Morus. atropurpurea* Roxb.	A : A A (99.5:0.5)	100	20	13.5 ± 0.8 mg/GAE g dw	[120]
ASE (2)	*Morus. atropurpurea* Roxb.	Met:W:AA (50:49.5:0.5)	100	20	15.1 ± 0.1 mg/GAE g dw	[120]
ASE (2)	*Morus. atropurpurea* Roxb.	A : W : A A (50:49.5:0.5)	100	20	14 ± 1 mg/GAE g dw	[120]
Mac	*Cassia alata*	Ethanol	60	120	70.13 ± 4.43 mg QE/g dw	[121]
Mac (2)	*Brassica oleracea* L var. *botrytis* L subvar *cymosa* (flower)	A:W (70:30) Methanol	4	1440	0.32 ± 0.05 mg GAE/g fw	[122]
Mac (2)	*B. oleracea* L var. *botrytis* L (flower)	A:W (70:30) Methanol	4	1440	0.18 ± 0.01 mg GAE/g fw	[122]
Mac (2)	*B. oleracea* L var. *capitata* L (leaves)	A:W (70:30) Methanol	4	1440	0.1 ± 0.01 mg GAE/g fw	[122]
Mac (2)	*Lactuca sativa* L (leaves)	A:W (70:30) Methanol	4	1440	0.1 ± 0.01 mg GAE/g fw	[122]

Type (cycles)	Substrate	Solvent (%)	Temperature (°C)	Time (min)	Yields	Ref.
Mac (2)	*Brassica chinensis* L (leaves)	A:W (70:30) Methanol	4	1440	0.94 ± 0.7 mg GAE/g fw	[122]
Mac (2)	*Artemisia vulgaris* Cantley (leaves)	A:W (70:30) Methanol	4	1440	0.44 ± 0.07 mg GAE/g fw	[122]
Mac (2)	*Daucus carot* L subsp. *sativus* (Hoffm) *Arcang* (root)	A:W (70:30) Methanol	4	1440	0.045 ± 0.002 mg GAE/g fw	[122]
Mac (2)	*Allium cepa* L (bulb)	A:W (70:30) Methanol	4	1440	0.51 ± 0.04 mg GAE/g fw	[122]
Mac (2)	*Lycopersicon esculentum* Mill (fruit)	A:W (70:30) Methanol	4	1440	0.08 ± 0.01 mg GAE/g fw	[122]
Mac (2)	*Capsicum annum* L (fruit)	A:W (70:30) Methanol	4	1440	0.32 ± 0.02 mg GAE/g fw	[122]

SBE: sequential batch extraction; HRE: heat reflux extraction; SAE: stirring-assisted extraction; QE: quercetin equivalent; RE: rutin equivalent; Mac: maceration; AA: acetic acid; A: acetone; MA: maceration with agitation; GAE: gallic acid equivalent; ASE: accelerated solvent extraction; dw: dry weight; and fw: fresh weight.

3.2 Novel techniques

Non-conventional extraction techniques put their effort in concentrate the energy to extract the bioactive compounds in a more efficient and/or selective way than in conventional extractions. Nowadays, methods that employ microwaves, ultrasounds, high pressure, supercritical fluids or digestive enzymes can extract more compounds of interest at a lower cost. Moreover, these novel techniques decrease the extraction time, increase the compounds' selectivity and reduce the amount of solvent per extraction. In addition to solvent reduction, some of these techniques allow the use of solvents less harmful to the environment and human health. Therefore, some of these techniques are green methods that can be used with green solvents. This fact has prompted companies to optimize these techniques for subsequent implementation on an industrial scale [101, 123].

3.2.1 Ultrasound-assisted extraction (UAE)

The parameter most characteristic of ultrasound is the frequency. The frequency of ultrasound is between 20 kHz and 10 MHz, while the frequency of sound is between 16 Hz and 20 kHz. The ultrasound-assisted extraction (UAE) uses one of the two types of ultrasound, power ultrasound (low frequency and high intensity), to extract different compounds from a wide variety of sources [124]. The mechanism of action of UAE consists in the formation of cavitation bubbles in the medium or solvent used. The appearance of voids by the compression and rarefaction cycle, which subjects the liquid to points above its critical molecular distance, creates cavitation bubbles in the medium. The compression and rarefaction cycle produce the enlargement of the bubbles until the bubbles collapse. This collapsing liberates a considerable amount of energy and subject the medium to high temperatures (4726.85°C) and pressures (2000 atm) now of the collapse.

This extreme condition produces microjets that can break solid surfaces like vegetable cells favoring intracellular compounds' extraction. Moreover, microjets also benefit the solvent-substrate interaction by reducing the particle size [124, 125, 126].

This method has been used in the food and pharmaceutical industries for several purposes [125]. The yield of the flavonoid extraction depends on diverse parameters: frequency, solvent, solid-solvent ratio. Table 2 shows numerous studies about the optimization of flavonoid extraction from different raw materials. Although UAE is considered a green extraction technique for their reduction of energy and time consuming, the use of green solvents with UAE is currently a new trend. In the case of flavonoids and other phenolic compounds, deep eutectic solvents (DES) are becoming a viable alternative to traditional polluting solvents (Table 2) [127, 128, 129, 130, 131]. The use of UAE has numerous benefits compared with other conventional and novel techniques. UAE obtains higher yields and productivity with lower extraction times and solvent consumption than the conventional techniques. Moreover, it is more ecofriendly and a wider variety of solvents can be used. Furthermore, the extraction can be carried out at low temperatures reducing the risk of thermal degradation of flavonoids. Nevertheless, UAE has some drawbacks. Before the extraction, a filtration step is required, and the unstable compounds are not suitable for this method [132].

Table 2. Parameters that affect novel extraction techniques of flavonoids from different sources.

Type (cycles)	Substrate	Solvent	Temperature (°C)	Time (min)	Yield	Ref.0
UAE	*Celastrus hindsii* (leaves)	Eth:W (65:35)	40	29	23.60 ± 0.31 mg QE/g dw	[72]
UADESE	*Lycium barbarum* (fruit)	CC:1,2-Propanediol (33:67)	Rt	90	4.9 mg/g fw	[35]
UADESE	*L. barbarum* (fruit)	CC:Glycerol (33:67)	Rt	90	2.9 mg/g fw	[35]
UADESE	*L. barbarum* (fruit)	CC:EG (33:67)	Rt	90	8.7 mg/g fw	[35]
UADESE	*L. barbarum* (fruit)	CC:Malic a. (50:50)	Rt	90	11.1 mg/g fw	[35]
UADESE	*L. barbarum* (fruit)	CC:Malonic a. (50:50)	Rt	90	8 mg/g fw	[35]
UADESE	*L. barbarum* (fruit)	CC:p-Ta (33:67)	Rt	90	88.9 mg/g fw	[35]
UADESE	*L. barbarum* (fruit)	CC:La. (33:67)	Rt	90	16.5 mg/g fw	[35]
UADESE	*L. barbarum* (fruit)	CC: Oxalic a. (33:67)	Rt	90	11 mg/g fw	[35]
UADESE	*L. barbarum* (fruit)	CC:Resorcinol (25:75)	Rt	90	6.5 mg/g fw	[35]
UADESE	*L. barbarum* (fruit)	CC:Xylitol (50:50)	Rt	90	3.2 mg/g fw	[35]
UADESE	*L. barbarum* (fruit)	CC: Urea (33:67)	Rt	90	5.5 mg/g fw	[35]

Type (cycles)	Substrate	Solvent	Temperature (°C)	Time (min)	Yield	Ref.0
UADESE	*L. barbarum* (fruit)	Water	Rt	90	3.6 mg/g fw	[35]
UADESE	*L. barbarum* (fruit)	Methanol	Rt	90	3.2 mg/g fw	[35]
UADESE	*L. barbarum* (fruit)	Ethanol	Rt	90	4.2 mg/g fw	[35]
UAE	*Citrus* sp. (peel)	Water	40	30	19.595 ± 2.114 mg GAE/g dw	[73]
UAE	*Crataegus* sp. (seeds)	Eth:W (72:28)	65	37	16.45 ± 0.2 mg/g dw	[74]
UAE	*Crinum asiaticum*	Eth:W (60:40)	64	47	16.4 mg/g dw	[75]
UAE	*Prunella vulgaris* L. (fruit)	Eth:W (41:59)	79	30.5	36.2 mg/g dw	[76]
HPAE	*Momordica cochinchinensis* (leaves)	Eth:Water (50:50)	25	3	5.8 ± 0.1 mg QE/g dw	[77]
UAE	*M. cochinchinensis* (leaves)	Eth:Water (50:50)	25	20	5.9 ± 0.2 mg QE g dw	[77]
HHPE	*Solanum lycopersicum* (pulp)	Hexane:W (60:40)	20	10	21.5 ± 0.1 mg QE g dw	[78]
HPAE	*Agave americana* (leaves)	Methanol	150	240	15.5 mg/g dw	[79]
HPAE	*Flos sphorae*	Eth:W (75:25)	80	120	200 ± 8.63 mg/g dw	[80]
MAE	*Aegle marmelos*	Eth:W (80:20)	50	1	77.26 mg RE/g dw	[16]
MAE	*Trigonella foenum-graecum* (seeds)	Eth:W (80:20)	80	4	37.18 mg QE/g dw	[81]
MAE	*Cassia alata*	Ethanol	Rt	4	135 ± 3 mg QE/g dw	[82]
MAE	*Epimedium sagittatum* (leaves)	Eth:E (60:20)	65	25	97.19 mg/g dw	[83]
MAE	*Astragalus mongolicus* (root)	Eth:W (90:10)	110	25	1.19 ± 0.04 mg dw	[84]
SFE (CO_2)	*Ziziphus jujuba* Mill. (leaves)	Eth:W (90:10)	52.5	113	29.011 mg/g dw	[85]
SFE (CO_2)	*Odontonema strictum* (leaves)	Eth:W (95:05)	65	270	18.92 mg QE g dw	[86]
SFE (CO_2)	*Maydis stigma* (flowers)	Eth:W (20:80)	51	120	4.24 mg/d dw	[87]
SFE (CO_2)	*Pueraria lobata* (Roots)	Ethanol	50	90	16.95 ± 0.43 mg/g dw	[88]
SFE (CO_2)	*Vaccinium myrtillus* (fruit)	Eth:W (10:80)	Rt	30	72.18 ± 1.13 mg/g dw	[89]
SFE (CO_2)	*Lepidium sativum*	Eth:W (96:4)	50	70	58 mg RuE/g dw	[90]

Type (cycles)	Substrate	Solvent	Temperature (°C)	Time (min)	Yield	Ref.0
UAE: ultrasound-assisted extraction; CC: choline chloride; EG: Ethylene glicol; a.: acid; p-Ta: p-toluenesulfonic acid; La.: levulinic acid; UADESE: ultrasound-assist deep eutectic solvent extraction; HPAE: high pressure assisted extraction; HHPE: high hydrostatic, pressure extraction; Rt: room temperature; SFC-CO_2: supercritical CO_2 fluid extraction; dw: dry weight; and fw: fresh weight.						

3.2.2 High pressure assisted extraction (HPAE)

Le Chatelier's principle states that if a system in equilibrium is perturbed, it restores the balance changing other parameters [133]. Therefore, if a system (solvent – raw material) is subjected to an increase in pressure, it will suffer a decrease in volume that will result in a more efficient extraction [134]. The volume changes produce variations in the cellular membrane and other big molecules that can cause the cell membrane and organelles' rupture, thus facilitating the transfer of bioactive compounds to the solvent [135]. The process of high pressure-assisted extraction (HPAE) has three stages. Firstly, the sample is mixed with the solvent in the pressure vessel at ambient pressure. The sample is subjected to a sudden pressure change up to 100–1000 MPa. At this point, the plant cell wall, the cell membrane, or any other barriers are subjected to a large differential pressure between the inside and outside of the barrier producing deformations and ruptures. The solvent penetrates the barriers through the ruptures and deformations, accessing the cell interior. Once the solvent is in the cell interior, the mass transfer of soluble compounds is favored. Moreover, the differential pressure could exceed the cell's deformation limit (cell wall and/or membrane). This will collapse, resulting in the liberation of all the compounds which will flow to the outside and dissolved in the solvent. Finally, all pressure is quickly released to atmospheric pressure, which produces cell expansion deforming the cell wall and membrane again [136]. The parameters that are usually considered at the time of the extraction are: temperature, pressure, type of solvent and concentration, holding pressure time, the ratio of solvent to raw material and the number of cycles [137]. Table 2 shows how some of these parameters affect the extraction of flavonoids.

Regarding their advantages, the use of HPAE has demonstrated that the extraction could be performed at low temperatures without damaging heat-sensitive compounds or other compounds. Moreover, HPAE is considered an environment-friendly process, so it is a suitable alternative [138]. Other advantages are: the possible combination of more than one solvent to extract more than one type of compound, short periods of extraction, low use of energy or high cell penetration, resulting in higher mass transfer and extraction performance [137]. Nevertheless, in most cases, this technique uses some contaminant solvents, and after the extraction process, a filtration step is mandatory [139].

3.2.3 Microwave-assisted extraction (MAE)

The microwave-assisted extraction (MAE) consists of applying electromagnetic waves to produce changes in the cell wall and membrane. Microwaves have a frequency between 300 MHz and 300 GHz and belong to the electromagnetic field [140]. The MAE process's main advantage is the synergetic combination of heat and mass gradients flowing in the same direction [141]. The electromagnetic waves interact with the polar components inside the cells producing heat through ionic conduction and dipole rotation only in the compounds with an adequate dielectric constant [142]. Depending on the interaction between the compounds and the microwaves the compounds can be classified into three categories: opaque, transparent and absorbing materials. Microwaves heat only absorbing materials by the absorption of the energy of the electromagnetic waves. The mass transfer of flavonoids is produced because of the capacity to heat the cell's intracellular volume, causing an increase of the intracellular pressure producing the collapse of the cell wall and membrane. Then, the compounds can flow out of the cell and the gradient of heat flows [143].

The yield of the flavonoid extraction will depend on the raw material and selected parameters, such as temperature and time of the extraction, composition of the solvent, solvent-to-feed ratio, microwave power, the water content of the matrix and the number of cycles for optimal extraction of flavonoids [141]. Table 2 shows the yields of some flavonoid extractions by MAE and how some parameters affect flavonoid recovery. In comparison with conventional extractions, MAE has demonstrated better time of the extraction, yield, selectivity and quality of the flavonoid extracted. Moreover, the amount of solvent required is lower than in other techniques [144]. Nevertheless, the solvent and target compounds must fulfill some characteristics, compounds must be polar, and solvent must be not too viscous and absorb microwave energy. However, thermally labile compounds cannot be extracted with this method and after the extraction, extract filtration is required [132].

3.2.4 Supercritical fluid extraction (SFE)

A supercritical fluid (SF) is a homogeneous liquid in which the liquid and gas state's demarcation surface disappears. This homogeneous state is caused by exceeding the critical point of temperature and pressure [145]. The diffusivity and density of a SF are between what is expected in a gas and a liquid. As the same as gases, SFs experience a change of density when temperature or pressure are altered, which can produce variations in the density affecting the solvating power [146]. Therefore, these phenomena can improve the solubility of the compounds in the SFs. Supercritical fluid extraction (SFE) is a complex process widely studied along the literature [145, 146, 147, 148]. Nowadays, CO_2 is the most SF used for SFE. CO_2 has some very advantageous characteristics for SFE, low critical temperature (32°C) and pressure (704 MPa). Moreover, CO_2 in low concentrations is non-explosive, non-toxic, non-inflammable and is easy to purchase at a low price with a high degree of purity. Besides, CO_2 has more than double the diffusivity of other fluids with lower surface tension and viscosity. Nevertheless, CO_2 is more suitable for nonpolar compounds than for polar compounds [149, 150].

The main limiting parameters are temperature, pressure and time of extraction [151]. Table 2 shows how these parameters affect the yield of flavonoid SFE. Moreover, other factors like flow rate, modifiers and fractionation can affect the yield of the extraction [146]. The main advantages of SFE are rapidity, low amount of solvent, high selectivity and yield. On the other hand, SFE is a complex process with many parameters to optimize. High investment is needed, and specific alterations such as adding modifiers when extracting polar compounds are necessary [132].

3.2.5 Enzyme assisted extraction (EAE)

The enzyme assisted extraction (EAE) consists of the disruption of the plant cell wall and membrane by the enzymatic digestion of the polysaccharides that conform these two barriers. The plant cell wall comprises a complex structural mixture of polysaccharides, such as hemicellulose, cellulose and pectin, together with other molecules such as structural proteins and lignin [152]. Pectin is composed of a chine of α-D-galacturonate and L-rhamnose units linked by glycosidic bonds in α-1,4 or 1,2 that create the structure called pectic elbows [153]. For the hydrolysis of pectin, several types of pectinases (protopectinases, esterases, depolymerases) are used in the juice industry but also in the extraction of polyphenols [154, 155]. Cellulose is a polymer consisting of glucose β-1,4, which linked to other molecules, gives protection and stability to the cell wall [156]. Cellulases catalyze the breakdown of cellulose. Although its mechanism of action is not fully established, the most accepted theory affirms that three different types of proteins work synergistically during cellulose catalysis. Endonucleases act first, followed by the cellobiohydrolases and, finally, exoglucanases, resulting in free glucose molecules [157]. Hemicellulose is a heterogeneous mixture of carbohydrates homologous to cellulose, such as xyloglucans and mannans. Hemicellulases are a big group of enzymes with several enzymatic activities to break down all hemicellulose forms [158]. Lignins refer to aromatic polymers resulting from the oxidative combinatorial coupling of 4-hydroxyphenylpropanoids [159]. Nowadays, enzymes kits for digestion of the cell wall are prepared to carry out the functions previously mentioned and thus liberate flavonoids in the cell interior and improve the solvent's mass transfer [160]. Besides, EAE could be used alone or combined with other techniques (MAE, UAE, SFE or HPAE) [161].

Parameters like temperature and pH are essential when working with enzymes. Moreover, selected enzymes, mode of action and time are other parameters to consider [161]. Table 3 shows several studies of the extraction of flavonoids from different sources and the yield variation depending on some parameters that affect extraction efficiency. In terms of environmental pollution, this method is one of the most environmentally friendly. Besides, EAE could be performed at low temperature, valid for many different raw materials, and different enzymes can be selected depending on the targets of the extraction [160, 162, 163, 164, 165].

Table 3. Parameters that affect enzyme assisted extraction (EAE) of flavonoids.

Substrate	Enzymes	Solvent	Temperature (°C)	Time (min)	Compound	Yield	Ref.
Ginko biloba (leaves)	Cellulase and pectinase	Eth:W (50:50)	60	1800	Flavonoids	28.3 mg/g dw	[91]
Grape skins	Lallzyme EX-V (commercial) cellulase and hemicellulose, polygalacturonase, pectin lyase, pectin methylesterase	Water	45	179	Flavonoid glycoside and flavan-3-oil	4 mg/g dw	[92]
Grape skins	Lallzyme HC (commercial) cellulase, polygalacturonase, pectin lyase, pectin methylesterase	Water	31	162	Flavonoid glycoside and flavan-3-oil	3.7 mg/g dw	[92]
Grape skins	Endozym Rouge (commercial) cellulase and hemicellulose, polygalacturonase, pectin lyase, pectin methylesterase	Water	39	85	Flavonoid glycoside and flavan-3-oil	3.7 mg/g dw	[92]
Grape skins	Endozym Contact Pelliculaire (commercial) cellulase and hemicellulose, polygalacturonase, pectin lyase, pectin methylesterase	Water	36	128	Flavonoid glycoside and flavan-3-oil	3.7 mg/g dw	[92]
Cajanus cajan (L.) Millsp.	Cellulase, beta-glucosidase, pectinase	Water	32.5	1080	Luteolin and apigenin	0.4 mg/g dw	[93]
Larix gmelina (Rupr) Rupr.	Cellulase, pectinase	Water	32	1080	Flavonoids	4.96 ± 0.29 mg/g dw	[94]
Citrus x *paradisi* (peel)	Cellulose ® MX, Kleerase ® AFP	Water	50	180	Phenolics	1.62 mg GAE/g fw	[95]

SBE: sequential batch extraction; HRE: heat reflux extraction; SAE: stirring-assisted extraction; QE: quercetin equivalent; RE: rutin equivalent; dw: dry weight; and fw: fresh weight.

4. Bioavailability of Flavonoids

Bioavailability refers to the concentration of a molecule or related like-molecules that become absorbed and available for exerting their biological activity in the site of drug action of the target tissue, organ or system [166]. The term bioavailability is strongly related to the concept of bioaccessibility and to bioactivity. Bioaccessibility refers to the number of compounds that, after digestion, becomes available and absorbable through the intestinal epithelium. This definition is linked to bioactivity, which involves the physiological effects that biomolecules trigger in the organism and includes their transport through systemic circulation to the target receptor and their interaction with other biomolecules [167].

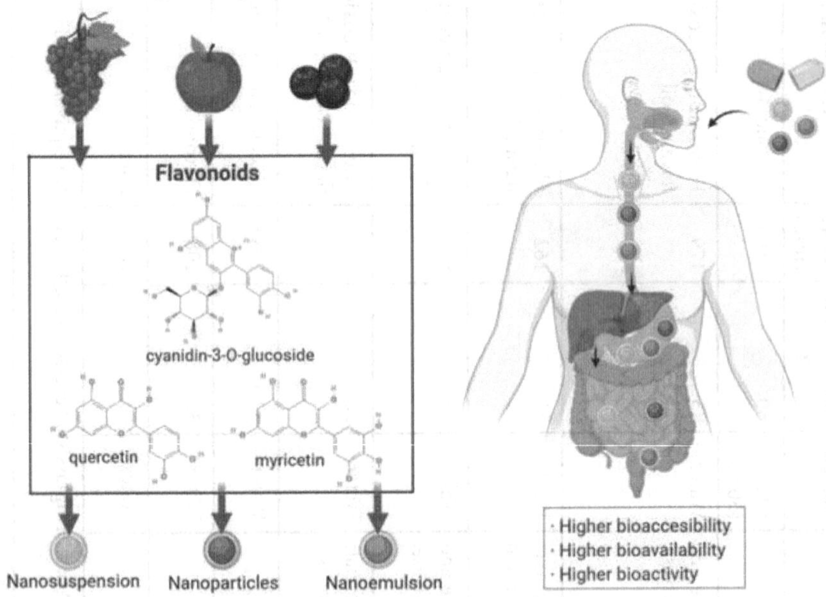

Figure 7. Strategies to enhance flavonoids bioavailability. Nanosuspension, nanoencapsulation or nanoemulsions have been proved as successful approaches to improve flavonoids solubility and enhance their bioavailability, bioaccesibility and, bioactivity.

The bioavailability of polyphenolic compounds has been described to be poor since they hardly reach bioaccessibility rates higher than 30–50% [167]. Among the parameters involved in this low bioavailability, there are several physicochemical properties of flavonoids which include their chemical structure, polymerization degree, solubility, variability of attached saccharides or potential interactions they established with other compounds, or flavonoids stability, both during storage and along the digestion process [168]. Different approaches have been developed to enhance the accessibility to the final number of flavonoids or to blur their metabolism through digestion and improve and extend their chemical stability. These intend to maximize the bioavailability of flavonoids. To increase the available concentration of flavonoids in food, different treatments have been applied to food matrixes. The main purpose is to alter the matrix's structural organization in which biomolecules are embedded so they can get easily released. Both heating and freezing approaches have been tested and demonstrated to positively affect the bioaccessibility of polyphenols [167]. Nevertheless, other techniques requiring more technological development have demonstrated a better performance to improve flavonoids bioaccessibility (Figure 7). In fact, the pharmaceutical industry has established alternative approaches to improve the oral bioavailability of flavonoids with clinical applications. Some of the most utilized strategies are the use of absorption enhancers (nonionic surfactants, myo-inositol hexaphosphate, chitosan or pectin), the induction of structural transformations which include the introduction of functional groups with higher polarity (sulfuric acids, amino acids, carbamoyls, glycosides, etc.), or the complexation with a carrier (such as cyclodextrins, phospholipids or polymeric carriers) [169].

Among these approaches, nanosuspension, in which pure drug particles are combined with stabilizers, has been demonstrated as a promising strategy to enhance the bioavailability of flavonoids. This system facilitates the delivery of flavonoids using particles in the nanometers range, which allows reaching a higher concentration quickly by increasing solubility and dissolution rates. For instance, in a recently published work in which different nanosuspension formulae were applied, the solubility of myricetin was increased from 43 to nearly 75 times. This increment was accompanied by an improved bioavailability in the relative range of 161–357% [169]. Another flavonol, quercetin, was also submitted to nanosuspension. This strategy improved its saturation solubility about eleven times which also provided a much better bioaccessibility. The bioaccesibility increased slowly, reaching its maximum peak between 2 and 3 h, while the pure molecule reached this maximum at 1.5–2 h. The amount of quercetin released from the nanosuspension was higher than the pure, duplicating its bioaccessibility even at the last measured times [170]. The bioactivity of the orally administrated flavanone, naringenin, was also tested using rats. It was shown that the nanosuspension of naringenin was nearly 4 times higher when compared against the control [171].

A common methodology applied in both food and pharmacological industries for increasing flavonoids bioavailability and bioaccesibility, which ultimately enhances their potential bioactivity, relies on their encapsulation [169, 172]. Different techniques, with diverse complexity degrees, permit the encapsulation of a considerable variability of core ingredients using different shell materials for obtaining capsules with various physical properties. Some of the most used encapsulation methods include spray or freeze-drying, spray chilling and cooling, coacervation, fluidized bed coating, liposome entrapment, rotational suspension separation, extrusion and inclusion complexation, (micro) emulsions, etc. [173]. The main aim of the encapsulation process is to prevent biological and physicochemical degradation of bioactive ingredients. Encapsulation permits to extend the chemical stability of the target molecules and thus their bioactivities. Besides, encapsulation may also allow the controlled release of the compounds delivered using concentrations. Scientific literature provides several examples of flavonoids that have been encapsulated. Among the benefits of encapsulation, bioaccesibility and bioavailability are two parameters that can be improved using this approach. Besides, encapsulation permits to embed flavonoids in the most appropriate matrices to reinforce their stability [172]. The flavonoid subclass of anthocyanins has been extensively used to evaluate the performance of different encapsulation techniques and materials. For instance, anthocyanins from grape peels have been submitted to encapsulation by emulsification/internal gelation using both spray and freeze-drying techniques. The former provided smaller microcapsules (0.6 μm) with higher encapsulation efficiency and better microcapsule and anthocyanins stability, which extended their release in simulated gastrointestinal digestion and improved their bioaccessibility [174]. Two major anthocyanins were identified in extracts from *Rubus idaeus* as cyanidin-3-*O*-glucoside and 3-*O*-sophoroside and encapsulated using β-lactoglobulin. This encapsulation strategy was demonstrated to increase the bioavailability of these anthocyanins after submitting them to simulated gastrointestinal digestion up to nearly double (19%) than its free presentation (11%) [175]. Different prototypes of nanoparticles were created for encapsulating commercial anthocyanins, containing as major representatives cyanidin-3-*O*-glucoside and peonidin-3-*O*-glucoside. The improvement of free anthocyanins' bioavailability was determined to get increased from 17 to 27% when loaded in chitosan-based nanoparticles and to 40% when anthocyanins were encapsulated using a mixture of chitosan and β-lactoglobulin [176]. Other flavonoids have also been encapsulated with a demonstrated enhancement of their bioavailability. Quercetin, naringenin, and hesperetin were nanoencapsulated, thus preventing their degradation under the hydrolytic conditions of the small intestine. It also reinforced their bioaccesibility, especially in the case of quercetin which reached a relative bioaccesibility after simulated digestion of around 80% when encapsulated against the scarcely 10% when present as a free molecule. Similarly, free naringenin and hesperetin showed low bioaccesibility (about 15–20%) that reach about 70% when encapsulated [177]. The encapsulation of the glycoside flavonoid, naringin, prompted a progressive release in both gastric and intestinal environments when compared to the free molecule. In the intestine, the liberation was faster, probably due to the presence of bile salts and pancreatin. However, this poorer release at the gastric level and higher at the intestinal level is desirable. It would allow the arrival of higher concentrations of naringin to the intestine, and thus more amounts of this biocompound would be available for its adsorption [178].

Another technique tested for enhancing the bioavailability of flavonoids is based on their emulsion. This emulsion can be created by utilizing emulsifying agents or mixtures of oil-(co)surfactants-water, which permit the self-emulsion with a simple process, agitation. In fact, different alternatives of this approach have been proved successful for different kinds of flavonoids. Anthocyanins from blueberry fruits were micro-emulsified and their bioavailability and bioactivity were evaluated against the control (without vehicle). Non-purified and purified anthocyanins, especially malvidin-3-*O*-glucoside, showed hypoglycemic activity when emulsified while those without vehicle did not [179]. Quercetin was also self-emulsified in an oily matrix using Tween 20 as surfactant and ethanol as co-surfactant. The oral bioavailability of this specific emulsified flavonol was increased up to 5 times [180]. An even better bioavailability was obtained through the quercetin's nano-emulsion using an oil-water phase and different surfactants and co-surfactants. The optimized emulsified presentation of this flavonol was based on an equimolar mixture of Capryol and Labrafil as oily phase representing a 17% while water phase was around 33%, and remaining fractions corresponding to an equimolar combination of Labrasol and Tween 80 as surfactant and an equimolar mixture of Cremophor EL:polyethylene glycol 400 as co-surfactant. The emulsion obtained following this protocol improved its gastrointestinal permeability, accompanied by enhanced bioactivity (higher mice weight reduction) and bioavailability in rats, 33 times higher than its aqueous dispersion [181]. Similarly, the flavonol, myricetin, was micro-emulsified using 6% of Tween 80, 12% of

Cremophor RH40, 9% of Transcutol HP, 18% of WL 1349, and 55% of distilled water. This formula allowed the increment of its aqueous solubility more than 1000 times, increased its bioavailability 14 times and hence its bioactivities, both antioxidant and anti-proliferative [182]. From the sub-group of flavones, tangeretin was also emulsified using as part of the oily phase medium-chain triacylglycerol and lecithin to which aqueous phase was added following previously optimized temperature, pressure and stirring conditions. This emulsion doubled the bioavailability of tangeretin, which ultimately improved its tested bioactivities, especially the anti-proliferative [183, 184]. Also the flavone baicalein has been nano-emulsified using a composed oil (isopropyl myristate) and water, Cremophor EL35 as surfactant and propylene glycol as co-surfactant. This preparation improved intestinal absorption of baicalein along all the different parts of the intestine, especially in duodenum and jejunum. This kind of nanoemulsions protect the core ingredient from enzymatic damage and can prolong drug retention in the intestine. In fact, its pharmacokinetic behavior was improved 7 times, and its oral exposure increased nearly 15 times compared to the suspension of baicalein [185].

Therefore, very different techniques have been proved to improve the poor solubility of flavonoids and to enhance their bioavailability, bioaccesibility and, hence, their bioactivity. Among these techniques, the most successful ones are based on the application of nanosuspensions, encapsulations or emulsions of the flavonoids.

5. Development of Flavonoids Based Natural Additives: Reported and Future Applications

Currently, consumers are increasingly aware of their healthier food choices, associating them with their health and well-being. In this way, there is a global demand on the part of the food industry to develop innovative natural products and health promoters that contain bioactive components [186].

Different bioactive ingredients, namely flavonoids, have been studied to adapt organoleptic, sensory and conservation properties. They have also been explored as functional ingredients with bioactive properties, such as antioxidant, anti-inflammatory and immunomodulatory referred to earlier in this manuscript [187, 188, 189]. Thus, bioactive compounds are considered valuable options to be explored in the design of innovative food formulations with health benefits.

Different flavonoids have been studied, and their bioactive properties have been proven by several authors, which has arisen the high interest of the food industry in their application in functional foods. Foods and beverages such as dairy products, bakery and confectionery products, meat products, juices and energy drinks, snacks, pasta, gums and sweets are some of the products explored the most in the addition of bioactive compounds [190].

In a recent study, the stability of anthocyanins from grape residues was evaluated when applied as a food coloring in carbonated water and proved that the degradation of the incorporated anthocyanins followed the kinetic behavior during storage, when exposed to light or dark [191]. The anthocyanin malvidin-3-glycoside showed the greatest stability when added to the water. Additionally, it was found that the light had adverse effects on the color of the carbonated water. Bakery and pastry products, recognized for providing consumers of all ages with pleasure and fun, have been explored exponentially in an attempt to find functional natural ingredients and/or colors with potential for application in a highly competitive area [192]. A recent study intended to explore the bioaccesibility and bioavailability of phenolic compounds (namely flavonoids) obtained from green tea in wheat bread. The results showed an increase in the nutraceutical potential and the protection of lipids against oxidation. Also, *in vitro* studies showed that digestion induced the release of phenolic compounds in bread, proving bioaccesibility and bioavailability [193]. In its turn, another study optimized the obtaining of an anthocyanin-rich extract (33.58 mg anthocyanins per g of extract where cyanidin-3-*O*-glucoside appears as the major anthocyanin compound) from *Rubus ulmifolius* Schott. The authors described an extract with bioactive properties (antioxidant, anti-tumor and antimicrobial potential) with excellent coloring ability when applied to a pastry product, the donut, proving to be a great approach for replacing artificial colorants [194].

Dairy products have been extensively tested and explored due to the industry's high interest to supplement functional ingredients [195]. Aqueous extracts of *Foeniculum vulgare* Mill. and *Matricaria recutita* L. rich in phenolic compounds namely flavonoids (11.52 ± 0.11 mg/g and 17.89 ± 0.91 mg/g flavonoids, respectively) were explored as health-promoting natural ingredients in cottage cheese [196, 197]. The results obtained demonstrate that the tested natural extracts improved the natural conservation of the cottage cheese by increasing the shelf life and adding antioxidant properties to the product. Since yogurt is considered one of the most traditionally consumed dairy snacks globally, several studies explore functional ingredients with preservative and coloring properties. The coloring ability of different anthocyanin extracts from edible flowers (rose, cornflower and dahlia) was tested in yoghurts as a replacement for artificial colors (specifically E163, anthocyanin extract). The results showed the hydrophilic rose extract as the most appropriate natural ingredient to replace E163 since, in addition to not altering the nutritional composition of the product, it presented close scores in the color parameters achieved by the artificial colorant [198].

Some fruits have also been explored as natural ingredients. In a recent study, extracts from *Vaccinium myrtillus* L. fruits (blueberry) revealed a high concentration of anthocyanins (21.1 ± 0.2 mg/g), highlighting malvidin glycoside and delphinidin glycoside derivatives as the majority [199]. The high content of these compounds was responsible for their bioactive properties that arouse interest for incorporation. In the same study, the coloring potential of the natural ingredient in yoghurts was tested. The results showed that although the blueberry extract has a lower coloring capacity when compared to the artificial additive E163, it presented greater stability over the storage time.

However, incorporating these compounds in this type of food product has represented a challenge about the quality of the final product and the stability of bioactive compounds. The high-water content and low pH value of yogurt as well as the low solubility of polyphenols have represented a great challenge for the use of herbal extracts, especially hydrophobic extracts [200]. The use of bioactive compounds as natural ingredients in food products has been characterized in several studies as limited due to their stability and bioavailability. Storage conditions, thermal and non-thermal processes and extraction treatments are some of the parameters identified as responsible for affecting these compounds' effectiveness [201, 202]. Some bioactive compounds are susceptible to environmental factors, namely pH, temperature, oxygen, enzymes, light, metal ions, sulfur dioxide and ascorbic acid [203]. The molecular interactions between bioactive compounds with other food ingredients can also affect some properties of these compounds, such as bioavailability, bioactivity and organoleptic properties [204]. After oral consumption, the chemical structure and bioactivities of the components are altered in intestinal metabolism. Thus, it is necessary to ensure that the bioactive compounds in the gastrointestinal tract are stable and allow controlled release at target points [205].

This type of limitations has been a concern for the food industry since it can hamper its industrial application. For example, quercetin is a flavonol recognized for its anti-diabetic properties; however, its low solubility and aqueous permeability limit its application. Anthocyanins, which are very attractive due to their ability to provide color and potential health benefits, have also represented a major industrial challenge in controlling their deterioration and increasing their bioavailability in food systems [206, 207].

For this reason, different microencapsulation and delivery systems have been explored to guarantee the production of functional foods with acceptable organoleptic characteristics and the controlled release of flavonoids, thus preventing interactions with other food components, and overcoming problems encountered during food processing and gastrointestinal transit [208, 209]. Some examples will be mentioned as follows. A blueberry-derived mixture of anthocyanins was encapsulated into chitosan nanoparticles, and its stability in a drink was evaluated. The results suggested that the chitosan nanoparticles delayed the anthocyanin degradation in the simulated gastrointestinal fluid and increased the anthocyanin storage stability in the drink [201]. In other work, an encapsulated polyphenolic extract (rich in anthocyanins) from Artemide black rice obtained through the atomization process with maltodextrins and gum arabic (50:50, w/w) was incorporated into biscuits. The results showed that the encapsulated ingredient emerged as the most stable during storage and cooking and with the most significant antioxidant capacity than the control biscuit [210].

Also, *in vitro* assays such as those presented above make it possible to understand the potential beneficial health effects these compounds may have after ingestion [211]. However, studies on *in vitro* release of microencapsulated phenolic compounds are still relatively scarce. A study demonstrated a better solubility of microencapsulated curcumin and quercetin in niosomes [212]. Also, an anthocyanin extract obtained from fruits of *V. myrtillus* L. was tested in a gastrointestinal model *in vitro*, and the results demonstrated an improvement in stability in adverse pH conditions during digestion, being released only in the intestinal mucosa [213].

The exploitation of natural ingredients with antioxidant and antimicrobial properties in combination with natural polymers has also been ceased by the scientific community in the development of edible films that allow to reduce the dependence on synthetic polymers and offer viable solutions for industrial application [214]. Polyamide, polyethylene terephthalate, ethylene vinyl alcohol, polyvinylidene chloride, polypropylene and polyethylene are some of the most widely used polymer materials in the food industry for food preservation. However, some authors report that polymers in direct contact with food allow the migration of the additives and other components to food, causing some adverse effects for consumers [215]. In this sense, studies on plastics and plasticizers and non-toxic bio-based food coatings to replace their synthetic counterparts have been increasing [216]. These coatings make it possible to coordinate natural polymers with bioactive ingredients from plant extracts with preservative and antimicrobial properties that improve the organoleptic and functional properties of food [217].

6. Concluding Remarks and Future Research Directions

Although many bioactive compounds are currently tested in different food matrices to improve their organoleptic properties, to fortify and functionalize these same products, it is considered that the protection of such functional ingredients in the food matrix during processing, storage and passage the gastrointestinal tract has been little explored. Several flavonoids have been extensively studied and have shown to be highly promising bioactive compounds, capable of improving the physical-chemical, sensory and health properties of food products. However, studies on the effectiveness and interactions of these bioactive compounds for the development of new innovative products are still scarce and there are some gaps between digestion, metabolism and bioactive substance delivery approaches across biological barriers that must be explored. This type of studies' transition to a commercial scale is an essential future step in innovation to provide more practical information that can be transposed to industry. Thus, and to meet consumers preferences and requirements, there is a great need for new and more complete *in vivo* studies capable of verifying the appropriate dosage for different ingredients to be incorporated in different food matrices and, consequently, build the proper design of food products ensuring the desired safety and functionality. The production of functional foods is a current trend in the growing exploitation of the food industry. Therefore, the exploration of new bioactive compounds from different sources to deliver and modulate the properties of foods will be the objective of analysis in the scientific community to respond to the industry's needs.

References

1. Colla L, Moraes F. Alimentos funcionais e nutracêuticos: Definições, legislação e benefícios à saúde. Revista Eletronica de Farmacia. 2006;3:109-122. DOI: 10.5216/ref.v3i2.2082

2. Valls J, Millán S, Martí MP, Borràs E, Arola L. Advanced separation methods of food anthocyanins, isoflavones and flavanols. Journal of Chromatography A. 2009;1216:7143-7172. DOI: 10.1016/j.chroma.2009.07.030

3. Wen W, Alseekh S, Fernie AR. Conservation and diversification of flavonoid metabolism in the plant kingdom. Current Opinion in Plant Biology. 2020;55:100-108. DOI: 10.1016/j.pbi.2020.04.004

4. Engin AB, Tsatsakis AM, Tsoukalas D, Engin A. Do flavanols-rich natural products relieve obesity-related insulin resistance? Food and Chemical Toxicology. 2018;112:157-167. DOI: 10.1016/j.fct.2017.12.055

5. Akhlaghi M, Ghobadi S, Mohammad Hosseini M, Gholami Z, Mohammadian F. Flavanols are potential anti-obesity agents, a systematic review and meta-analysis of controlled clinical trials. Nutrition, Metabolism, and Cardiovascular Diseases. 2018;28:675-690. DOI: 10.1016/j.numecd.2018.04.001

6. Šeruga M, Tomac I. Influence of chemical structure of some flavonols on their electrochemical behaviour. International Journal of Electrochemical Science. 2017;12:7616-7637. DOI: 10.20964/2017.08.79

7. Zhang N, Bi F, Xu F, Yong H, Bao Y, Jin C, et al. Structure and functional properties of active packaging films prepared by incorporating different flavonols into chitosan based matrix. International Journal of Biological Macromolecules. 2020;165:625-634. DOI: 10.1016/j.ijbiomac.2020.09.209

8. Mouffouk C, Mouffouk S, Mouffouk S, Hambaba L, Haba H. Flavonols as potential antiviral drugs targeting SARS-CoV-2 proteases (3CLpro and PLpro), spike protein, RNA-dependent RNA polymerase (RdRp) and angiotensin-converting enzyme II receptor (ACE2). European Journal of Pharmacology. 2021;891:173759. DOI: 10.1016/j.ejphar.2020.173759

9. Seleem D, Pardi V, Murata RM. Review of flavonoids: A diverse group of natural compounds with anti-Candida albicans activity in vitro. Archives of Oral Biology. 2017;76:76-83. DOI: 10.1016/j.archoralbio.2016.08.030

10. Jiang X, Shi Y, Fu Z, Li WW, Lai S, Wu Y, et al. Functional characterization of three flavonol synthase genes from Camellia sinensis: Roles in flavonol accumulation. Plant Science. 2020;300:110632. DOI: 10.1016/j.plantsci.2020.110632

11. Duan Y, Eduardo Melo Santiago F, Rodrigues dos Reis A, de Figueiredo MA, Zhou S, Thannhauser TW, et al. Genotypic variation of flavonols and antioxidant capacity in broccoli. Food Chemistry. 2021;338:127997. DOI: 10.1016/j.foodchem.2020.127997

12. Aherne SA, O'Brien NM. Dietary flavonols: Chemistry, food content, and metabolism. Proceedings of the Nutrition. 2002;18:75-81

13. Jones P, Messner B, Nakajima JI, Schäffner AR, Saito K. UGT73C6 and UGT78D1, glycosyltransferases Involved in flavonol glycoside biosynthesis in Arabidopsis thaliana. The Journal of Biological Chemistry. 2003;278:43910-43918. DOI: 10.1074/jbc.M303523200

14. Company A, Lloret-Fillol J, Costas M. Small molecule models for nonporphyrinic iron and manganese oxygenases. In: Comprehensive Inorganic Chemistry II (Second Edition): From Elements to Applications. Vol. 3. Ed. Jan Reedijk: Northwestern University, US; 2013. pp. 487-564. ISBN 9780080965291

15. Ottaviani JI, Momma TY, Heiss C, Kwik-Uribe C, Schroeter H, Keen CL. The stereochemical configuration of flavanols influences the level and metabolism of flavanols in humans and their biological activity in vivo. Free Radical Biology & Medicine. 2011;50:237-244. DOI: 10.1016/j.freeradbiomed.2010.11.005

16. Robbins RJ, Kwik-Uribe C, Hammerstone JF, Schmitz HH. Analysis of flavanols in foods: What methods are required to enable meaningful health recommendations? Journal of Cardiovascular Pharmacology. 2006;47:S110-S118. DOI: 10.1097/00005344-200606001-00004

17. Aron PM, Kennedy JA. Flavan-3-ols: Nature, occurrence and biological activity. Molecular Nutrition & Food Research. 2008;52:79-104. DOI: 10.1002/mnfr.200700137

18. Serra A, Macià A, Romero MP, Salvadó MJ, Bustos M, Fernández-Larrea J, et al. Determination of procyanidins and their metabolites in plasma samples by improved liquid chromatography-tandem mass spectrometry. Journal of Chromatography B. 2009;877:1169-1176. DOI: 10.1016/j.jchromb.2009.03.005

19. Singh M, Kaur M, Silakari O. Flavones: An important scaffold for medicinal chemistry. European Journal of Medicinal Chemistry. 2014;84:206-239. DOI: 10.1016/j.ejmech.2014.07.013

20. Maleki SJ, Crespo JF, Cabanillas B. Anti-inflammatory effects of flavonoids. Food Chemistry. 2019;299:125124. DOI: 10.1016/j.foodchem.2019.125124

21. Jiang N, Doseff AI, Grotewold E. Flavones: From biosynthesis to health benefits. Plants. 2016;5:27

22. Verma AK, Pratap R. The biological potential of flavones. Natural Product Reports. 2010;27:1571-1593. DOI: 10.1039/c004698c

23. Hariri BM, McMahon DB, Chen B, Freund JR, Mansfield CJ, Doghramji LJ, et al. Flavones modulate respiratory epithelial innate immunity: Anti-inflammatory effects and activation of the T2R14 receptor. The Journal of Biological Chemistry. 2017;292:8484-8497

24. Hariri BM, McMahon DB, Chen B, Adappa ND, Palmer JN, Kennedy DW, et al. Plant flavones enhance antimicrobial activity of respiratory epithelial cell secretions against Pseudomonas aeruginosa. PLoS One. 2017;12:e0185203

25. Ivanov M, Kannan A, Stojković DS, Glamočlija J, Calhelha RC, Ferreira ICFR, et al. Flavones, flavonols, and glycosylated derivatives—Impact on Candida albicans growth and virulence, expression of cdr1 and erg11, cytotoxicity. Pharmaceuticals. 2021;14:27

26. Barreca D, Mandalari G, Calderaro A, Smeriglio A, Trombetta D, Felice MR, et al. Citrus flavones: An update on sources, biological functions, and health promoting properties. Plants. 2020;9:288

27. Shankar E, Goel A, Gupta K, Gupta S. Plant flavone apigenin: An emerging anticancer agent. Current Pharmacology Reports. 2017;3:423-446

28. Zaragozá C, Villaescusa L, Monserrat J, Zaragozá F, Álvarez-Mon M. Potential therapeutic anti-inflammatory and immunomodulatory effects of dihydroflavones, flavones, and flavonols. Molecules. 2020;25:1017

29. Hostetler GL, Ralston RA, Schwartz SJ. Flavones: Food sources, bioavailability, metabolism, and bioactivity. Advances in Nutrition. 2017;8:423-435

30. Martens S, Mithöfer A. Flavones and flavone synthases. Phytochemistry. 2005;66:2399-2407. DOI: 10.1016/j.phytochem.2005.07.013

31. Woo Y, Shin SY, Hyun J, Lee SD, Lee YH, Lim Y. Flavanones inhibit the clonogenicity of HCT116 cololectal cancer cells. International Journal of Molecular Medicine. 2012;29:403-408. DOI: 10.3892/ijmm.2011.857

32. Khan MK, Zill-E-Huma; Dangles, O. A comprehensive review on flavanones, the major citrus polyphenols. Journal of Food Composition and Analysis. 2014;33:85-104. DOI: 10.1016/j.jfca.2013.11.004

33. Tomás-Barberán FA, Clifford MN. Flavanones, chalcones and dihydrochalcones—Nature, occurrence and dietary burden. Journal of the Science of Food and Agriculture. 2000;80:1073-1080. DOI: 10.1002/(SICI)1097-0010(20000515)80:7<1073::AID-JSFA568>3.0.CO;2-B
34. Khan A, Ikram M, Hahm JR, Kim MO. Antioxidant and anti-inflammatory effects of citrus flavonoid hesperetin: Special focus on neurological disorders. Antioxidants. 2020;9:609
35. Tutunchi H, Naeini F, Ostadrahimi A, Hosseinzadeh-Attar MJ. Naringenin, a flavanone with antiviral and anti-inflammatory effects: A promising treatment strategy against COVID-19. Phytotherapy Research. 2020;34:3137-3147
36. Scoditti E. Neuroinflammation and neurodegeneration: The promising protective role of the citrus flavanone hesperetin. Nutrients. 2020;12:2336
37. Mendes LA d O, Ponciano CS, Cataneo AHD, Wowk PF, Bordignon J, Silva H, et al. The anti-Zika virus and anti-tumoral activity of the citrus flavanone lipophilic naringenin-based compounds. Chemico-Biological Interactions. 2020;331:109218
38. Barreca D, Gattuso G, Bellocco E, Calderaro A, Trombetta D, Smeriglio A, et al. Flavanones: Citrus phytochemical with health-promoting properties. BioFactors. 2017;43:495-506
39. Stevens Y, Rymenant EV, Grootaert C, Camp JV, Possemiers S, Masclee A, et al. The intestinal fate of citrus flavanones and their effects on gastrointestinal health. Nutrients. 2019;11:1464. DOI: 10.3390/nu11071464
40. Testai L, Calderone V. Nutraceutical value of citrus flavanones and their implications in cardiovascular disease. Nutrients. 2017;9:502
41. Křížová L, Dadáková K, Kašparovská J, Kašparovský T. Isoflavones. Molecules. 2019;24:1076. DOI: 10.3390/molecules24061076
42. Ko KP. Isoflavones: Chemistry, analysis, functions and effects on health and cancer. Asian Pacific Journal of Cancer Prevention. 2014;15:7001-7010. DOI: 10.7314/APJCP.2014.15.17.7001
43. Jheng HF, Hayashi K, Matsumura Y, Kawada T, Seno S, Matsuda H, et al. Anti-inflammatory and antioxidative properties of isoflavones provide renal protective effects distinct from those of dietary soy proteins against diabetic nephropathy. Molecular Nutrition & Food Research. 2020;64:2000015
44. Wang S, Wang Y, Pan MH, Ho CT. Anti-obesity molecular mechanism of soy isoflavones: Weaving the way to new road of therapeutic. Food & Function. 2017;8:3831-3846
45. Bilir B, Sharma NV, Lee J, Hammarstrom B, Svindland A, Kucuk O, et al. Effects of genistein supplementation on genome-wide DNA methylation and gene expression in patients with localized prostate cancer. International Journal of Oncology. 2017;51:223-234
46. Gómez-Zorita S, González-Arceo M, Fernández-Quintela A, Eseberri I, Trepiana J, Portillo MP. Scientific evidence supporting the beneficial effects of isoflavones on human health. Nutrients. 2020;12:3853

47. Alipour MR, Karimi-Sales E. Molecular mechanisms of protective roles of isoflavones against chemicals-induced liver injuries. Chemico-Biological Interactions. 2020;329:109213

48. Ma L, Liu G, Ding M, Zong G, Hu FB, Willett WC, et al. Isoflavone intake and the risk of coronary heart disease in US men and women: Results from 3 prospective cohort studies. Circulation. 2020;141:1127-1137

49. Duru KC, Kovaleva EG, Danilova IG, Bijl P v d, Belousova AV. The potential beneficial role of isoflavones in type 2 diabetes mellitus. Nutrition Research. 2018;59:1-15

50. Pabich M, Materska M. Biological effect of soy isoflavones in the prevention of civilization diseases. Nutrients. 2019;11:1660

51. Vázquez L, Flórez AB, Guadamuro L, Mayo B. Effect of soy isoflavones on growth of representative bacterial species from the human gut. Nutrients. 2017;2017:9

52. Pessoa E d A, Convento MB, Castino B, Leme AM, Oliveira AS d, Aragão A, et al. Beneficial effects of isoflavones in the kidney of obese rats are mediated by PPAR-gamma expression. Nutrients. 2020;12:1624

53. Mojica L, Berhow M, Gonzalez de Mejia E. Black bean anthocyanin-rich extracts as food colorants: Physicochemical stability and antidiabetes potential. Food Chemistry. 2017;229:628-639. DOI: 10.1016/j.foodchem.2017.02.124

54. Schiozer AL, Barata LES. Estabilidade de Corantes e Pigmentos de Origem Vegetal. Revista Fitos. 2013;3:6-24

55. Khoo HE, Azlan A, Tang ST, Lim SM. Anthocyanidins and anthocyanins: Colored pigments as food, pharmaceutical ingredients, and the potential health benefits. Food & Nutrition Research. 2017;61:1361779. DOI: 10.1080/16546628.2017.1361779

56. Mazza G, Cacace JE, Kay CD. Methods of analysis for anthocyanins in plants and biological fluids. Journal of AOAC International. 2004;87:129-145. DOI: 10.1093/jaoac/87.1.129

57. Horbowicz, M.; Grzesiuk, A.; DĘBski, H.; Kosson, R. Anthocyanins of fruits and vegetables—Their occurrence, analysis and role in human. Vegetable Crops Research Bulletin 2008, 68, 5–22, doi:10.2478/v10032-008-0001-8

58. Bakowska-Barczak A. Acylated anthocyanins as stable, natural food colorants : A review. Polish Journal of Food and Nutrition Sciences. 2005;14:107-116

59. Kaume L, Howard LR, Devareddy L. The blackberry fruit: A review on its composition and chemistry, metabolism and bioavailability, and health benefits. Journal of Agricultural and Food Chemistry. 2012;60:5716-5727. DOI: 10.1021/jf203318p

60. Gürbüz N, Uluişik S, Frary A, Frary A, Doğanlar S. Health benefits and bioactive compounds of eggplant. Food Chemistry. 2018;268:602-610. DOI: 10.1016/j.foodchem.2018.06.093

61. Dixon RA, Pasinetti GM. Flavonoids and isoflavonoids: From plant biology to agriculture and neuroscience. Plant Physiology. 2010;154:453-457. DOI: 10.1104/pp.110.161430

62. Havsteen BH. The biochemistry and medical significance of the flavonoids. Pharmacology & Therapeutics. 2002;96:67-202

63. Marella S. Flavonoids—The most potent poly-phenols as antidiabetic agents: An overview. Modern Approaches in Drug Designing. 2017;1:2-6. DOI: 10.31031/madd.2017.01.000513

64. Robbins RJ. Phenolic acids in foods: An overview of analytical methodology. Journal of Agricultural and Food Chemistry. 2003;51:2866-2887

65. Shahidi F, Ambigaipalan P. Phenolics and polyphenolics in foods, beverages and spices: Antioxidant activity and health effects—A review. Journal of Functional Foods. 2015;18:820-897. DOI: 10.1016/j.jff.2015.06.018

66. Wang T, Li Q, Bi K. Bioactive flavonoids in medicinal plants: Structure, activity and biological fate. Asian Journal of Pharmaceutical Sciences. 2018;13:12-23. DOI: 10.1016/j.ajps.2017.08.004

67. Heim KE, Tagliaferro AR, Bobilya DJ. Flavonoid antioxidants: Chemistry, metabolism and structure-activity relationships. The Journal of Nutritional Biochemistry. 2002;13:572-584

68. Vo QV, Nam PC, Thong NM, Trung NT, Phan CTD, Mechler A. Antioxidant motifs in flavonoids: O–H versus C–H Bond Dissociation. ACS. Omega. 2019;4:8935-8942. DOI: 10.1021/acsomega.9b00677

69. Novaes P, Torres PB, Cornu TA, de Lopes JC, Ferreira MJP, Santos do DYAC. Comparing antioxidant activities of flavonols from Annona coriacea by four approaches. South African Journal of Botany. 2019;123:253-258. DOI: 10.1016/j.sajb.2019.03.011

70. Hidalgo M, Sánchez-Moreno C, de Pascual-Teresa S. Flavonoid-flavonoid interaction and its effect on their antioxidant activity. Food Chemistry. 2010;121. DOI: 10.1016/j.foodchem.2009.12.097

71. Ragab FA, Yahya TAA, El-Naa MM, Arafa RK. Design, synthesis and structure-activity relationship of novel semi-synthetic flavonoids as antiproliferative agents. European Journal of Medicinal Chemistry. 2014;82. DOI: 10.1016/j.ejmech.2014.06.007

72. Cushnie TPT, Lamb AJ. Antimicrobial activity of flavonoids. International Journal of Antimicrobial Agents. 2005;26

73. Cheesman MJ, Ilanko A, Blonk B, Cock IE. Developing new antimicrobial therapies: Are synergistic combinations of plant extracts/compounds with conventional antibiotics the solution? Pharmacognosy Reviews. 2017;11

74. Biharee A, Sharma A, Kumar A, Jaitak V. Antimicrobial flavonoids as a potential substitute for overcoming antimicrobial resistance. Fitoterapia. 2020;146

75. Alves-Santos AM, Sugizaki CSA, Lima GC, Naves MMV. Prebiotic effect of dietary polyphenols: A systematic review. Journal of Functional Foods. 2020;74

76. Gibson GR, Hutkins R, Sanders ME, Prescott SL, Reimer RA, Salminen SJ, et al. Expert consensus document: The International Scientific Association for Probiotics and Prebiotics (ISAPP) consensus statement on the definition and scope of prebiotics. Nature Reviews. Gastroenterology & Hepatology. 2017;14

77. Sanders ME, Merenstein DJ, Reid G, Gibson GR, Rastall RA. Probiotics and prebiotics in intestinal health and disease: From biology to the clinic. Nature Reviews. Gastroenterology & Hepatology. 2019;16

78. Vendrame S, Guglielmetti S, Riso P, Arioli S, Klimis-Zacas D, Porrini M. Six-week consumption of a wild blueberry powder drink increases Bifidobacteria in the human gut. Journal of Agricultural and Food Chemistry. 2011;59. DOI: 10.1021/jf2028686

79. González-Sarrías A, Romo-Vaquero M, García-Villalba R, Cortés-Martín A, Selma MV, Espín JC. The endotoxemia marker lipopolysaccharide-binding protein is reduced in overweight-obese subjects consuming pomegranate extract by modulating the gut microbiota: A randomized clinical trial. Molecular Nutrition & Food Research. 2018;62. DOI: 10.1002/mnfr.201800160

80. Spence C. On the psychological impact of food colour. Flavour. 2015;4:4-21. DOI: 10.1186/s13411-015-0031-3

81. Oplatowska-Stachowiak M, Elliott CT. Food colors: Existing and emerging food safety concerns. Critical Reviews in Food Science and Nutrition. 2017;57:524-548. DOI: 10.1080/10408398.2014.889652

82. Vinha AF, Rodrigues F, Nunes MA, Oliveira MBPP. Natural pigments and colorants in foods and beverages. In: Galanakis CM, editor. Polyphenols: Properties, Recovery, and Application. Woodhead Publishing/Elsevier; 2018. pp. 363-392. DOI: 10.1016/B978-0-12-813572-3.00011-7

83. EFSA. EFSA panel on food additives and nutrient sources added to food scientific opinion on the re-evaluation of anthocyanins (E 163) as a food additive. EFSA Journal. 2013;11:3145. DOI: 10.2903/j.efsa.2013.3145

84. Berli FJ, Moreno D, Piccoli P, Hespanhol-Viana L, Silva MF, Bressan-Smith R, et al. Abscisic acid is involved in the response of grape (Vitis vinifera L.) cv. Malbec leaf tissues to ultraviolet-B radiation by enhancing ultraviolet-absorbing compounds, antioxidant enzymes and membrane sterols. Plant, Cell and Environment. 2010;33:1-10. DOI: 10.1111/j.1365-3040.2009.02044.x

85. Hariram Nile S, Hwan Kim D, Keum Y-S. Determination of anthocyanin content and antioxidant capacity of different grape varieties. Ciência e Técnica Vitivinícola. 2015;30:60-68. DOI: 10.1051/ctv/20153002060

86. Sun J, Cao X, Bai W, Liao X, Hu X. Comparative analyses of copigmentation of cyanidin 3-glucoside and cyanidin 3-sophoroside from red raspberry fruits. Food Chemistry. 2010;120:553-557. DOI: 10.1016/j.foodchem.2009.11.031

87. Pan YZ, Guan Y, Wei ZF, Peng X, Li TT, Qi XL, et al. Flavonoid C-glycosides from pigeon pea leaves as color and anthocyanin stabilizing agent in blueberry juice. Industrial Crops and Products. 2014;58:142-147. DOI: 10.1016/j.indcrop.2014.04.029

88. Mohan S, Nandhakumar L. Role of various flavonoids: Hypotheses on novel approach to treat diabetes. Journal of Medical Hypotheses and Ideas. 2014;8;1-6. DOI: 10.1016/j.jmhi.2013.06.001

89. Sarian MN, Ahmed QU, Mat S'oAd SZ, Alhassan AM, Murugesu S, Perumal V, et al. Antioxidant and antidiabetic effects of flavonoids: A structure-activity relationship based study. BioMed Research International. 2017;2017:14. DOI: 10.1155/2017/8386065

90. Prasath GS, Sundaram CS, Subramanian SP. Fisetin averts oxidative stress in pancreatic tissues of streptozotocin-induced diabetic rats. Endocrine. 2013;44:359-368. DOI: 10.1007/s12020-012-9866-x

91. Teng H, Fang T, Lin Q, Song H, Liu B, Chen L. Red raspberry and its anthocyanins: Bioactivity beyond antioxidant capacity. Trends in Food Science and Technology. 2017;66:153-165

92. Yang B, Liu H, Yang J, Gupta VK, Jiang Y. New insights on bioactivities and biosynthesis of flavonoid glycosides. Trends in Food Science and Technology. 2018;79:116-124

93. Pontes M de CND, Torres-Rêgo M, de Aquino AKS, de Aquino NC, Teles YCF, Ximenes RM, et al. Harpalyce brasiliana Benth: A prolific source of bioactive flavonoids with antiophidic potential. Phytochemistry Letters. 2021;41:158-167. DOI: 10.1016/j.phytol.2020.09.025

94. Dimitrios B. Sources of natural phenolic antioxidants. Trends in Food Science and Technology. 2006;17:505-512. DOI: 10.1016/j.tifs.2006.04.004

95. Routray W, Orsat V. Microwave-assisted extraction of flavonoids: A review. Food and Bioprocess Technology. 2012;5:409-424. DOI: 10.1007/s11947-011-0573-z

96. Carlos ALE, Feliciano Priego-Capote; De Castro, M.D.L. Comparative study of the effect of sample pretreatment and extraction on the determination of flavonoids from lemon (Citrus limon). PLoS One. 2016;11:1-16. DOI: 10.1371/journal.pone.0148056

97. Molina-Calle M, Priego-Capote F, Luque De Castro MD. Development and application of a quantitative method for determination of flavonoids in orange peel: Influence of sample pretreatment on composition. Talanta. 2015;144:349-355. DOI: 10.1016/j.talanta.2015.05.054

98. Zhang X, Wang X, Wang M, Cao J, Xiao J, Wang Q. Effects of different pretreatments on flavonoids and antioxidant activity of Dryopteris erythrosora leave. bioRxiv. 2018;14:1-17

99. Nadeem M, Ubaid N, Qureshi TM, Munir M, Mehmood A. Effect of ultrasound and chemical treatment on total phenol, flavonoids and antioxidant properties on carrot-grape juice blend during storage. Ultrasonics Sonochemistry. 2018;45:1-6. DOI: 10.1016/j.ultsonch.2018.02.034

100. Ren F, Nian Y, Perussello CA. Effect of storage, food processing and novel extraction technologies on onions flavonoid content: A review. Food Research International. 2020;132:108953. DOI: 10.1016/j.foodres.2019.108953

101. Rocchetti G, Blasi F, Montesano D, Ghisoni S, Marcotullio MC, Sabatini S, et al. Impact of conventional/non-conventional extraction methods on the untargeted phenolic profile of Moringa oleifera leaves. Food Research International. 2019;115:319-327. DOI: 10.1016/j.foodres.2018.11.046

102. Barba FJ, Putnik P, Bursać Kovačević D, Poojary MM, Roohinejad S, Lorenzo JM, et al. Impact of conventional and non-conventional processing on prickly pear (Opuntia spp.) and their derived products: From preservation of beverages to valorization of by-products. Trends in Food Science and Technology. 2017;67:260-270. DOI: 10.1016/j.tifs.2017.07.012

103. Alara OR, Abdurahman NH, Ukaegbu CI. Soxhlet extraction of phenolic compounds from Vernonia cinerea leaves and its antioxidant activity. Journal of Applied Research on Medicinal and Aromatic Plants. 2018;11:12-17. DOI: 10.1016/j.jarmap.2018.07.003

104. Osorio-Tobón JF. Recent advances and comparisons of conventional and alternative extraction techniques of phenolic compounds. Journal of Food Science and Technology. 2020;57:4299-4315. DOI: 10.1007/s13197-020-04433-2

105. Farzaneh V, Carvalho IS. Modelling of microwave assisted extraction (MAE) of anthocyanins (TMA). Journal of Applied Research on Medicinal and Aromatic Plants. 2017;6:92-100. DOI: 10.1016/j.jarmap.2017.02.005

106. Chávez-González ML, Sepúlveda L, Verma DK, Luna-García HA, Rodríguez-Durán LV, Ilina A, et al. Conventional and emerging extraction processes of flavonoids. Processes. 2020;8:434. DOI: 10.3390/PR8040434

107. López-Bascón-Bascon MA, Luque de Castro MD. Soxhlet extraction. Liquid Extraction. In: Poole CF, editor. Handbooks in Separation Science, Liquid-Phase Extraction. Elsevier; 2020. 2019:327-354. DOI: 10.1016/B978-0-12-816911-7.00011-6. ISBN 9780128169117

108. Ye CL, Liu XG. Extraction of Flavonoids from Tetrastigma hemsleyanumDiels et Gilg and Their Antioxidant Activity. Journal of Food Processing & Preservation. 2015;39:2197-2205. DOI: 10.1111/jfpp.12464

109. Sonar MP, Rathod VK. Microwave assisted extraction (MAE) used as a tool for rapid extraction of Marmelosin from Aegle marmelos and evaluations of total phenolic and flavonoids content, antioxidant and anti-inflammatory activity. Chemical Data Collections. 2020;30:100545. DOI: 10.1016/j.cdc.2020.100545

110. Li C, Wang E, Elshikh MS, Alwahibi MS, Wang W, Wu G, et al. Extraction and purification of total flavonoids from Gnaphalium affine D. Don and their evaluation for free radicals' scavenging and oxidative damage inhabitation potential in mice liver. Arabian Journal of Chemistry. 2021;14:103006. DOI: 10.1016/j.arabjc.2021.103006

111. Zhu H, Wang Y, Liu Y, Xia Y, Tang T. Analysis of flavonoids in Portulaca oleracea L. by UV-vis spectrophotometry with comparative study on different extraction technologies. Food Analytical Methods. 2010;3:90-97. DOI: 10.1007/s12161-009-9091-2

112. Lin X, Wu L, Wang X, Yao L, Wang L. Ultrasonic-assisted extraction for flavonoid compounds content and antioxidant activities of India Moringa oleifera L. leaves: Simultaneous optimization, HPLC characterization and comparison with other methods. Journal of Applied Research on Medicinal and Aromatic Plants. 2021;20:100284. DOI: 10.1016/j.jarmap.2020.100284

113. Singh Chouhan KB, Tandey R, Sen KK, Mehta R, Mandal V. Microwave hydrodiffusion and gravity model with a blend of high and low power microwave firing for improved yield of phenolics and flavonoids from oyster mushroom. Sustainable Chemistry and Pharmacy. 2020;17:100311. DOI: 10.1016/j.scp.2020.100311

114. Veiga BA, Hamerski F, Clausen MP, Errico M, de Paula Scheer A, Corazza ML. Compressed fluids extraction methods, yields, antioxidant activities, total phenolics and flavonoids content for Brazilian Mantiqueira hops. Journal of Supercritical Fluids. 2021;170:105155. DOI: 10.1016/j.supflu.2020.105155

115. Hartonen K, Parshintsev J, Sandberg K, Bergelin E, Nisula L, Riekkola ML. Isolation of flavonoids from aspen knotwood by pressurized hot water extraction and comparison with other extraction techniques. Talanta. 2007;74:32-38. DOI: 10.1016/j.talanta.2007.05.040

116. Bimakr M, Rahman RA, Taip FS, Ganjloo A, Salleh LM, Selamat J, et al. Comparison of different extraction methods for the extraction of major bioactive flavonoid compounds from spearmint (Mentha spicata L.) leaves. Food and Bioproducts Processing. 2011;89:67-72. DOI: 10.1016/j.fbp.2010.03.002

117. Tan L, Ji T, Jiang G, Hu F. Simultaneous identification and quantification of five flavonoids in the seeds of Rheum palmatum L. by using accelerated solvent extraction and HPLC–PDA–ESI/MSn. Arabian Journal of Chemistry. 2019;12:1345-1352. DOI: 10.1016/j.arabjc.2014.10.038

118. Gomes SVF, Portugal LA, dos Anjos JP, de Jesus ON, de Oliveira EJ, David JP, et al. Accelerated solvent extraction of phenolic compounds exploiting a Box-Behnken design and quantification of five flavonoids by HPLC-DAD in Passiflora species. Microchemical Journal. 2017;132:28-35. DOI: 10.1016/j.microc.2016.12.021

119. Kam WYJ, Abas F, Hussain N, Mirhosseini H. Comparison of crude extract from durio zibethinus M. (durian) leaf waste via ultrasound-assisted extraction and accelerated solvent extraction: Antioxidant activity and cytotoxicity. Natural Product Research. 2020;34:1937-1941. DOI: 10.1080/14786419.2018.1564296

120. Yang J, Ou XQ, Zhang X, Zhou ZY, Ma LY. Effect of different solvents on the measurement of phenolics and the antioxidant activity of mulberry (Morus atropurpurea Roxb.) with accelerated solvent extraction. Journal of Food Science. 2017;82:605-612. DOI: 10.1111/1750-3841.13638

121. Yi Ling Y, Sook Fun P, Yeop A, Yusoff MM, Gimbun J. Assessment of maceration, ultrasonic and microwave assisted extraction for total phenolic content, total flavonoid content and kaempferol yield from cassia alata via microstructures analysis. Materials Today: Proceedings. 2019;19:1273-1279. DOI: 10.1016/j.matpr.2019.11.133

122. Bahorun T, Luximon-Ramma A, Crozier A, Aruoma OI. Total phenol, flavonoid, proanthocyanidin and vitamin C levels and antioxidant activities of Mauritian vegetables. Journal of the Science of Food and Agriculture. 2004;84:1553-1561. DOI: 10.1002/jsfa.1820

123. Bursać Kovačević D, Maras M, Barba FJ, Granato D, Roohinejad S, Mallikarjunan K, et al. Innovative technologies for the recovery of phytochemicals from Stevia rebaudiana Bertoni leaves: A review. Food Chemistry. 2018;268:513-521. DOI: 10.1016/j.foodchem.2018.06.091

124. Chemat F, Zill-E-Huma; Khan, M.K. Applications of ultrasound in food technology: Processing, preservation and extraction. Ultrasonics Sonochemistry. 2011;18:813-835. DOI: 10.1016/j.ultsonch.2010.11.023

125. Rutkowska M, Namieśnik J, Konieczka P. Ultrasound-assisted extraction. Application of Green Solvents in Separation Processes. 2017;1295:301-324. DOI: 10.1016/B978-0-12-805297-6.00010-3

126. Chemat F, Rombaut N, Sicaire AG, Meullemiestre A, Fabiano-Tixier AS, Abert-Vian M. Ultrasound assisted extraction of food and natural products. Mechanisms, techniques, combinations, protocols and applications. A review. Ultrasonics Sonochemistry. 2017;34:540-560

127. Mansur AR, Song NE, Jang HW, Lim TG, Yoo M, Nam TG. Optimizing the ultrasound-assisted deep eutectic solvent extraction of flavonoids in common buckwheat sprouts. Food Chemistry. 2019;293:438-445. DOI: 10.1016/j.foodchem.2019.05.003

128. Ali MC, Chen J, Zhang H, Li Z, Zhao L, Qiu H. Effective extraction of flavonoids from Lycium barbarum L. fruits by deep eutectic solvents-based ultrasound-assisted extraction. Talanta. 2019;203:16-22. DOI: 10.1016/j.talanta.2019.05.012

129. Xia GH, Li XH, Jiang Y. hang Deep eutectic solvents as green media for flavonoids extraction from the rhizomes of Polygonatum odoratum. Alexandria Engineering Journal. 2021;60:1991-2000. DOI: 10.1016/j.aej.2020.12.008

130. Guo H, Liu S, Li S, Feng Q, Ma C, Zhao J, et al. Deep eutectic solvent combined with ultrasound-assisted extraction as high efficient extractive media for extraction and quality evaluation of Herba Epimedii. Journal of Pharmaceutical and Biomedical Analysis. 2020;185:113228. DOI: 10.1016/j.jpba.2020.113228

131. Meng Z, Zhao J, Duan H, Guan Y, Zhao L. Green and efficient extraction of four bioactive flavonoids from Pollen Typhae by ultrasound-assisted deep eutectic solvents extraction. Journal of Pharmaceutical and Biomedical Analysis. 2018;161:246-253. DOI: 10.1016/j.jpba.2018.08.048

132. Easmin MS, Sarker MZI, Ferdosh S, Shamsudin SH, Yunus KB, Uddin MS, et al. Bioactive compounds and advanced processing technology: Phaleria macrocarpa (sheff.) Boerl—A review. Journal of Chemical Technology and Biotechnology. 2015;90:981-991

133. Smelt JPPM. Recent advances in the microbiology of high pressure processing. Trends in Food Science and Technology. 1998;9:152-158. DOI: 10.1016/S0924-2244(98)00030-2

134. Xi J, Shen D, Li Y, Zhang R. Ultrahigh pressure extraction as a tool to improve the antioxidant activities of green tea extracts. Food Research International. 2011;44:2783-2787. DOI: 10.1016/j.foodres.2011.06.001

135. Khan SA, Aslam R, Makroo HA. High pressure extraction and its application in the extraction of bio-active compounds: A review. Journal of Food Process Engineering. 2019;42:e12896. DOI: 10.1111/jfpe.12896

136. Kumar M, Tomar M, Potkule J, Verma R, Punia S, Mahapatra A, et al. Advances in the plant protein extraction: Mechanism and recommendations. Food Hydrocolloids. 2021;115:106595. DOI: 10.1016/j.foodhyd.2021.106595

137. Moreira SA, Pintado M, Saraiva JA. High Hydrostatic Pressure-Assisted Extraction: A Review on Its Effects on Bioactive Profile and Biological Activities of Extracts. Amsterdam, Netherlands: Elsevier Inc.; 2020

138. Xi J. Effect of high pressure processing on the extraction of lycopene in tomato paste waste. Chemical Engineering and Technology. 2006;29:736-739. DOI: 10.1002/ceat.200600024

139. Xi J. Ultrahigh pressure extraction of bioactive compounds from plants—A review. Critical Reviews in Food Science and Nutrition. 2017;57:1097-1106. DOI: 10.1080/10408398.2013.874327

140. Azmir J, Zaidul ISM, Rahman MM, Sharif KM, Mohamed A, Sahena F, et al. Techniques for extraction of bioactive compounds from plant materials: A review. Journal of Food Engineering. 2013;117:426-436

141. Chemat, F.; Abert Vian, M.; Zill EH Microwave Assisted—Separations: Green Chemistry in Action; Green Chemistry Research Trends. 2009:33-62

142. Chan CH, Yusoff R, Ngoh GC, Kung FWL. Microwave-assisted extractions of active ingredients from plants. Journal of Chromatography A. 2011;1218:6213-6225. DOI: 10.1016/j.chroma.2011.07.040

143. Vinatoru M, Mason TJ, Calinescu I. Ultrasonically assisted extraction (UAE) and microwave assisted extraction (MAE) of functional compounds from plant materials. Trends in Analytical Chemistry. 2017;97:159-178. DOI: 10.1016/j.trac.2017.09.002

144. Gharaati Jahromi S. Extraction techniques of phenolic compounds from plants. In: Soto-Hernández M, García-Mateos R, Palma-Tenango M, editors. Plant Physiological Aspects of Phenolic Compounds. 2019

145. Ibáñez, E.; Mendiola, J.A.; Castro-Puyana, M. Supercritical fluid extraction. In: Caballero B, Finglas PM, Toldrá F, editors. Encyclopedia of Food and Health; 2015

146. Yousefi M, Rahimi-Nasrabadi M, Pourmortazavi SM, Wysokowski M, Jesionowski T, Ehrlich H, et al. Supercritical fluid extraction of essential oils. Trends in Analytical Chemistry. 2019;118:182-193. DOI: 10.1016/j.trac.2019.05.038

147. Fabrowska J, Ibañez E, Łęska B, Herrero M. Supercritical fluid extraction as a tool to valorize underexploited freshwater green algae. Algal Research. 2016;19:237-245. DOI: 10.1016/j.algal.2016.09.008

148. Reverchon E, De Marco I. Supercritical fluid extraction and fractionation of natural matter. Journal of Supercritical Fluids. 2006;38:146-166. DOI: 10.1016/j.supflu.2006.03.020

149. Rostagno MA, Araújo JMA, Sandi D. Supercritical fluid extraction of isoflavones from soybean flour. Food Chemistry. 2002;78:111-117. DOI: 10.1016/S0308-8146(02)00106-1

150. Careri M, Furlattini L, Mangia A, Musci M, Anklam E, Theobald A, et al. Supercritical fluid extraction for liquid chromatographic determination of carotenoids in Spirulina Pacifica algae: A chemometric approach. Journal of Chromatography. A. 2001;912:61-71. DOI: 10.1016/S0021-9673(01)00545-3

151. Liza MS, Abdul Rahman R, Mandana B, Jinap S, Rahmat A, Zaidul ISM, et al. Supercritical carbon dioxide extraction of bioactive flavonoid from Strobilanthes crispus (Pecah Kaca). Food and Bioproducts Processing. 2010;88:319-326. DOI: 10.1016/j.fbp.2009.02.001

152. Puri M, Sharma D, Barrow CJ. Enzyme-assisted extraction of bioactives from plants. Trends in Biotechnology. 2012;30:37-44. DOI: 10.1016/j.tibtech.2011.06.014

153. Mohnen D. Pectin structure and biosynthesis. Curr Opin Plant Biol. 2008 Jun;11(3):266-77. DOI: 10.1016/j.pbi.2008.03.006. Epub 2008 May 15. PMID: 18486536

154. Danalache F, Mata P, Alves VD, Moldão-Martins M. Enzyme-assisted extraction of fruit juices. In: Rajauria G, Tiwari B, editors. Fruit Juices: Extraction, Composition, Quality and Analysis. London, United Kingdom 2018

155. Casas, P.; Dom, H. Enzyme-Assisted Aqueous Extraction Processes; In: Dominguez H, Munoz MJG, editors. Amsterdam, Netherlands: Elsevier; 2017

156. Brown RM. Cellulose structure and biosynthesis: What is in store for the 21st century? Journal of Polymer Science Part A Polymer Chemistry. 2004;42:487-495. DOI: 10.1002/pola.10877

157. Muniglia L, Claisse N, Baudelet PH, Ricochon G. Enzymatic Aqueous Extraction (EAE). In: Chemat F, Maryline Abert Vian MA, editors. Alternative Solvents for Natural Products Extraction. 2014. pp. 167-204

158. Scheller HV, Ulvskov P. Hemicelluloses. Annual Review of Plant Biology. 2010;61:263-289

159. Vanholme R, Demedts B, Morreel K, Ralph J, Boerjan W. Lignin biosynthesis and structure. Plant Physiology. 2010:153:895-905. DOI: 10.1104/pp.110.155119

160. Gligor O, Mocan A, Moldovan C, Locatelli M, Crişan G, Ferreira ICFR. Enzyme-assisted extractions of polyphenols—A comprehensive review. Trends in Food Science and Technology. 2019;88:302-315. DOI: 10.1016/j.tifs.2019.03.029

161. Nadar SS, Rao P, Rathod VK. Enzyme assisted extraction of biomolecules as an approach to novel extraction technology: A review. Food Research International. 2018;108:309-330. DOI: 10.1016/j.foodres.2018.03.006

162. Belwal T, Ezzat SM, Rastrelli L, Bhatt ID, Daglia M, Baldi A, et al. A critical analysis of extraction techniques used for botanicals: Trends, priorities, industrial uses and optimization strategies. Trends in Analytical Chemistry. 2018;100:82-102. DOI: 10.1016/j.trac.2017.12.018

163. Gullón B, Lú-Chau TA, Moreira MT, Lema JM, Eibes G. Rutin: A review on extraction, identification and purification methods, biological activities and approaches to enhance its bioavailability. Trends in Food Science and Technology. 2017;67:220-235. DOI: 10.1016/j.tifs.2017.07.008

164. Banerjee J, Singh R, Vijayaraghavan R, MacFarlane D, Patti AF, Arora A. Bioactives from fruit processing wastes: Green approaches to valuable chemicals. Food Chemistry. 2017;225:10-22. DOI: 10.1016/j.foodchem.2016.12.093

165. Pham DC, Nguyen HC, Nguyen THL, Ho HL, Trinh TK, Riyaphan J, et al. Optimization of ultrasound-assisted extraction of flavonoids from celastrus hindsii leaves using response surface methodology and evaluation of their antioxidant and antitumor activities. BioMed Research International. 2020;2020:1-9. DOI: 10.1155/2020/3497107

166. Ferreira ICFR, Martins N, Barros L. Phenolic compounds and its bioavailability: In vitro bioactive compounds or health promoters? Advances in Food and Nutrition Research. 2017;82:1-44. DOI: 10.1016/bs.afnr.2016.12.004

167. Thakur N, Raigond P, Singh Y, Mishra T, Singh B, Lal MK, et al. Recent updates on bioaccessibility of phytonutrients. Trends in Food Science and Technology. 2020;97:366-380. DOI: 10.1016/j.tifs.2020.01.019

168. Câmara JS, Albuquerque BR, Aguiar J, Corrêa RCG, Gonçalves JL, Granato D, et al. Food bioactive compounds and emerging techniques for their extraction: Polyphenols as a case study. Food. 2021;10:37-71. DOI: 10.3390/foods10010037

169. Zhao J, Yang J, Xie Y. Improvement strategies for the oral bioavailability of poorly water-soluble flavonoids: An overview. International Journal of Pharmaceutics. 2019;570:118642. DOI: 10.1016/j.ijpharm.2019.118642

170. Lu M, Ho C-T, Huang Q. Improving quercetin dissolution and bioaccessibility with reduced crystallite sizes through media milling technique. Journal of Functional Foods. 2017;37:138-146. DOI: 10.1016/j.jff.2017.07.047

171. Singh MK, Pooja D, Ravuri HG, Gunukula A, Kulhari H, Sistla R. Fabrication of surfactant-stabilized nanosuspension of naringenin to surpass its poor physiochemical properties and low oral bioavailability. Phytomedicine. 2018;40:48-54. DOI: 10.1016/j.phymed.2017.12.021

172. Albuquerque BR, Heleno SA, Oliveira MBPP, Barros L, Ferreira ICFR. Phenolic compounds: current industrial applications, limitations and future challenges. Food & Function. 2021;12:14-29

173. Risch SJE. Overview of uses and techniques. In: Encapsulation and Controlled Release of Food Ingredients. Vol. 590. Washington, DC: American Chemical Society; 1995. pp. 1-2

174. Zhang R, Zhou L, Li J, Oliveira H, Yang N, Jin W, et al. Microencapsulation of anthocyanins extracted from grape skin by emulsification/internal gelation followed by spray/freeze-drying techniques: Characterization, stability and bioaccessibility. LWT. 2020;123:109097. DOI: 10.1016/j.lwt.2020.109097

175. Salah M, Mansour M, Zogona D, Xu X. Nanoencapsulation of anthocyanins-loaded β-lactoglobulin nanoparticles: Characterization, stability, and bioavailability in vitro. Food Research International. 2020;137:109635. DOI: 10.1016/j.foodres.2020.109635

176. Ge J, Yue X, Wang S, Chi J, Liang J, Sun Y, et al. Nanocomplexes composed of chitosan derivatives and β-Lactoglobulin as a carrier for anthocyanins: Preparation, stability and bioavailability in vitro. Food Research International. 2019;116:336-345. DOI: 10.1016/j.foodres.2018.08.045

177. Ban C, Park SJ, Lim S, Choi SJ, Choi YJ. Improving flavonoid bioaccessibility using an edible oil-based lipid nanoparticle for oral delivery. Journal of Agricultural and Food Chemistry. 2015;63:5266-5272. DOI: 10.1021/acs.jafc.5b01495

178. Feng T, Wang K, Liu F, Ye R, Zhu X, Zhuang H, et al. Structural characterization and bioavailability of ternary nanoparticles consisting of amylose, α-linoleic acid and β-lactoglobulin complexed with naringin. International Journal of Biological Macromolecules. 2017;99:365-374. DOI: 10.1016/j.ijbiomac.2017.03.005

179. Grace MH, Ribnicky DM, Kuhn P, Poulev A, Logendra S, Yousef GG, et al. Hypoglycemic activity of a novel anthocyanin-rich formulation from lowbush blueberry, Vaccinium angustifolium Aiton. Phytomedicine. 2009;16:406-415. DOI: 10.1016/j.phymed.2009.02.018

180. Jain S, Jain AK, Pohekar M, Thanki K. Novel self-emulsifying formulation of quercetin for improved in vivo antioxidant potential: Implications for drug-induced cardiotoxicity and nephrotoxicity. Free Radical Biology & Medicine. 2013;65:117-130. DOI: 10.1016/j.freeradbiomed.2013.05.041

181. Pangeni R, Kang S-W, Oak M, Park EY, Park JW. Oral delivery of quercetin in oil-in-water nanoemulsion: In vitro characterization and in vivo anti-obesity efficacy in mice. Journal of Functional Foods. 2017;38:571-581. DOI: 10.1016/j.jff.2017.09.059

182. Guo RX, Fu X, Chen J, Zhou L, Chen G. Preparation and characterization of microemulsions of myricetin for improving its antiproliferative and antioxidative activities and oral bioavailability. Journal of Agricultural and Food Chemistry. 2016;64:6286-6294. DOI: 10.1021/acs.jafc.6b02184

183. Ting Y, Jiang Y, Lan Y, Xia C, Lin Z, Rogers MA, et al. Viscoelastic emulsion improved the bioaccessibility and oral bioavailability of crystalline compound: A mechanistic study using in vitro and in vivo models. Molecular Pharmaceutics. 2015;12:2229-2236. DOI: 10.1021/mp5007322

184. Ting Y, Chiou Y-S, Pan M-H, Ho C-T, Huang Q. In vitro and in vivo anti-cancer activity of tangeretin against colorectal cancer was enhanced by emulsion-based delivery system. Journal of Functional Foods. 2015;15:264-273. DOI: 10.1016/j.jff.2015.03.034

185. Wu L, Bi Y, Wu H. Formulation optimization and the absorption mechanisms of nanoemulsion in improving baicalin oral exposure. Drug Development and Industrial Pharmacy. 2018;44:266-275. DOI: 10.1080/03639045.2017.1391831

186. Neri-Numa IA, Arruda HS, Geraldi MV, Maróstica Júnior MR, Pastore GM. Natural prebiotic carbohydrates, carotenoids and flavonoids as ingredients in food systems. Current Opinion in Food Science. 2020;33:2632

187. Chai J, Jiang P, Wang P, Jiang Y, Li D, Bao W, et al. The intelligent delivery systems for bioactive compounds in foods: Physicochemical and physiological conditions, absorption mechanisms, obstacles and responsive strategies. Trends in Food Science and Technology. 2018;78:144-154

188. Petzold G, Moreno J, Gianelli MP, Cerda F, Mella K, Zúñiga P, et al. Food technology approaches for improvement of organoleptic properties through preservation and enrichment of bioactive compounds. In: Grumezescu AM, Alina Maria Holban AM, editors. Food Processing for Increased Quality and Consumption. San Diego, United States; 2018

189. Artiga-Artigas, M.; Odriozola-Serrano, I.; Oms-Oliu, G.; Martín-Belloso, O. Nanostructured systems to increase bioavailability of food ingredients. In: Amparo LR, Rovira M, Sanz M, Gomez-Mascaraque LG, editors. Nanomaterials for Food Applications; Exeter, United Kingdom: 2018

190. Delfanian M, Sahari MA. Improving functionality, bioavailability, nutraceutical and sensory attributes of fortified foods using phenolics-loaded nanocarriers as natural ingredients. Food Research International. 2020;137:109-555

191. Montibeller MJ, de Lima Monteiro P, Tupuna-Yerovi DS, Rios A d O, Manfroi V. Stability assessment of anthocyanins obtained from skin grape applied in kefir and carbonated water as a natural colorant. Journal of Food Processing & Preservation. 2018;42:e13698. DOI: 10.1111/jfpp.13698

192. Doan CD, Tavernier I, Okuro PK, Dewettinck K. Internal and external factors affecting the crystallization, gelation and applicability of wax-based oleogels in food industry. Innovative Food Science and Emerging Technologies. 2018;45:42-52

193. Świeca M, Gawlik-Dziki U, Dziki D, Baraniak B. Wheat bread enriched with green coffee: In vitro bioaccessibility and bioavailability of phenolics and antioxidant activity. Food Chemistry. 2017;221:1451-1457. DOI: 10.1016/j.foodchem.2016.11.006

194. Da Silva LP, Pereira E, Prieto MA, Simal-Gandara J, Pires TCSP, Alves MJ, et al. Rubus ulmifolius Schott as a Novel Source of Food Colorant: Extraction optimization of coloring pigments and incorporation in a bakery product. Molecules. 2019;24:2181. DOI: 10.3390/molecules24112181

195. da Silva SC, Fernandes IP, Barros L, Fernandes Â, José Alves M, Calhelha RC, et al. Spray-dried Spirulina platensis as an effective ingredient to improve yogurt formulations: Testing different encapsulating solutions. Journal of Functional Foods. 2019;60:103427. DOI: 10.1016/j.jff.2019.103427

196. Caleja C, Barros L, Antonio AL, Ciric A, Soković M, Oliveira MBPP, et al. Foeniculum vulgare Mill. As natural conservation enhancer and health promoter by incorporation in cottage cheese. Journal of Functional Foods. 2015;12:428-438. DOI: 10.1016/j.jff.2014.12.016

197. Caleja C, Barros L, Antonio AL, Ciric A, Barreira JCM, Sokovic M, et al. Development of a functional dairy food: Exploring bioactive and preservation effects of chamomile (Matricaria recutita L.). Journal of Functional Foods. 2015;16:114-124. DOI: 10.1016/j.jff.2015.04.033

198. Pires TCSP, Dias MI, Barros L, Barreira JCM, Santos-Buelga C, Ferreira ICFR. Incorporation of natural colorants obtained from edible flowers in yogurts. LWT. 2018;97:668-675. DOI: 10.1016/j.lwt.2018.08.013

199. Pires TCSP, Dias MI, Carocho M, Barreira JCM, Santos-Buelga C, Barros L, et al. Extracts from: Vaccinium myrtillus L. Fruits as a source of natural colorants: Chemical characterization and incorporation in yogurts. Food & Function. 2020;11:3227-3234. DOI: 10.1039/c9fo02890k

200. Kwak HS, Mijan M, Al; Ganesan, P. Application of nanomaterials, nano- and microencapsulation to milk and dairy products. In: Kwak HS, editor. Nano- and Microencapsulation for Foods. Vol. 9781118292334. New York, United States: 2014

201. He B, Ge J, Yue P, Yue XY, Fu R, Liang J, et al. Loading of anthocyanins on chitosan nanoparticles influences anthocyanin degradation in gastrointestinal fluids and stability in a beverage. Food Chemistry. 2017;221:1671-1677. DOI: 10.1016/j.foodchem.2016.10.120

202. Khan H, Jawad M, Kamal MA, Baldi A, Xiao J, Nabavi SM, et al. Evidence and prospective of plant derived flavonoids as antiplatelet agents: Strong candidates to be drugs of future. Food and Chemical Toxicology. 2018;119:355-367. DOI: 10.1016/j.fct.2018.02.014

203. Xiao J. Stability of dietary polyphenols: It's never too late to mend? Food and Chemical Toxicology. 2018;119:3-5. DOI: 10.1016/j.fct.2018.03.051

204. Jakobek L. Interactions of polyphenols with carbohydrates, lipids and proteins. Food Chemistry. 2015;175:556-567

205. Faridi Esfanjani A, Assadpour E, Jafari SM. Improving the bioavailability of phenolic compounds by loading them within lipid-based nanocarriers. Trends in Food Science and Technology. 2018;76:56-66. DOI: 10.1016/j.tifs.2018.04.002

206. Hribar U, Ulrih N. The Metabolism of Anthocyanins. Current Drug Metabolism. 2014;15:3-13. DOI: 10.2174/1389200214666131211160308

207. Chen L, Teng H, Xie Z, Cao H, Cheang WS, Skalicka-Woniak K, et al. Modifications of dietary flavonoids towards improved bioactivity: An update on structure–activity relationship. Critical Reviews in Food Science and Nutrition. 2018;58:513-527. DOI: 10.1080/10408398.2016.1196334

208. Bernardes AL, Moreira JA, Tostes M d GV, Costa NMB, Silva PI, Costa AGV. In vitro bioaccessibility of microencapsulated phenolic compounds of jussara (Euterpe edulis Martius) fruit and application in gelatine model-system. LWT. 2019;102:173-180. DOI: 10.1016/j.lwt.2018.12.009

209. Estrada JD, Boeneke C, Bechtel P, Sathivel S. Developing a strawberry yogurt fortified with marine fish oil 1. Journal of Dairy Science. 2011;94:5760-5769. DOI: 10.3168/jds.2011-4226

210. Papillo VA, Locatelli M, Travaglia F, Bordiga M, Garino C, Arlorio M, et al. Spray-dried polyphenolic extract from Italian black rice (Oryza sativa L., var. Artemide) as new ingredient for bakery products. Food Chemistry. 2018;269:8021-8030. DOI: 10.1016/j.foodchem.2018.07.059

211. Carbonell-Capella JM, Buniowska M, Barba FJ, Esteve MJ, Frígola A. Analytical methods for determining bioavailability and bioaccessibility of bioactive compounds from fruits and vegetables: A review. Comprehensive Reviews in Food Science and Food Safety. 2014;13:155-171. DOI: 10.1111/1541-4337.12049

212. Tavano L, Muzzalupo R, Picci N, De Cindio B. Co-encapsulation of antioxidants into niosomal carriers: Gastrointestinal release studies for nutraceutical applications. Colloids Surfaces B Biointerfaces. 2014;114:82-88. DOI: 10.1016/j.colsurfb.2013.09.058

213. Park KM, Sung H, Choi SJ, Choi YJ, Chang PS. Double-layered microparticles with enzyme-triggered release for the targeted delivery of water-soluble bioactive compounds to small intestine. Food Chemistry. 2014;161:53-59. DOI: 10.1016/j.foodchem.2014.03.125

214. Carocho M, Heleno S, Rodrigues P, Barreiro MF, Barros L, Ferreira ICFR. A novel natural coating for food preservation: Effectiveness on microbial growth and physicochemical parameters. LWT. 2019;104:76-83. DOI: 10.1016/j.lwt.2019.01.031

215. Stojanoska MM, Milosevic N, Milic N, Abenavoli L. The influence of phthalates and bisphenol A on the obesity development and glucose metabolism disorders. Endocrine. 2017;55:666-681

216. Brostow W, Lu X, Osmanson AT. Nontoxic bio-plasticizers for PVC as replacements for conventional toxic plasticizers. Polymer Testing. 2018;69:63-70. DOI: 10.1016/j.polymertesting.2018.03.007

217. Dehghani S, Hosseini SV, Regenstein JM. Edible films and coatings in seafood preservation: A review. Food Chemistry. 2018;240:505-513. DOI: 10.1016/J.FOODCHEM.2017.07.034

Chapter-5

Food Supplementation with Vitamins and Minerals: An Overview

1. Introduction

According to the US Food and Drug Administration (FDA), a dietary supplement is a product taken by mouth that contains a "dietary ingredient" intended to supplement the diet [1]. The use of dietary supplements is widespread in the United States, Canada as well as in several European countries [2, 3, 4, 5]. Up to 50% of adults and one-third of children in economically advanced economies consume these supplements [6]. Dietary supplements may be used to correct deficiencies or maintain adequate intake of some nutrients. However, they may be harmful with excess intake. Hence, it is necessary to control their levels for a safe intake [7, 8, 9]. Vitamins and minerals are the most frequently used dietary supplements among adults in United States [10]. Thirteen vitamins (A, B_1, B_2, B_3, B_5, B_6, B_8, B_9, B_{12}, C, D, E, K) and fifteen minerals (calcium, phosphorus, potassium, sodium, chloride, magnesium, iron, zinc, iodine, chromium, copper, fluoride, molybdenum, manganese, and selenium) have been recognized as essential for the maintenance of human health [11]. The first multivitamin/mineral (MVM) formulas was introduced in the 1930s and the most common definition of this term is a dietary supplement containing 3 or more vitamins and minerals [12]. MVM supplements are recommended to prevent or treat nutrition-related infectious diseases such as the Acquired Immune Deficiency Syndrome (AIDS) disease that arises from human immunodeficiency virus (HIV) infection and which is responsible for more than half of total deaths in some developing countries [13]. Tuberculosis is another chronic infectious disease that is correlated to the presence of vitamin A levels in human bodies. Indeed several epidemiological studies found that healthy population has significantly higher vitamin A serum levels than tuberculosis patients [14, 15]. MVM supplements are also widely used to promote health and prevent chronic diseases and cancer. However, controversial findings were reported on the beneficial effect of these supplements on cardiovascular diseases prevention [16, 17, 18]. The aim of this chapter was to give a global idea on the main vitamins and minerals that are essential in our life and to present the most important users of MVM supplements as well as their frequency of use and the related deficiency diseases.

2. Vitamins

Vitamins are organic molecules that are necessary for the organism and which humans cannot synthesize in a sufficient quantity. Unlike habitual nutrients that are introduced in large quantities in the body and that contribute to the production of energy, only small amounts of vitamins are required for a healthy metabolism (micrograms or milligrams per day). Vitamins are obtained naturally from a balanced and diversified diet or can be added to foods. Their deficiency can cause health disorders such as cardiovascular diseases or cancers while an overconsumption can lead to toxic effects in the medium or long term (Table 1) [1, 2, 3, 4, 5].

Table 1. Deficiency and toxic effects of selected vitamins.

Vitamin	Deficiency symptoms	Excess intake symptoms
Vitamin A	Measles Night blindness Respiratory infection Diarrhea *Xerophthalmia* disease	Nausea and vomiting Anorexia Abdominal pain Skin desquamation Blurred vision
Vitamin B_1	Anorexia and muscle weakness Gastrointestinal disturbances Peripheral and central neuropathy Cardiovascular irregularities *Beriberi* disease	Gastric upset
Vitamin B_2	Angular stomatitis and glossitis Seborrheic dermatitis Intense photophobia Superficial vascularization of the cornea	No toxic effects
Vitamin B_3	Fatigue and headache Diarrhea Dementia and dermatitis Pigmented rush *Pellagra* disease	Flushing of the skin Hyperuricemia Hepatic and ocular abnormalities Hyperglycemia Liver damage
Vitamin B_6	Epileptiform convulsions Stomatitis and cheilitis Glossitis Irritability Depression and confusion	No toxic effects
Vitamin B_9	Irritability and hostility Memory loss and paranoid behavior Poor growth Glossitis Megaloblastic anemia	No toxic effects

Vitamin	Deficiency symptoms	Excess intake symptoms
Vitamin B_{12}	Weakness and fatigue Shortness of breath and palpitations Memory loss and dementia Loss of appetite Flatulence and constipation	No toxic effects
Vitamin C	Fatigue Ecchymoses and coiled hairs Myocardial infractions Stroke *Scurvy* disease	Diarrhea Nausea Kidney stones Iron overabsorption Gastrointestinal disturbances
Vitamin D	Bow legs or knock knees Pelvic and thoracic deformities Spine curvature Epiphyseal abnormalities *Rickets* and *Osteomalacia* diseases	Nausea and vomiting Hypercalceamia and muscle weakness Polyuria and polydipsia Hypercalciuria Anorexia and thirst

Thirteen vitamins were discovered during the first half of the twentieth century and the last discovery was that of vitamin B_{12} in 1948. These compounds have diverse biochemical functions such as antioxidants, coenzymes, hormones or mediators of cell signaling. They are classified in two categories: fat-soluble (e.g. vitamins A and D) and water-soluble (e.g. vitamins B and C). While water-soluble vitamins should be consumed daily, fat-soluble vitamins can be consumed at less regular intervals due to their ability to be stored in the liver and in adipose tissue [1, 2, 6]. Among the thirteen known vitamins, six water-soluble and three fat-soluble ones were described, due to their high nutritive value and their frequent use as food supplements and nutritional additives.

2.1 Vitamin A

Vitamin A is a fat-soluble vitamin discovered in the early 1900s by McCollum and independently by Osborne and Mendel as an essential dietary factor for growth [2, 7]. Forms of vitamin A include provitamin A carotenoids (principally β-carotene, α-carotene, and β-cryptoxanthin), retinol (preformed vitamin A), retinal, retinoic acid and retinyl esters. The requirements for vitamin A are expressed in retinol activity equivalents (RAEs), such that 1 μg RAE = 1 μg performed retinol = 12 μg β-carotene = 24 μg α-carotene or β-cryptoxanthin [8, 9].

2.1.1 Dietary sources

Preformed vitamin A is found naturally only in animal-based products such as liver and fish liver oil, dairy products, eggs and fish. It is also used in food fortification of sugar, cereals, fats or condiments. The main dietary sources of provitamin A carotenoids are pigmented vegetables and fruits including spinach, parsley, amaranth, lettuce, carrot, papaya, mango, etc. [1, 9, 10].

2.1.2 Absorption, metabolism and storage

About 70-90% of preformed vitamin A is absorbed from the small intestine dissolved in lipid micelles. Carotenoids are absorbed into the small intestine by passive diffusion. Their biological availability varies between 5 and 60% [8].

After absorption, retinal esters and nonhydrolyzed carotenoids are transported to the liver in chylomicrons and chylomicron remnants. About 90% of retinol is stored in the liver as retinol palmitate. The remaining fraction is transported to other adipose tissues by Retinol Binding Protein (RBP). The total body vitamin A is excreted in the urine or the feces and to a lesser extent in the bile [1, 9, 11].

2.1.3 Function

Vitamin A is required for normal vision, embryonic development, reproduction, gene expression, growth and immune function [7, 9]. In the visual system, the dim-light vision is related to the presence of rhodopsin, a photosensitive pigment formed in rod cells, after the binding of the 11-*cis*-retinaldehyde to the optin protein. Following an equivalent mechanism, the vision of shapes and colors is carried out in the presence of iodopsin pigments formed in the cones after the binding of the 11-*cis*-retinaldehyde to the photopsin protein [4, 10, 11]. As regards embryonic development, retinoic acid as well as endogenous retinoids play an important role in the anterior–posterior development of the central body axis and in limb development. Their concentrations are highly regulated, both spatially and temporally [3]. Vitamin A is essential for reproduction too, through the biosynthesis of glycoproteins specific to spermatogenesis [4, 11]. It is also required to regulate the expression of a number of genes through interactions of retinoic acid with various cytosolic and nuclear receptors [11, 12]. These genes include the growth hormone gene, and therefore the normal growth and development of children are affected by vitamin A deficiency. Finally, vitamin A enhances cellular and humoral immunity through all-trans-retinoic acid. It stimulates the synthesis of immunoglobulins and the proliferation of lymphocytes and thymocytes [4, 11].

2.2 Vitamin B_1 (thiamin)

Historically, vitamin B_1 has been known as the preventive and curative agent of the human disease "beriberi" [2, 7]. Its structure determination and synthesis were carried out in 1936. This water-soluble vitamin includes substituted pyrimidine and thiazole moieties linked by a methylene bridge. Three phosphorylated forms of thiamin occur in nature: thiamin monophosphate, thiamin diphosphate (the active coenzyme also known as thiamin pyrophosphate) and thiamin triphosphate. Thiamin is particularly sensitive to sulfites and polyphenols, which destroy its biological activity [8].

2.2.1 Dietary sources

The most abundant *sources* of vitamin B_1 include yeast and yeast extract, whole cereal grains, nuts, lean pork, wheat bran, heart, kidney, liver and fortified food such as bread and breakfast cereals. Fresh food, such as fruits, vegetables and dairy products except butter, could protect against vitamin B_1 deficiency if it is consumed regularly and in sufficient amount [1, 3, 7].

2.2.2 Absorption, metabolism and storage

Dietary thiamin phosphates are hydrolyzed to free thiamin by intestinal phosphatases. Then, absorption of free thiamin takes place in the duodenum and proximal jejunum [3, 4]. Two parallel mechanisms are involved in thiamin absorption: a saturable active transport at low physiological concentrations of the vitamin and a passive diffusion at higher concentrations [3, 7, 9]. Transport of thiamin is modulated by various biological factors such as age, diabetic state and alcohol exposure [7]. Some thiamin is phosphorylated to thiamin monophosphate in the intestinal mucosa and both free thiamin and thiamin monophosphate circulate in the bloodstream, the former bound to plasma proteins. Approximately 30 mg of total thiamin is found in human body, of which about 80% is thiamin diphosphate, 10% is thiamin triphosphate, and the remainder is free thiamin and thiamin monophosphate. Due to the low storage capacity and the relatively high turnover rate of thiamin, a regular intake of this vitamin is then necessary [1, 3].

2.2.3 Function

Thiamin diphosphate is a coenzyme for three multienzyme complexes involved in the oxidative decarboxylation of oxoacids: pyruvate dehydrogenase, α-ketoglutarate dehydrogenase and branched-chain ketoacid dehydrogenase. Thiamin triphosphate plays a role in nerve transmission for sodium and potassium transport [1, 3].

2.3 Vitamin B_2 (riboflavin)

The water-soluble vitamin riboflavin was synthesized independently by Kuhn's and Karrer's groups in 1935 [3]. This fluorescent, yellow crystalline compound was discovered during the 1920s as a preventive factor from human "pellagra" as well as "beriberi" disease. Riboflavin is a flavin with the flavin ring attached to an alcohol related to ribose [2]. It is a relatively heat–stable molecule but it is rapidly inactivated in ultraviolet (UV) light [7].

2.3.1 Dietary sources

Major food sources of vitamin B_2 include milk and dairy products as well as meat and meat products. Green vegetables such as collard greens, turnip greens and broccoli are good sources of riboflavin. Natural grain products contain low amounts of riboflavin but enrichment of these food items with the vitamin has improved its intake. Vitamin B_2 is also widely used as food color due to its intense yellow color [2, 3, 7].

2.3.2 Absorption, metabolism and storage

Flavin coenzymes are hydrolyzed by alkaline phosphatase in the upper small intestine to free riboflavin, which is then absorbed. Absorption is enhanced when riboflavin is ingested along with other foods and in the presence of bile salts. Much of the absorbed riboflavin is phosphorylated in the intestinal mucosa and enters the bloodstream as riboflavin phosphate. About 50% of plasma riboflavin in plasma is free riboflavin, with somewhat less flavin adenine dinucleotide (FAD) and less than 10% flavin mononucleotide (FMN). The average concentration of riboflavin in plasma is about 0.03 μM [3, 4, 7]. The liver is the main storage organ of vitamin B_2. Other storage sites are the spleen, kidney and cardiac muscle. Urinary excretion of vitamin B_2 occurs predominantly in the form of free riboflavin with a small amount as a variety of its glycosides and metabolites [4].

2.3.3 Function

Riboflavin functions as the precursor of the flavin coenzymes, FMN and FAD, and of covalently bound flavins. These coenzymes catalyze numerous oxidation–reduction reactions in several metabolic pathways including the mitochondrial electron transport chain. The majority of flavoproteins require FAD as the prosthetic group rather than FMN. Other major functions of riboflavin include drug, lipid and steroid metabolisms [3, 7, 8].

2.4 Vitamin B_3 (niacin)

Niacin is a generic term referring to two vitamers nicotinic acid and nicotinamide. Nicotinic acid was discovered by Huber in 1867 as a product of nicotine oxidation, but its curative and preventive role against "pellagra" disease was not recognized until much later, in 1938. Niacin is a whitish water-soluble crystalline compound. It can be synthesized in the body from the essential amino acid tryptophan, which constitutes an important route for meeting the body's niacin requirement [1, 2, 7].

2.4.1 Dietary sources

Niacin requirements are generally expressed as mg niacin equivalents; 1 mg niacin equivalent = 1 mg preformed niacin + 1/60 × mg tryptophan. Due to the contribution of tryptophan, foods rich in proteins are important sources of the vitamin. Lean red meat, fish, liver and poultry contain high amounts of both niacin and tryptophan. Other contributors to niacin intake include ready-to-eat cereals supplemented with nicotinic acid. Vegetables and fruits provide useful amounts of the vitamin, depending upon the dietary intake [1, 3, 7].

2.4.2 Absorption, metabolism and storage

Niacin is present in food largely in the form of the nicotinamide nucleotides, nicotinamide adenine dinucleotide (NAD) and nicotinamide adenine dinucleotide phosphate (NADP). These nucleotides are hydrolyzed to free nicotinamide in the intestinal lumen. At low concentrations, absorption from the small intestine is mediated by a sodium-dependent facilitated diffusion. At higher amounts, absorption is by passive diffusion [7, 9]. Niacin compounds entering the portal circulation are either transported to the liver or internalized by erythrocytes. In the liver, nicotinic acid and nicotinamide, together with tryptophan, are converted to (NAD). On reaching the tissues, the niacin vitamers are used for the intracellular synthesis of NAD and NADP. The liver plays an important role in the preparation of niacin for urinary excretion, converting excess niacin to methylated derivatives [3, 7].

2.4.3 Function

The nicotinamide nucleotides NAD and NADP play a major role in a wide variety of oxidation–reduction reactions. The NAD coenzyme is involved in catabolic reactions while NADP is more often required in synthetic mechanisms such as the synthesis of steroids and fatty acids [1]. The coenzyme NAD is easily convertible to NADP and vice versa, while both molecules can also exchange their oxidation state. This conversion maintains a balance between the energy-consuming synthetic reactions and the catabolic reactions that produce energy [1].

2.5 Vitamin B_6

The pure crystalline water-soluble vitamin B_6 was first isolated in 1938 by Lepkovsky, than synthesized by Harris and Folkers in 1939. There are six nutritionally active B_6 vitamers, namely pyridoxine (PN), pyridoxal (PL), pyridoxamine (PM), and the corresponding 5'-phosphate esters pyridoxine phosphate (PNP), pyridoxal phosphate (PLP) and pyridoxamine phosphate (PMP) [2, 3]. Aqueous vitamin B_6 solutions are colorless, except for PLP solutions, which are yellow and not very stable in neutral or alkaline medium [11].

2.5.1 Dietary sources

High quantities of vitamin B_6 are found in yeast, liver, fortified cereals and grain germs. Other important sources include nuts, pulses, meat, fish, avocado, banana and potatoes. Eggs, fruit and milk, contain relatively low concentrations of the vitamin [1, 3, 9].

2.5.2 Absorption, metabolism and storage

The absorption of vitamin B_6 occurs following the hydrolysis of the phosphorylated forms in the lumen of intestine through a sodium-dependent carrier-mediated system [7]. Once absorbed, the different forms of the vitamin B_6 vitamers are interconverted and conveyed to the liver in the form of free non-phosphorylated vitamers. In the liver, the vitamers are accumulated by diffusion and converted to phosphorylated forms, under the action of pyridoxal kinase enzyme. The PNP and PMP are oxidized to PLP by pyridoxine oxidase. A proportion of PLP is released into the bloodstream bound to plasma albumin. The remaining portion of PLP is dephosphorylated and oxidized to pyridoxic acid, which is released into the plasma and excreted in the urine [3].

2.5.3 Function

The metabolically active vitamer is PLP, which is involved in a wide variety of amino acids reactions, including transamination, transulphuration, desulphuration and decarboxylation. PLP is also required in the regulation of steroid hormone actions. In addition, PLP vitamer acts as the cofactor of glycogen phosphorylase in muscle and liver and has an essential role in lipid metabolism and immune function [3, 4, 8].

2.6 Vitamin B_9 (folates)

The water-soluble vitamin B_9 is represented by the group of folates (from *folium* the Latin word for leaf). The isolation, chemical structure determination and synthesis of vitamin B_9 was carried out in 1946 by Angier's group [3]. Since the early 1990s, the associations between folate intake and birth outcome or chronic disease risk were investigated. Research studies established a reduction in the risk of neutral tube defects after a daily periconceptional supplementation with folic acid. These important results led to the implementation of public health policies that include mandatory folic acid fortification in North America [7].

2.6.1 Dietary sources

Good sources of folate include fortified grain products, orange juice, fresh dark green leafy vegetables, lentils, kidney beans, lima beans, avocado, peas and peanut products. Meat in *general* is not a good source of folate, except liver [1, 3, 7].

2.6.2 Absorption, metabolism and storage

About 80% of dietary folate occurs in a polyglutamate form, which is hydrolyzed to the monoglutamate form before absorption by enterocytes primarily in the jejunum. Monoglutamyl folate is then transported across the intestinal mucosa via a saturable, pH dependent active transport process. Nevertheless, at high concentrations, the intestinal uptake takes place by a nonsaturable passive diffusion. Plasma folate, primarily 5-methyl THF, is distributed in two major forms: free folate and folate bound to albumin. A small proportion of plasma folate (less than 5%) is bound to high-affinity binders [3, 4, 8]. The majority of reduced monoglutamates arriving at the liver from the intestine is metabolized and retained or released into the blood or bile [3]. Folate is excreted in the urine as folate derivatives and only 1%-2% of the vitamin is emitted as intact urinary folate [7].

2.6.3 Function

Due to their chemical structure, folates play a major role in the mobilization of one-carbon units in intermediary metabolism. Such reactions are required in the synthesis of the purines guanine and adenine and the pyrimidine thymine, which are used in the formation of nucleoproteins deoxyribonucleic acid (DNA) and ribonucleic acid (RNA) [2, 3]. Folates are also essential in the metabolism of amino acids, by means of THF derivatives that serve as donors of one-carbon units in a variety of synthetic reactions [3].

2.7 Vitamin B_{12}

Vitamin B_{12} was discovered after a series of important contributions from different fields including medicine, chemistry and microbiology, which led to the awarding of Nobel Prizes. The chemical structure of vitamin B_{12} was elucidated by the x-ray crystallographer Dorothy Hodgkin, that was awarded the Nobel Prize for chemistry in 1964. The total chemical synthesis of vitamin B_{12} took 11 years and was led by Robert Woodward, who received the Nobel Prize in Chemistry in 1965 [7]. The term vitamin B_{12} is used as a generic descriptor of cobalamins, which have a structure similar to heme but the central iron atom is replaced by a cobalt atom. Cyanocobalamin, a water-soluble purplish red powder, is the most stable of the vitamin B_{12}-active cobalamins. It is widely used in food supplementation and pharmaceutical preparations. However, it is a light sensitive molecule [3, 11].

2.7.1 Dietary sources

Vitamin B_{12} is found almost exclusively in foods of animal origin and liver is the most abundant dietary source of the vitamin followed by kidney and heart [2, 3]. Other important sources include milk, fish, eggs, shellfish and muscle meat. Different forms of vitamin B_{12} exist in foods such as adenosyl- and hydroxocobalamins in meat and fish, Sulphitocobalamin in canned meats and fish and small quantities of Cyancobalamin in egg white and cheeses [3].

2.7.2 Absorption, metabolism and storage

Vitamin B_{12} is absorbed by two different mechanisms: one active and the other passive. The major route of vitamin B_{12} active absorption is by attachment in the intestinal lumen to a small glycoprotein secreted by the parietal cells of the gastric mucosa named intrinsic factor. In the stomach, vitamin B_{12} binds to cobalophilin protein, which is hydrolyzed in the duodenum, releasing vitamin B_{12} to bind to intrinsic factor. In plasma, vitamin B_{12} circulates bound to transcobalamin I and II proteins, which prevent the vitamin from urine excretion as it passes through the kidney [3, 8]. With over 60% of the total vitamin content present in the body, the liver is the richest organ in vitamin B_{12}.

2.7.3 Function

Viamin B_{12} is involved in the metabolism of proteins, fats and carbohydrates. It is essential in the regeneration of the 5-methyl-folic acid into folic acid, which prevents from its direct urine excretion. In addition, vitamin B_{12} contributes to the transport and storage of folic acid in cells and plays a crucial role in the synthesis of myelin lipoprotein [1]. Vitamin B_{12} is also an important cofactor in the maintenance of normal DNA synthesis, the regeneration of methionine for adequate protein synthesis and methylation capacity, and the avoidance of homocysteine accumulation leading to various degenerative diseases [7].

2.8 Vitamin C (ascorbic acid)

The scurvy disease, due to vitamin C deficiency, was described in the Ebers papyrus of 1500 B.C. and by Hippocrates [4]. It was endemic in many areas, especially in the earliest explorations of the New World where over 2 million sailors died of scurvy during the era, often called the "Age of Sail" [7]. The structure of vitamin C, was elucidated in 1933 by Walter Haworth and his associates, at the University of Birmingham in England. It was synthesized in the same year by the same research team [3]. Vitamin C is the generic descriptor for all compounds with the biological activity of L-ascorbic acid. It is a water-soluble white crystalline powder, that is stable at solid state and that serves as a good reducing agent [2, 11].

2.8.1 Dietary sources

Vitamin C is synthesized by almost all living organisms, except humans and other primates, guinea pigs, fish, fruit-eating bats and some exotic birds [3, 8]. It occurs in significant amounts in green vegetables, fresh fruits, especially citrus fruits and blackcurrants and in animal organs such as liver, kidney and brain. Other good sources of vitamin C include potatoes, tomatoes, strawberries and cabbage. Muscle meat and cereal grains are poor dietary sources of the vitamin. Cooking causes significant losses of vitamin C through leaching into the cooking water and also atmospheric oxidation [3, 7].

2.8.2 Absorption, metabolism and storage

The biologically active forms of vitamin C are ascorbic acid and dehydroascorbic acid. Both vitamers are absorbed across the buccal mucosa by carrier-mediated passive processes. At physiological intakes, the intestinal absorption of vitamin C is by active transport mechanism, while at high doses, there is a less efficient passive diffusion, which proceeds at a very low rate [11]. In plasma, vitamin C occurs mainly as free ascorbate, and dehydroascorbic acid is present at undetectable concentrations. Liver has the greatest store of vitamin C in the body by virtue of its size. Other organs containing ascorbate include kidney, pancreas and brain. Excretion of vitamin C is mainly in urines with negligible amounts of ascorbic acid or its catabolites excreted in feces [3].

2.8.3 Function

Vitamin C acts as a cofactor in various hydroxylation reactions catalyzed by oxygenases. For example, vitamin C plays an important role in collagen and catecholamines biosynthesis. It is also involved in carnitine synthesis, which takes place in liver from the amino acid lysine. Both vitamin C and copper act as cofactors in the catabolism of phenylanaline and thyrosine enzymes, which is impaired with vitamin C deficiency. In addition, some peptide hormones like peptidylglycine are activated in the presence of copper, oxygen and ascorbate. By promoting the incorporation of iron into ferritin, vitamin C influences the distribution of this mineral in the body. It is also required in the metabolism of histamine and nitrosamine. Moreover, vitamin C increases the mobility of leukocytes and protects their membranes from oxidative attacks. Due to its high reducing power, vitamin C has also an important biochemical function of antioxidant. The powerful antioxidant action of vitamin C allows it to protect the sperm from AND oxidative attacks or the eye from radical attacks [11].

2.9 Vitamin D

Rickets, a disease due to vitamin D deficiency, occurred in ancient times about 50,000 B.C. [7]. Due to air pollution and little sunlight, Rickets has spread in northern Europe, England, and the United States during the Industrial Revolution. There are two common forms of vitamin D: cholecalciferol (vitamin D_3) and ergocalciferol (vitamin D_2). The structures of vitamin D_2 and vitamin D_3 were determined by Windaus and his associates respectively in 1932 and in 1936, in Germany. Dorothy Hodgkin, the Nobel laureate x-ray crystallographer, was the first to develop a three-dimensional model of vitamin D_3. While vitamin D_2 is synthesized in plants, fungi and yeasts by the solar irradiation of ergosterol, vitamin D_3 is produced *in vivo* by the action of sunlight on skin [3, 7].

2.9.1 Dietary sources

There are relatively few sources of vitamin D. The richest dietary sources are fish-liver oils, especially halibut-liver oil. Other major sources of the vitamin are fatty fish such as tuna, sardines, pilchards or herring. Eggs, mammalian liver and dairy products are good sources of vitamin D. However, vegetables, fruit and cereals are exempt of the vitamin. Fortified products include margarine, breakfast cereals and milk. Vitamin D can also be synthesized by the skin after exposure to sunlight. This endogenous synthesis constitutes a much more important source of vitamin D than foodstuff [2, 3, 8].

2.9.2 Absorption, metabolism and storage

Vitamin D is absorbed in lipid micelles and incorporated into chylomicrons through the lymphatic system. Following intestinal absorption, vitamin D is rapidly taken up by the liver, where it is hydroxylated to form the 25-hydroxy derivative calcidiol before entering the circulation bound to a vitamin D binding globulin. On arrival at the kidney, the calcidiol undergoes 1-hydroxylation to yield the active metabolite 1,25-dihydroxyvitamin D (calcitriol) or 24-hydroxylation to yield an apparently inactive metabolite, 24,25-dihydroxyvitamin D (24-hydroxycalcidiol) [3, 8]. These conversions are regulated in response to the concentrations of calcium and phosphorous minerals in plasma. Vitamin D is mainly excreted in bile after catabolism into highly polar inactivation products by the liver [3].

2.9.3 Function

The primary physiologic role of vitamin D is to maintain the plasma concentrations of calcium and phosphorous in the body. Calcitriol acts to rise intestinal absorption of calcium, to decrease its excretion and to mobilize the mineral from bone [4]. It plays an important role in immune muscle functions, nervous system and cell differentiation. In addition, it regulates more than 50 genes by activating nuclear receptors that modulate gene expression. 24-hydroxycalcidiol has also a biological activity in cartilage or in bone formation during the development of knockout mice [4].

3. Minerals

Minerals are inorganic substances that are required for normal metabolism, development, growth, body structure, cell function regulation, and electrolyte balance [13]. Minerals are divided in two groups, the major (macro) minerals, which are present in the human body in concentrations greater than 100 milligrams per kilogram and trace (micro) minerals occurring at much lower amounts (micrograms or milligrams per kilogram) [8]. Minerals are considered as essential when deficiency symptoms are noted with depletion or removal. Typically, calcium constitutes about 46% and phosphorous about 29% of total body minerals. Chloride, potassium, sulfur, sodium and magnesium together account for about 25%, while essential trace elements represent less than 0.3% of the total. Bone is the primary storage site for many essential elements (99% of the total calcium, 80 to 85% of phosphorous and some 70% of magnesium), while the thyroid gland contains more than 80% of the total body iodine. Generally, mineral distribution within the body's tissue is not uniform and each organ has a specific mineral composition [14]. Table 2 shows the benefits and potential harmful effects associated with deficiency or excess intake of four frequently used minerals as food additives and/or supplements [1, 2, 9, 10, 15].

Table 2. Deficiency and toxic effects of selected minerals.

Mineral	Deficiency symptoms	Excess intake symptoms
Calcium	Memory loss Muscle cramps Increased risk of fractures Osteopenia Osteoporosis	Hypercalcemia Renal insufficiency Kidney stones
Magnesium	Ischemic heart disease Hypertension Osteoporosis Diabetes Stroke	Diarrhea Muscular paralysis Central nervous system depression

Mineral	Deficiency symptoms	Excess intake symptoms
Iron	Fatigue and weakness Anemia Mental retardation Brain damage Behavioral impairment	Vomiting and diarrhea Liver, kidney and heart diseases Neurological defects Gastrointestinal effects Hormonal abnormalities
Iodine	Goiter and cretinism Hypothyroidism Growth retardation Neurological defects Reproductive disturbance	Goiter Hypothyroidism Hyperthyroidism Thyroiditis

3.1 Calcium

Calcium is a soft, silvery white metal ranked fifth in order of abundance among the elements in the earth crust. It has an atomic weight of 40.08, and an atomic number of 20. Calcium is found in nature only in compounds, mainly as carbonate, fluoride, sulfate and phosphate [8, 14]. Metallic calcium or calcium compounds are used in steel industry, plastics, cement manufacture, paper industry, alloys production, etc. Calcium is the most abundant mineral in the organism and 99% of this macro mineral is found in bone and teeth [14].

3.1.1 Dietary sources

Dairy products are the most important dietary sources of calcium including milk, yoghurt and cheese. Tinned fish such as sardine, oilseeds like almonds and dried fruit are also good sources of calcium. Generally, foods of plant origin are poor sources of calcium. However, due to their high level of consumption they make a significant contribution to the total calcium intake. Contributions from calcium-enriched foods such as breakfast cereals and dietary supplements containing calcium salts may be also significant [8].

3.1.2 Absorption, metabolism and storage

The absorption of calcium takes place by two mechanisms. The first is active and regulated by vitamin D and the second is by passive diffusion. At low levels of dietary calcium, most of the absorption is by active transport and is done principally in the duodenum. Passive diffusion is more important at high calcium intakes and ileum is the most active absorptive site. The large intestine contributes also to calcium absorption. Generally, only 30 to 50% of dietary calcium is normally absorbed [14]. Calcium absorption is largely dependent on other nutrients in the diet, hormonal status and physiological conditions such as pregnancy and breastfeeding [1]. Lactose and vitamin C supplementation increase calcium absorption. Dietary sources rich in oxalic acid or phytic acid could inhibit calcium absorption from foodstuff [9]. Calcium homeostasis is regulated by calcitonin and parathyroid hormones that are associated to the active form of vitamin D, 1,25-dihydroxyvitamin D. Bone is the primary storage site of calcium and the decrease of calcium levels in blood leads to a quick metal mobilization from the bone to bring the blood levels back to normal. Plasma contains from 9 to 12 mg of calcium per 100 mL, which is distributed in ionized, protein bound and complexed fractions. Excretion of calcium is done principally in feces.

3.1.3 Function

The primary role of calcium in the body is to form the structures of bone and teeth. During skeletal growth and maturation, calcium accumulates in the skeleton at an average rate of 150 mg/day. During maturity, the skeleton is in calcium equilibrium until the age of fifty, from which the bone is lost from all skeletal sites. This bone loss increases the risk of hip fracture with aging. Extraskeletal calcium, which represent around 1% of the total body calcium, play important roles in various essential functions in body metabolism such as enzymatic activation, cell division, muscular contraction, nerve transmission and vesicular secretion [1, 8].

3.2 Magnesium

Magnesium is a silvery white metal ranked seventh in order of abundance in the crust. It has an atomic number of 12 and an atomic weight of 24.31. Magnesium occurs in nature in compounds such as dolomite, magnetite, epsomite and carnallite. Due to its relatively low density (1.74 g/cm^3), magnesium is widely used as a construction material. Other applications include production of alloys, steel industry, automotive construction, aviation and space technology, glass and cement manufacture, etc. [16]. In green plants, magnesium plays a central role, as photosynthesis does not proceed, when the metal has been removed from the chlorophyll molecule.

3.2.1 Dietary sources

Dietary sources rich in magnesium include whole grains, green leafy vegetables, tofu and nuts. Milk, legumes and potatoes are good sources of the metal. Plants are generally well endowed with magnesium due to its central role in photosynthesis [8, 9].

3.2.2 Absorption, metabolism and storage

The main site of magnesium absorption is the small intestine. Active transport, solvent drag and passive diffusion are the three mechanisms of magnesium absorption. Generally, 20–70% of ingested magnesium is absorbed in normal healthy humans [8]. Magnesium absorption could be affected by many dietary and physiological factors. While protein intake seems to increase magnesium absorption, fat supplementation has been shown to generate the opposite effect. In addition, high levels of phosphorous and calcium inhibit magnesium absorption. Magnesium is primarily stored in the bones (60–70%) and the remainder is evenly distributed between muscle and other soft tissue. Magnesium is excreted mainly in the urine or in feces [14].

3.2.3 Function

Magnesium has various physiological functions. Its is involved in more than 300 enzymatic processes in the body and plays an important role in skeletal mineralization and development, fatty acids synthesis, phosphorylation and dephosphorylation mechanisms, DNA and protein synthesis, insulin action in the liver, metabolic pathways for energy production, sodium and phosphorous metabolism and maintenance of transmembrane electrical potentials in nerves and muscle [1, 14].

3.3 Iron

Iron is a silvery-white metal classified second in the order of abundance of the elements in the earth crust after aluminum. It has an atomic number of 26 and an atomic weight of 55.8. It is mainly combined with oxygen in the crust forming iron oxide ores such as hematite, magnetite or limonite. Due to its interesting physico-chemical properties, iron is used in various domains such as steel industry, alloys, food fortification, environmental protection, bio-medical applications, etc. Iron is an essential element for almost all living organisms. It exists in the body in complex forms bound to protein as heme compounds (hemoglobin and myoglobin), heme enzymes, or nonheme compounds (transferrin, ferritin, and hemosiderin [14]. Iron deficiency anemia is the most common deficiency disorder in the world [8, 14].

3.3.1 Dietary sources

Organ meats such as liver and kidney, egg yolk, dried legumes, cocoa, cane molasses, and parsley are among the richest dietary sources of iron. Meat meals and fish meals contain lower iron amounts than blood meals. Generally, the bioavailability of heme iron found in animal products is better than that of nonheme iron found in vegetables or chlorophyll plants. Poor sources of iron include dairy products, fresh fruits and vegetables, white floor, white sugar and unenriched bread [1, 14].

3.3.2 Absorption, metabolism and storage

Generally, iron absorption increases with higher dietary intake of the metal. Only 5 to 15% of ingested iron is absorbed by human adults. However, this level may increase to twice or more in children and in deficient adults. The primary sites of iron absorption are the duodenum and the jejunum in the small intestine. Iron absorption is enhanced by ascorbic acid and cysteine dietary intake. However, tannins, phytates, egg yolk, tea, coffee, milk and soy proteins have an inhibitory effect on iron absorption [1]. Upon entering plasma, ferrous iron is oxidized to ferric form, then most of it binds to transferrin, and the rest to ferritin. Transferrin regulates iron body distribution by transporting more than 70% of plasma iron to the bone marrow for hemoglobin synthesis. The synthesized hemoglobin is then destroyed by phagocytes and the released iron is either returned to the circulation via plasma transferrin or stored as ferritin or hemosiderin. The liver, spleen, and bone marrow are the main sites of iron storage in the body [9]. Excretion of iron is done primary in feces and urine in addition to losses through sweat, hair, and nails [14].

3.3.3 Function

Almost *two*-thirds of body's iron is found in hemoglobin, a quarter is stored and most of the remaining 15% is in the muscle protein myoglobin. The primary function of iron is the transport of oxygen and carbon dioxide in the organism through red cells [1]. Iron is also a component of various enzymes required for energy production, immune system functioning or adenosine triphosphate (ATP) production.

3.4 Iodine

Iodine is a bluish black solid that belongs to halogen family including fluorine, chlorine and bromine. It has an atomic number of 53 and an atomic weight of 126.9. It is a relatively rare element in the earth's crust, which occurs in the dispersed state in soil, water, air, and living organisms. There are many regions in the world low in iodine. This could be due to distance from the sea, low annual rainfall or recent glaciation [14]. Iodine is used in many industrial fields such as pharmaceutics, medicine, metallurgy, lasers, herbicides, animal feeds, colorants, photographic equipment, etc. [8]. Iodine deficiency is one of the main cause of impaired cognitive development in children.

3.4.1 Dietary sources

The iodine content of food is highly dependent on the type of soil, climatic conditions and fertilizers. Hence, it is useless to give the concentration of dietary iodine in food composition tables since it is highly variable. Nevertheless, seafood, fish, seaweed and plants of the seaside are very rich in iodine. Fortified foods are also good iodine sources such as breakfast cereals or milk and milk products. The salt iodized with potassium iodate remains the most reliable source of iodine worldwide [1, 9].

3.4.2 Absorption, metabolism and storage

Iodine in water and food occurs usually as iodide or iodate compound. In these forms, it is rapidly absorbed in the gastrointestinal tract and circulates in the blood to all body tissues. Over 90% of ingested iodine is concentrated in the thyroid gland for adequate thyroid hormone synthesis or excreted in urine. If iodine supply is abundant, only 10% of iodine absorbed by the gut appears in the thyroid. However, this fraction may reach 80% or more, with long-standing iodine deficiency [14]. Iodine is stored in thyroid in the form of thyroglobulin, an iodinated glycoprotein that represent 90% of the total thyroid iodine. It is excreted either as organic iodine in the feces or as free iodine in the urine. Important goitrogen-containing foods include cassava, broccoli, cabbage, sprouts, kohlrabi, turnips, swedes, rapeseed, and mustard [8].

3.4.3 Function

Iodine is an essential constituent of the thyroid hormones thyroxine, and triiodothyronine. These hormones have multiple biological roles in thermoregulation, reproduction, growth and development, intermediary metabolism, cell activity, and muscle function. Thyroid hormones also affect heart rate, respiratory rate, metabolism of carbohydrates and lipogenesis. In addition, they regulate proteins associated with cartilage metabolism, skin epidermis and hair production [8, 14].

4. Multivitamins and Minerals as Food Supplements

4.1 Prevalence of use and associated factors

Vitamins and minerals are found in diets rich in fruits and vegetables. In the early 20th century, these micronutrients have been isolated, purified and manufactured as dietary supplements. Multivitamin/mineral (MVM) supplements, defined by the National Institutes of Health (NIH) as dietary supplements containing 3 or more vitamins and minerals, are the most commonly used type of dietary supplements among adults in the United States (US) [17]. According to the National Health and Nutrition Examination Survey (NHANES) 2011–2014 data (n = 11,024), 52% of US adults took at least one dietary supplement in a 30-day period and MVM supplements account for the vast majority of total dietary supplements use [18]. MVM use was higher among women (34.0%) than men (28.3%) and the use raised linearly with age. Supplementation with MVM was particularly higher among older women (44.0%), non-Hispanic whites (35.7%), higher educated adults (36.3%), overweight adults (34.0%), former smokers (37.8%), moderate alcohol consumers (1 drink/day) (39.0%), adults with excellent or very good self-reported health status (36.7%) and those with private health coverage (35.1%) compared to their counterparts [18]. Perlitz et al. reported data on the use of vitamins and mineral supplements by adolescent living in Germany from the second wave of the German Health Interview and Examination Survey for Children and Adolescents (EsKiMoII), conducted from June 2015 to September 2017 [19]. They found that 16.4% of the adolescents (girls: 18.8%, boys: 14.0%) aged 12 to 17 years had consumed vitamin or mineral supplements in the previous four week. The use of micronutrients was higher among girls (18.8%) than boys (14.0%), normal weight adolescent (17.0%) than overweight ones (9.3%) and among adolescents with a high level of physical exercise (19.9%) than those with a low level of physical exercise (11.5%). Only one supplement, with both vitamins and minerals nutrients, was consumed by the majority of users. The most frequently used vitamin supplements were vitamin C (43.9%), followed by vitamin D (41.1%) and vitamin B_{12} (30.4%). As regards mineral supplements, those containing magnesium (45.9%), zinc (28.1%), and iron (24.1%) were the most commonly consumed [19]. In Iran, the prevalence of dietary supplements use among children and adolescents was 34.1% [20]. Iron supplements were the most frequently used dietary supplements, with a prevalence of use of 12.9%, followed by multivitamins (8.1%) and vitamin D supplement (2.9%). Results showed that boys, children with excess weight and those with high-educated parents used less supplements compared to their counterparts [20].

Pregnant women are particularly vulnerable population groups and the MVM supplementation especially with iron and folic acid in pregnancy is highly recommended to improve birth outcome and to reduce low birth weight [21]. Data on MVM supplementation of pregnant women living in European countries revealed the presence of iron, iodine, vitamin A and zinc deficiencies, not only in low-income countries but also to an extent in Europe [22]. The inadequate intake among pregnant women may be due to poor knowledge about adequate nutrition, healthy vegetarian or vegan diets, special diets that avoid excessive weight gain, etc. [22]. Data on the dietary intake and mineral status of Polish pregnant women showed that 53.7% of pregnant women used supplements during pregnancy. Of supplement users, 93% took folic acid and only 16% consumed iron. Composite vitamin and mineral supplements were used by only 17.6% of Polish pregnant women [23]. Results of the NHANES 2007-2014 cross-sectional study revealed that 61.6% of American pregnant women used MVM supplements during pregnancy. However, the majority of this group do not consume the recommended amount of fruits and vegetables (five servings per day) [24].

4.2 Recommended intakes

Table 3 displays the recommended intakes of selected vitamins and minerals for men and women according to AFSSA (French Food Safety Agency), EFSA (European Food Safety Authority), IOM (Institute of Medicine) and WHO (World Health Organization) [25]. Table 3 shows that several terms, referring to the population's nutritional references, are used by the different organisms. Hence, it is necessary to harmonize these terms.

Table 3. Population's nutritional references of selected vitamins and minerals (mg/d). [1]

Organism	Gender	VitA 2	$VitB_1$	$VitB_2$	$VitB_3$	$VitB_6$	$VitB_9$	$VitB_{12}$	VitC	VitD	Ca	Mg	Fe	I
AFSSA (RNI) 3	Men	0.80	1.50	1.80	17.40	1.80	0.33	0.0024	110	10	900	420	9	0.15
	Women	0.80	1.20	1.50	14.40	1.50	0.30	0.0024	110	10	900	360	16	0.15
EFSA (PRI) 4	Men	0.75	ND 5	ND	17.40	ND	0.33	0.0040	110	15	920	350	11	0.15
	Women	0.65	ND	ND	14.40	ND	0.33	0.0040	95	15	950	300	16	0.15
IOM (RDA) 6	Men	0.90	1.20	1.30	17.40	1.30	0.40	0.0024	90	15	1000	420	8	0.15
	Women	0.70	1.10	1.10	14.40	1.30	0.40	0.0024	75	15	1000	320	18	0.15
WHO (RNI)	Men	0.60	1.20	1.30	14.40	1.30	0.40	0.0024	45	5	1000	260	9.1-27.4	0.15
	Women	0.50	1.20	1.10	11.40	1.30	0.40	0.0024	45	5	1000	220	19.6-58.8	0.15

[1] mg/day.

[2] Expressed in retinol equivalent (RE): 1 mg retinol = 1 mg RE; 1 mg β-carotene = 1/12 mg RE.

[3] Recommended Nutrient Intake.

[4] Population Reference Intake.

[5] Not Defined.

[6] Recommended Dietary Allowance.

4.3 Availability and regulatory status

MVM supplements can be obtained directly by the health care providers or purchased from retail stores such as pharmacies, grocery stores or health food stores. Many forms of vitamins and minerals and many routes of administration are available [13]. For example, vitamin D is manufactured as cholecalciferol or ergocalciferol and iodine is available as iodide or iodate [26]. Vitamin and mineral dietary supplements can be compounded in various ways and sold in capsules, pills, tablets, sachets of powder, ampoules of liquids and other similar forms of powder and liquids [13].

According to the Codex Alimentarius Guidelines for Vitamin and Mineral Food Supplements, the sources of vitamin and mineral supplements may be natural or synthetic and their selection should be based on considerations such as safety and bioavailability [27]. The minimum level of each vitamin and/or mineral contained in a vitamin and mineral supplement per daily portion of consumption should be 15% of the WHO recommended dietary intake. Moreover, the vitamin and mineral supplements must be packed in containers that are safe and suitable for their intended use.

4.4 Health effects

The use of MVM supplements is recommended in disease states, to correct deficiencies, or to promote health and protect from future chronic diseases in healthy subjects [28, 29, 30, 31]. The main nutrient deficiency diseases prevalent in developing countries include *Xerophthalmia,* which is responsible of the blindness of over five million children suffering from visual impairment in the world [32]. According to Beaton et al. (1993), vitamin A supplementation reduces the mortality of children aged from six months to five years by 23% [33]. In addition, data showed a decrease in the prevalence of *Xerophthalmia* in children who regularly received vitamin A supplements, reaching 70% [32, 33]. In 1999, over 700 million people living in developing countries suffered from goiter and cretinism with severe brain damage and mental retardation [8]. An important progress was made with universal salt iodization including an approximate 13-point increase in intelligence quotient [32]. Iron deficiency anemia have adverse effects on the development of children as well as on the mortality and morbidity of infant and mother during pregnancy [8]. Research studies established that the use of iron supplements allowed a 20% reduction in mothers mortality [32]. In Nepal, the results of a demographic and health survey carried out in 2006 showed that in five years, the coverage of iron supplementation increased from 23–59%. These improvements, associated to additional measures, led to a reduction in iron deficiency anemia in pregnant women from 75–42% [34]. Zinc deficiency is associated to the incidence of serious childhood infectious diseases such as pneumonia and diarrhea, which causes 18% of deaths in children under five [35, 36, 37]. Zinc supplementation of children during acute diarrhea allowed the decrease of the illness duration by 9-23% [36]. In a community-based, double-blind, randomized trial conducted in India, zinc supplementation resulted in a 33% lower diarrheal incidence in children with low plasma concentrations [38]. During pregnancy, there is a high risk of fetal neural tube defects with folate deficiency. Research studies proved that folate intake by pregnant women reduced the risk of neural tube defects by 50% [32, 39].

5. Addition of Vitamins and Minerals to Food

Vitamins and minerals are added to foods mainly to improve or maintain their nutritional quality. However, these nutrients are also used for other additive functions (Table 4) [40]. Various forms of vitamin and mineral additives are commercially available such as powders, oily suspensions or emulsions [2, 40]. The most important factors affecting their stability include heat, oxygen, pH, moisture and light. For example, the stability of β-carotene and other provitamin A carotenoids is enhanced by ascorbic acid antioxidant, both in liquid or powder forms [41]. In addition, vitamin D is more stable when prepared in edible oils than in powder form. Preparations of vitamin D are usually provided in lightproof containers with inert gas flushing. The detrimental interactions between vitamins and minerals are also taken in consideration in the manufacture of these micronutrients, especially in liquid preparations. Some interactions between vitamins are advantageous such as niacinamide that act as solubilizer for riboflavin and folic acid [41].

Table 4. Examples of use of vitamins and minerals as food additives.

Function	Vitamins and minerals
Antioxidants	Ascorbic acid, β-carotene, calcium ascorbate, ascorbyl palmitate, sodium ascorbate, tocopherols (mixed α, γ and δ), tocopherol (mixed natural concentrate)
Colourings	β-carotene, riboflavin, riboflavin-5⊠-phosphate, iron oxide, calcium carbonate
Flavor enhancers	Calcium chloride, magnesium sulfate
Floor additives	Calcium carbonate
Nutritive additives	Ascorbic acid, β-carotene, ascorbyl palmitate, calcium ascorbate, calcium carbonate, calcium lactate, cholecalciferol, ergocalciferol, ferrous sulphate, folic acid, magnesium oxide, nicotinic acid, retinol, riboflavin, retinyl acetate, thiamin, thiamin hydrochloride, thiamin mononitrate, α-tocopherol
Sequestrants	Calcium chloride, calcium citrate, calcium sulfate

6. Conclusion

Despite the essential role of vitamins and minerals in human's health, these micronutrients are not consumed in sufficient amounts in developing countries where millions of children die each year from micronutrients deficiency due to malnutrition. Therefore, additional efforts should be made worldwide to minimize disparities between developing and developed countries in the quality of nutrition.

References

1. Mouray, A. Nutrition manual for humanitarian intervention (in french); Comité international de la Croix-Rouge: Geneva, switzerland, 2004
2. Food additives, Second edition ed.; CRC Press: Boca Raton, USA, 2001
3. Ball, G.F.M. Vitamins. Their Role in the Human Body; Blackwell Publishing: London, UK, 2004
4. Bender, D.A. Nutritional Biochemistry of the Vitamins, Second edition ed.; Cambridge University Press: London, UK, 2003

5. Sinbad, O.O.; Folorunsho, A.A.; Olabisi, O.L.; Ayoola, O.A.; Temitope, E.J. Vitamins as Antioxidants. J Food Sci Nutr Res 2019, 2 214-235
6. Lukaski, H.C. Vitamin and mineral status: effects on physical performance. Nutrition 2004, 20, 632-644, doi:10.1016/j.nut.2004.04.001
7. Handbook of vitamins, Fourth edition ed.; CRC Press. Taylor & Francis Group: Boca Raton, USA, 2007
8. The Nutrition Society. Introduction to Human Nutrition, Second edition ed.; Wiley-Blackwell: Oxford, UK, 2009
9. Intitute of Medicine of the National Academies. Dietary Reference Intakes: The Essential Guide to Nutrient Requirements; The National Academies Press: Washington, DC, USA, 2006
10. WHO/FAO. Vitamin and mineral requirements in human nutrition: report of a joint FAO/WHO expert consultation; Bangkok, Thailand, 21-30 September 1998, 1998
11. Société Francophone Vitamines et Biofacteurs. Vitamins (in french). Bioforma: Paris, France, 2007
12. Polcz, M.E.; Barbul, A. The Role of Vitamin A in Wound Healing. Nutr Clin Pract 2019, 34, 695-700, doi:10.1002/ncp.10376
13. Gagner, J.J. Nutritional, Herbal, and Homeopathic Supplements. In Evidence-Based Management of Low Back Pain, Dagenais, S., Haldeman, S., Eds. Mosby: Maryland Heights, Missouri, USA, 2012; pp. 258-268
14. McDowell, L.R. Minerals in Animal and Human Nutrition, Second edition ed.; Elsevier: Amsterdam, Netherlands, 2003
15. WHO. Iodine. In Trace elements in human nutrition and health, WHO, Ed. World Health Organization: Geneva, Switzerland, 1996; pp. 49-71
16. Enghag, P. Encyclopedia of the Elements; Technical Data · History · Processing · Applications; WILEY-VCH Verlag GmbH & Co. KGaA: Weinheim, 2004
17. NIH State-of-the-Science Conference: Multivitamin/Mineral Supplements and Chronic Disease Prevention, May 15-17, 2006, Bethesda, Maryland, USA. Am J Clin Nutr 2007, 85, 251S-327S
18. Cowan, A.E.; Jun, S.; Gahche, J.J.; Tooze, J.A.; Dwyer, J.T.; Eicher-Miller, H.A.; Bhadra, A.; Guenther, P.M.; Potischman, N.; Dodd, K.W., et al. Dietary Supplement Use Differs by Socioeconomic and Health-Related Characteristics among U.S. Adults, NHANES 2011(−)2014. Nutrients 2018, 10, doi:10.3390/nu10081114
19. Perlitz, H.; Mensink, G.B.M.; Lage Barbosa, C.; Richter, A.; Brettschneider, A.K.; Lehmann, F.; Patelakis, E.; Frank, M.; Heide, K.; Haftenberger, M. Use of vitamin and mineral supplements among adolescents living in Germany-Results from EsKiMo II. Nutrients 2019, 11, doi:10.3390/nu11061208
20. Namazi, N.; Kelishadi, R.; Heshmat, R.; Motlagh, M.E.; Sanaei, M.; Shafiee, G.; Ziaodini, H.; Beshtar, S.; Taheri, M.; Aminaee, T., et al. Determinants of taking dietary supplements in Iranian children and adolescents: the CASPIAN-V study. J Diabetes Metab Disord 2019, 18, 409-417, doi:10.1007/s40200-019-00432-z

21. Biesalski, H.K.; Tinz, J. Multivitamin/mineral supplements: Rationale and safety - A systematic review. Nutrition 2017, 33, 76-82, doi:10.1016/j.nut.2016.02.013

22. Biesalski, H.K. Sustainable Micronutrients in Europe: Is There Cause for Concern? In Sustainable Nutrition in a Changing World, Biesalski, H., Drewnowski, A., Dwyer, J., Strain, J., Weber, P., Eggersdorfer, M., Eds. Springer, Cham: New York, USA, 2017; pp. 143-166

23. Kocylowski, R.; Lewicka, I.; Grzesiak, M.; Gaj, Z.; Sobanska, A.; Poznaniak, J.; von Kaisenberg, C.; Suliburska, J. Assessment of dietary intake and mineral status in pregnant women. Arch Gynecol Obstet 2018, 297, 1433-1440, doi:10.1007/s00404-018-4744-2

24. Francis, E.C.; Zhang, L.; Witrick, B.; Chen, L. Health behaviors of American pregnant women: a cross-sectional analysis of NHANES 2007-2014. J Public Health (Oxf) 2019, 10.1093/pubmed/fdz117, doi:10.1093/pubmed/fdz117

25. ANSES. Actualisation des repères du PNNS : élaboration des références nutritionnelles (in french); Agence nationale de sécurité sanitaire de l'alimentation, de l'environnement et du travail.: December 2016

26. Anadón, A.; Martínez-Larrañaga, M.R.; Ares, I.; Aránzazu Martínez, M. Evaluation and Regulation of Food Supplements: European Perspective. In Nutraceuticals. Efficacy, Safety and Toxicity, Gupta, R.C., Ed. Academic Press: Cambridge, Massachusetts, 2016

27. FAO. Codex Guidelines for Vitamin and Mineral Food Supplements. CAC/GL55 - 2005; The Food and Agriculture Organization: Rome, Italy, 2005

28. Combet, E.; Buckton, C. Micronutrient deficiencies, vitamin pills and nutritional supplements. Medicine 2015, 43, 66-72

29. Lentjes, M.A.H. The balance between food and dietary supplements in the general population. Proc Nutr Soc 2019, 78, 97-109, doi:10.1017/S0029665118002525

30. Fortmann, S.P.; Burda, B.U.; Senger, C.A.; Lin, J.S.; Whitlock, E.P. Vitamin and mineral supplements in the primary prevention of cardiovascular disease and cancer: An updated systematic evidence review for the U.S. Preventive Services Task Force. Ann Intern Med 2013, 159, 824-834, doi:10.7326/0003-4819-159-12-201312170-00729

31. Jenkins, D.J.A.; Spence, J.D.; Giovannucci, E.L.; Kim, Y.I.; Josse, R.; Vieth, R.; Blanco Mejia, S.; Viguiliouk, E.; Nishi, S.; Sahye-Pudaruth, S., et al. Supplemental Vitamins and Minerals for CVD Prevention and Treatment. J Am Coll Cardiol 2018, 71, 2570-2584, doi:10.1016/j.jacc.2018.04.020

32. FFI; GAIN; MI; USAID; Bank, T.W.; UNICEF. Investing in the future. A united call to action on vitamin and mineral deficiencies: Global Report 2009; 2009

33. Beaton, G.H.; Martorell, R.; Aronson, K.J.; Edmonston, B.; McCabe, G.; Ross, A.C.; Harve, B. Effectiveness of Vitamin A Supplementation in the Control of Young Child Morbidity and Mortality in Developing Countries In Nutrition policy discussion paper No. 13, NUTRITION, A.C.O.C.S.O., Ed. Geneva, Unites States, 1993

34. Ministry of Health and Population from Nepal. Nepal Demographic and Health Survey. 2006; Ministry Of Health and Population from Nepal,: Katmandou, 2006

35. Black, R.E. Therapeutic and preventive effects of zinc on serious childhood infectious diseases in developing countries. Am J Clin Nutr 1998, 68, 476S-479S, doi:10.1093/ajcn/68.2.476S

36. Bryce, J.; Boschi-Pinto, C.; Shibuya, K.; Black, R.E.; Group, W.H.O.C.H.E.R. WHO estimates of the causes of death in children. Lancet 2005, 365, 1147-1152, doi:10.1016/S0140-6736(05)71877-8

37. Bahl, R.; Bhandari, N.; Hambidge, K.M.; Bhan, M.K. Plasma zinc as a predictor of diarrheal and respiratory morbidity in children in an urban slum setting. Am J Clin Nutr 1998, 68, 414S-417S, doi:10.1093/ajcn/68.2.414S

38. Sazawal, S.; Black, R.E.; Bhan, M.K.; Jalla, S.; Sinha, A.; Bhandari, N. Efficacy of zinc supplementation in reducing the incidence and prevalence of acute diarrhea--a community-based, double-blind, controlled trial. Am J Clin Nutr 1997, 66, 413-418, doi:10.1093/ajcn/66.2.413

39. Christianson, A.; Howson, C.P.; Modell., Y.B. March of Dimes Global Report on Birth Defects: The Hidden Toll of Dying and Disabled Children; March of Dimes Birth Defects Foundation: 2006

40. Food additives data book. Second edition; John Wiley & Sons Ltd: West Sussex, UK, 2011

41. Food fortification and supplementation. Technological, safety and regulatory aspects; CRC Press: Boca Raton, USA, 2008

Chapter-6

New Trends in Natural Emulsifiers and Emulsion Technology for The Food Industry

1. Introduction

Food emulsions are produced from two immiscible liquids (usually oil and water), which in the presence of an emulsifier and by applying an emulsification method, can be dispersed one into another. Some typical examples include mayonnaise, salad dressings, sauces, milk, ice cream, and sausages. These systems can be used to encapsulate, protect, and deliver biocompounds, including vitamins, flavors, colorants, and nutraceuticals [1]. Emulsifiers are food additives acting by forming a physical barrier between the oil and water, enabling their compatibilization. Effective emulsifiers must be quickly adsorbed at the oil–water interfaces leading to a rapid decrease in the interfacial tension, preventing droplets aggregation. Moreover, they must generate strong repulsive interactions promoting emulsion stability [2, 3].

Synthetic emulsifiers (e.g., Tweens and Spans) are well-known for their ability to form highly stable emulsions. However, consumers' preferences for healthy, sustainable and natural lifestyle habits have increased worldwide. Moreover, some studies have reported intestinal dysfunctions caused by synthetic emulsifiers [4, 5]. In this context, natural emulsifiers have emerged as great alternatives to replace their conventional counterparts, namely proteins [6], polysaccharides [7], phospholipids [8] and saponins [9]. Concerning protein-based natural emulsifiers, the most use ones come from animal sources (e.g., whey proteins, caseins, egg protein, gelatin) [10]. However, plant-based proteins have demonstrated to be good alternatives for their replacement in products with dietary restrictions (e.g., lactose-free) and in vegetarian and vegan foods. Moreover, plant-based proteins are more sustainable as they have a lower carbon footprint [11, 12]. Examples include pea [13, 14] and soy proteins [15], which have been reported for emulsions production.

Aligned with natural emulsifiers, Pickering stabilizers (in particular organic-based colloidal particles) are emerging as promising solutions. Pickering emulsions or particle-stabilized emulsions present high resistance to coalescence and Oswald ripening due to the tight fixation of the particles to the droplets surface [16]. Several food-grade particles have been studied, namely particles based on proteins [17], polysaccharides [18], and protein/polysaccharide complexes [19]. Furthermore, natural emulsifiers from microbial origin such as biosurfactants and bioemulsifiers are also potential alternatives to be explored in food emulsions [20, 21].

This chapter covers a bibliographic review focused on the last 10-years on natural emulsifiers and emulsion technology field. Research and market trends are also highlighted, showing the most relevant natural emulsifier families. Basic concepts concerning emulsion production, classification, and stabilization methods are introduced. A special emphasis is given to Pickering emulsions regarding novel trends in food emulsion systems.

1.1 Bibliometric and market trend analysis

According to the Research and Markets report, amidst the Covid-19 crisis, the global emulsifiers' market is projected to reach US$ 6.1 Billion by 2027, growing at a Compound Annual Growth Rate (CAGR) of 4.8% over the forecast period (2020–2027). Particularly, natural emulsifiers' area is estimated to get US$ 3.3 Billion, recording a 5.4% CAGR [22]. In agreement, the "Global Food Emulsifiers Market 2020-2027 report" from MarketResearch, foresees a high potential for the plant-based emulsifiers in the global food emulsifiers market [23].

Concurrently, scientific literature corroborates the global food emulsifiers report's projections. More than 8,000 documents were found using the terms "natural emulsifier*" OR "bioemulsifier*" OR "bio-emulsifier*" OR "biosurfactant*" OR "bio-surfactant*" OR "Pickering emulsion*" searched in title, abstract, keywords and Keywords plus sections using the Web of Science Core Collection (SCI-EXPANDED), in the 2010–2020 period. Excluding documents with early publication and applying the "Food Science and Technology" filter from WOS, 792 documents were found. By removing 4 documents from 2021 in a final manual screening, 788 documents were analyzed using Biblioshiny app from the Bibliometrix-R package (RStudio) [24] and VosViewer software [25]. The survey was performed on April 25th, 2021.

Table 1 presents some of the retrieved 788 documents concerning the application of natural emulsifiers or Pickering stabilizers in emulsion formation/stability, including their use in biocompound delivery systems. Some works regarding the production of bioemulsifiers or biosurfactants by microorganisms were also found [31, 32]. Several studies addressing Pickering emulsions and the use of high-pressure homogenization were identified.

Table 1. Studies reporting the use of natural molecules and Pickering stabilizers selected from the retrieved 788 documents of the bibliometric search.

Natural emulsifiers or Pickering stabilizers	Emulsification method	Main target	Ref.
Zein-Chitosan complex particles	High-shear homogenization	Delivery system (Curcumin)	[26]
Mannoprotein	High-shear homogenization	Formation/Stability	[27]
Whey protein Gum arabic *Quillaja* saponin Lecithin	Dual-channel microfluidization	Formation/Stability	[28]

Natural emulsifiers or Pickering stabilizers	Emulsification method	Main target	Ref.
Gum arabic Beet pectin Corn fiber gum	High-shear homogenization Microfluidizer	Formation/Stability	[7]
Ginseng saponins	High-shear/ High-pressure homogenization	Delivery system (Astaxanthin)	[9]
Wheat gluten nanoparticles (WPN) WPN-xanthan gum nanoparticles	High-shear homogenization	Delivery system (β-carotene)	[29]
Pea protein microgel particles	High-pressure homogenization	Formation/Stability	[30]

Figure 1a shows the wordcloud from Author's Keyword. The higher font size indicates an increased frequency of the keyword. Figure 1b also illustrates keyword co-occurrence network analysis; the terms distributed in the same cluster present the higher similarity, in comparison with the terms distributed in different clusters.

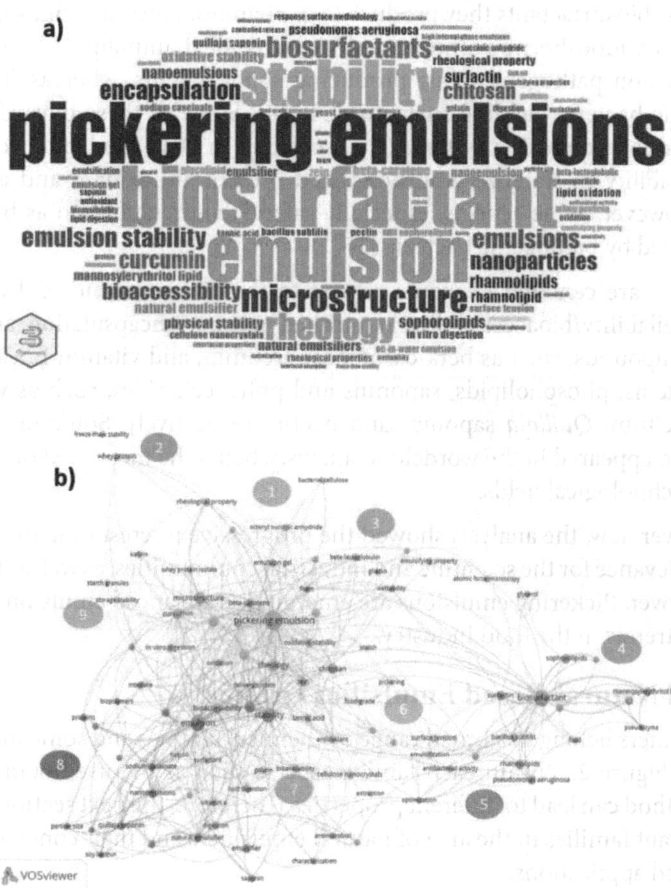

Figure 1. (a) Wordcloud from Author's keywords (100 keywords; minimum frequency of 5); (b) keyword co-occurrence network (9 clusters; Author's keywords; number of occurrences 5).

"Pickering emulsions" is the most frequent keyword, followed by biosurfactant (Figure 1). Other keywords (e.g., whey protein, sodium caseinate, glycolipid, sophorolipids, rhamnolipids, *Quillaja* saponin) appeared in the wordcloud.

These findings substantiate the keyword co-occurrence analysis (Figure 1). 93 keywords (Author's keywords) were organized in 9 clusters. The number of occurrences indicates the number of documents where the keyword appears. Each circle represents a keyword with at least 5 occurrences, being their areas proportional to the number of occurrences. The clusters are characterized by different colors and their words can be related.

Some clusters present words associated to recent trends in the area of natural emulsifiers. Clusters 1, 6, 8 and 9 refer to "Pickering emulsions" and other inter-related words, including nanoparticles, Pickering stabilization, and some commonly used Pickering stabilizers such as starch granules, cellulose nanocrystals and kafirin. Clusters 1 and 2 comprise terms related to the rheological properties of emulsions, an important parameter in food applications. The words included in clusters 4 and 5 are associated with microorganisms (e.g., *Pseudomonas aeruginosa; Starmerella bombicola; Bacillus subtilis)* and the biosurfactants they produce (e.g., rhamnolipids; sufactin; sophorolipids). *P. aeruginosa* is a food-borne pathogen and a source of rhamnolipids [33]. *Starmerella bombicola* is a non-pathogenic yeast producing sophorolipids, whereas *B. subtilis* is a non-pathogenic bacteria yielding surfactin [31, 34]. Some of these biosurfactants have been applied in food emulsion systems (e.g., surfactin in O/W food emulsions [20]) due to their high ability to stabilize emulsions, and present antioxidant and antimicrobial properties. However, some aspects, especially safety, require attention as biosurfactants may be produced by pathogenic bacteria [35].

Cluster 7 and 9 are centered in words related to the biocompounds delivery systems, namely bioavailability/bioaccessibility, controlled release, encapsulation and examples of used biocompounds, such as beta-carotene, curcumin, and vitamin E. Clusters 8 and 9 refer to proteins, phospholipids, saponins and polysaccharides, such as whey protein isolate, soy lecithin, *Quillaja* saponin, and pectin, respectively. Some of these natural emulsifiers also appeared in the wordcloud analysis, being the most used ones in the food science and technological fields.

In a general overview, the analysis showed the progressive interest in natural emulsifiers due to their relevance for the scientific and industrial communities, as well as for the global market. Moreover, Pickering emulsions are emerging as advanced emulsion technologies within future trends in the food industry.

2. Principal Natural-Based Emulsifier Groups

Natural emulsifiers belong to a broad range of chemical families and some main examples are shown in Figure 2. Within each family, aspects such as the used natural source or extraction method can lead to different properties. Therefore, the next sections summarize the most relevant families in the area of natural emulsifiers and their contextualization in the field of food applications.

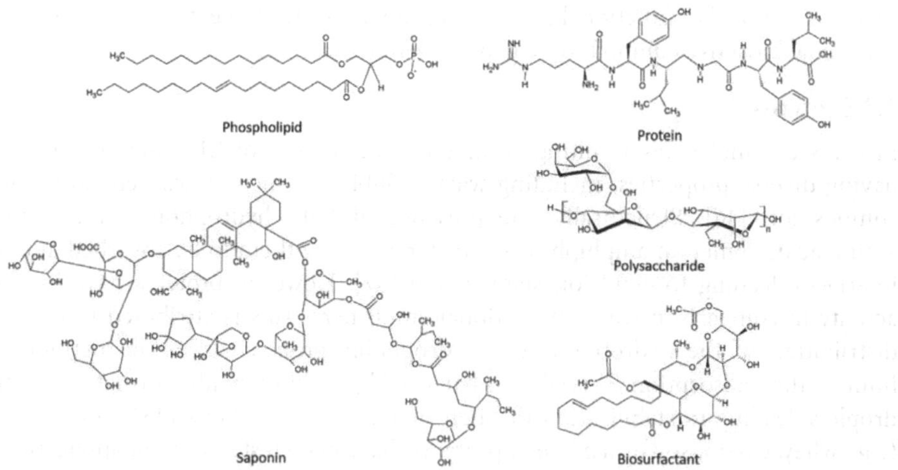

Figure 2. Representative chemical structures for each emulsifier family.

2.1 Phospholipids

Phospholipids are amphiphilic molecules, and a main constituent of natural membranes. Their structure comprises a hydrophilic head holding a phosphoric acid (H_3PO_4), combined with a hydrophobic tail composed by one or two non-polar fatty acids. They comprise groups as glycerophospholipids or sphingolipids, with lecithins (glycerophospholipid) assuming an important role. Phospholipids can be obtained from diverse natural sources, including milk, vegetable oils (soybean, rapeseed or sunflower), egg yolk, meat and fish [36, 37]. Specifically, lecithins are known to be good stabilizers for food emulsions, for example the ones derived from soy or egg yolk are applied in mayonnaise, creams, or sauces [38]. Other phospholipid examples include phosphatidylcholine, phosphatidylglycerol, phosphatidylethanolamine, phosphatidylinositol, phosphatidylserine, phosphatidic acid, sphingomyelin. The amphiphilic character of these compounds supports their capacity to stabilize emulsions. Concurrently to their ability to stabilize emulsions they can act as texturizing agents, thus influencing the organoleptic attributes of the final product [39].

2.2 Saponins

Saponins are a complex family derived from plants, constituted by triterpenes or steroid aglycones linked to glycosyl derived sugar structures [40]. Usually the aglycones involve pentacyclic triterpenoids with oleanolic acid and the sugars moieties comprise rhamnose, xylose, glucose or galactose [41]. Factors conditioning the composition of saponins are their botanical origin and extraction method. *Quillaja saponaria* Molina is the principal source of saponins, named *Quillaja* saponins, characterized by high contents of quillaic acid groups (hydrophobic) and rhamnose, galactose or glucuronic acids (hydrophylic) [42]. Saponins produce highly stable emulsions, including at the nanoscale, and at relatively low surfactant contents, with promising stability in terms of pH, ionic strength or temperature conditions [40]. Their promising properties avail their use in diverse applications, with examples of food-grade saponins applied in beverages added with flavors or bioactives such as vitamins [43]. More recently, the utilization of saponins was extended to the use of saponin-rich extracts obtained from plant sources [44], by-products and food wastes [40]. Besides the increased costs of using highly pure saponins, they provide weaker functional properties comparatively with extracts, due to the lack of additional bioactive compounds, e.g. polyphenols [45].

2.3 Proteins

Proteins are molecules resulting from the combination of 21 different amino acids, having diverse properties, including water solubility, which varies depending on their composition [46]. Structurally, the presence of both hydrophobic and hydrophilic amino acids confer an amphiphilic character, allowing them to be absorbed at oil/water interfaces, leading to emulsion stabilization [47]. However, proteins have low surface activity in comparison with conventional emulsifiers. This is attributed to the random distribution of the hydrophilic and hydrophobic groups within the peptide chains, limiting their adsorption. This effect is balanced by the protein film formation around the droplets, leading to stabilization through molecular interactions [48]. Diverse proteins (e.g., whey, casein, soy or faba bean proteins) have been tested in food applications, e.g., emulsions for the controlled release of lutein [49], w-3 oil [50], bioactive hydrophobic compounds [51], fish oil [52, 53] or β-carotene [54]. Their application in final products is still hindered by environmental conditions such as pH, temperature and ionic strength [48]. However, these drawbacks can be surpassed by using more complex formulations, namely by combining proteins with polysaccharides [48] or by chemically modifying the proteins trough grafting with other compounds such as polyphenols [54].

2.4 Polysaccharides

Polysaccharides are biopolymers composed of monosaccharide units such as glucose, fructose, mannose or galactose, bonded by glycosidic bonds. Their structural rearrangement, i.e., type and number of monosaccharides, type of glycosidic bonds, molecular weight, electrical charge, branching degree, hydrophobicity and the presence of other groups (carboxylate, sulfate or phosphate), rule the polysaccharides functional properties such as solubility, rheology, and amphiphilic character, among others [10]. Their amphiphilicity depends on the presence of hydrophobic (glycolipids) and hydrophilic (hydroxyls) groups, being adsorbed at the interface, forming a thick stabilizing layer (e.g., pectins, gum Arabic) [55]. Moreover, non-amphiphilic polysaccharides can contribute to emulsion stabilization due to their thickener role, increasing the viscosity and decreasing oil droplets' motion (e.g., alginates, carrageenan) [56]. Despite the high number of polysaccharides available in nature, only few are authorized as food emulsifiers in EU, namely alginic acid (E400), gum Arabic (E414), pectin (E440), cellulose and chemically modified celluloses (E460 to E469) [57]. Polysaccharides can be obtained from animal, vegetal, microbial fermentation or marine sources (algae), being their properties mostly dependent on the source and extraction process [10].

2.5 Natural based emulsifiers from microbial sources

Microbial synthetic routes are emerging as valuable sustainable and green alternatives to produce emulsifiers. They generate compounds with low ecotoxicity, biodegradability, stability (pH and salinity) and low critical micellar concentration (CMC), in addition to biological activity, biocompatibility and digestibility [58]. Emulsifiers produced by microorganisms are classified according to their molecular weight. Low molecular weight family includes glycolipids (e.g., rhamnolipids, sophorolipids, trehalose lipids) and lipopeptides (e.g., surfactin, iturin, fengycin) and are referred as biosurfactants. Polysaccharides, proteins, lipoproteins, and lipopolysaccharides belong to the high molecular weight family and are referred as bioemulsifiers [59, 60]. Glycolipids like rhamnolipids and trehalose lipids are mostly produced by bacterial strains like *Pseudomonas aeruginosa*, *Pseudomonas fluorescens*, *Rhodococcus erythropolis*, *Nocardia erythropolis*, *Arthrobacter sp.* or *Mycobacterium sp.* while sophorolipids are generally produced by yeasts (e.g. *Candida bombicola* or *Candida antartica*) or by filamentous fungi like *Aspergillus flavus* or *Rhizopus oryzae* [58, 61]. Lipopeptides can be produced by *Bacillus sp. Serratia marcescens* and *P. fluorescens* [58]. Bioemulsifiers such as Emulsan/Biodispersan are commercially available products, being produced by *Acinetobacter spp.*, while mannoproteins are commonly obtained from the yeasts *Saccharomyces cerevisiae* or *Kluyveromyces marxianus* [61]. The excellent properties of both microbial derived biosurfactants and bioemulsifiers make them appealing as natural based emulsifiers for foods. Several studies reported the use of glycolipids for fat emulsions stabilization [62, 63], and glycolipids and lipopeptides as rheology modifiers in cookies and muffins dough [64, 65, 66]. Other works refer bioemulsifiers (e.g., exopolysaccharides, mannoproteins) as having high potential in aromas emulsification [67]. Nevertheless, the practical application in foods is still limited due to two main factors, their high production costs narrowing the commercial profit, together with the legal regulations that limit the use of compounds produced by microbial strains classified as pathogenic in food applications. Examples include the bacteria genera like *Pseudomonas* and *Bacillus*. In opposition, yeasts like *S. cerevisiae* and *Kluyveromyces lactis*, which are classified as GRAS organisms, are authorized for food applications [68].

3. Emulsion Technology

Emulsions are colloidal systems constituted by two immiscible liquids (oil and water), formed in the presence of an emulsifier, and, usually, by applying an energy input. The emulsifier selection is therefore an important step to reach stability. They can be classified based on the hydrophilic region that correspond to ionic structures (anionic or cationic surfactants), change charge with pH (amphoteric surfactants) or present no charged centers (nonionic surfactants) [69]. Among them, nonionic surfactants are often used in food applications because they are less toxic and less affected by pH and ionic strength changes [70, 71]. The choice of a nonionic surfactant can be based on the hydrophilic–lipophilic balance (HLB) index [72]. This scale (0–20), reflects the changing from hydrophobic to hydrophilic character, that is, a lower HLB value corresponds to a lipophilic surfactant being appropriate to stabilize water-in-oil (W/O) emulsions, whereas a high HLB indicates the ability to stabilize oil-in-water (O/W) emulsions, due to the strong hydrophilic balance [72].

3.1 Emulsion classification

Emulsions can be classified according to their typology and structure. The first refers to the relative distribution of the immiscible phases (oil and water), and the latter refers to the arrangement of the emulsified entities [73]. Considering the typology, they can be classified as simple (O/W and W/O) or double (oil-in-water-in-oil (O/W/O), and water-in-oil-in-water (W/O/W)) emulsions (Figure 3). Examples of O/W emulsions in food systems include products such as milk, sauces, beverages, yogurts, ice-creams, and mayonnaise [74]. W/O emulsions are not so frequent but can be found in butter and margarine [73, 75]. For double emulsions, W/O/W are the most used systems due to their ability to generate reduced-fat products, when compared to O/W emulsions. Moreover, they can serve as base systems to encapsulate and control the release of sensitive water-soluble compounds, such as flavors or bioactive ingredients [16, 75, 76].

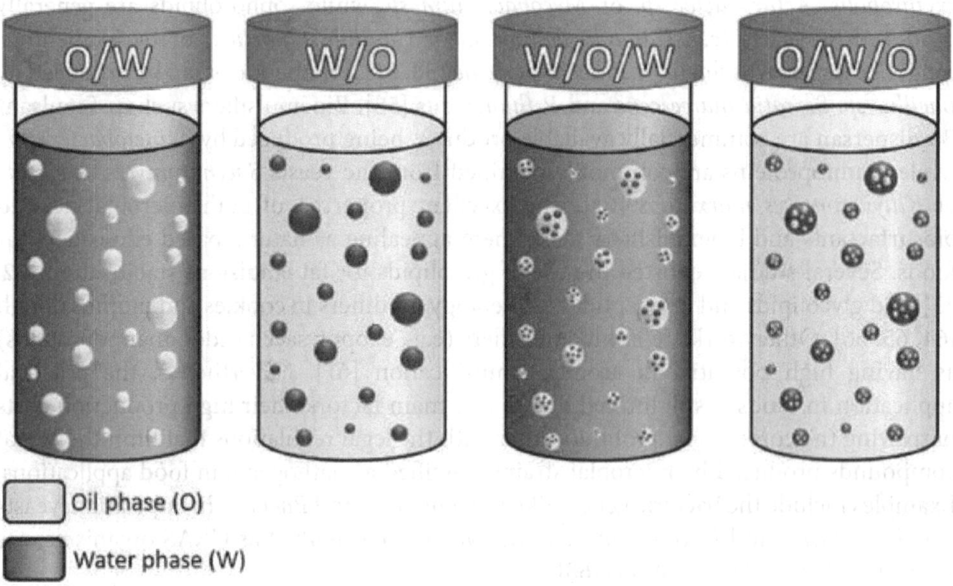

Figure 3. Typology of simple and double emulsions.

Regarding structure, emulsions can be classified as macroemulsions (usually called emulsions), nanoemulsions, or microemulsions. These systems present specific physicochemical properties that influence their range of applications [71]. Emulsions and nanoemulsions are thermodynamically unstable systems because their free energy is higher than the one of the individual phases [74, 77]. Thus, considering that all systems tend to their lowest energy state, phase separation will occur. However, due to their kinetic stability, they may remain in a metastable state for a considerable period of time, delaying the phase separation phenomenon. The kinetic stability is governed by two mechanisms, namely the energy barriers between the two states (emulsified and separated phases) and mass transfer between the phases. Therefore, high energy barriers and slow mass transfer processes delay phase separation [78]. By contrast, microemulsions are thermodynamically stable systems because their free energy is lower than the one of separate phases. Thus, they can be formed spontaneously under particular compositions and temperature conditions. In practice, some energy input is needed due to the existence of kinetic energy barriers [71]. Regarding the droplet size, nanoemulsions and microemulsions present droplet sizes <200 nm, whereas emulsions hold sizes between 200 nm and 100 μm [16, 71].

Nanoemulsions and microemulsions are optically transparent or slightly turbid due to their small droplet size, being valuable for applications requiring transparency, such as soft drinks [79]. Comparatively with nanoemulsions, microemulsions require a higher emulsifier content, have a lower particle size, and droplets can assume a non-spherical shape, feature that can be used to differentiate the two systems. Emulsions are typically turbid to opaque and are used in creamy systems such as dairy products [80]. Table 2 provides some application examples for each system addressing natural emulsifiers.

Table 2. Food applications of emulsions, nanoemulsions and microemulsions using natural emulsifiers.

System	Type	Natural emulsifiers	Final applications	Ref.
Emulsions	O/W	Whey protein	Ice-cream	[81]
	O/W	*Quillaja* saponin and soy lecithin	Coffee creamers	[82]
	W/O/W	Whey, rice and pumpkin seed proteins	Cheese	[83]
	O/W	Faba bean protein	Tofu and yogurts	[84]
	O/W	Pectin	Functional foods	[85]
	O/W	Rhamnolipid	Beverages	[86]
	O/W	Whey protein, locust bean gum, and iota-carrageenan	Mayonnaise, salad dressings, and sauces	[87]
	O/W	Modified starch	Dairy products	[88]
Nanoemulsions	O/W	Soy lecithin	Fruit juices	[89]
	O/W	*Quillaja* saponin, whey protein, and soy lecithin	Soft foods, creams, sauces, and salad dressings	[90]
	O/W	*Quillaja* saponin and soy lecithin	Functional foods and beverages	[91]
	O/W	Ginseng saponin		[9]
	O/W	Whey protein, gum Arabic, and soy lecithin		[92]
	O/W	Modified starch, whey and casein proteins		[93]
Microemulsions	W/O	Soy lecithin	Extraction of edible vegetable oils	[94]
	W/O	Soy lecithin	Functional foods and soft drinks	[95]
	O/W	Soy lecithin		[96]
	O/W	*Quillaja* saponin, Rhamnolipid, and soy lecithin	Soft drinks and minced meat	[97]

3.2 Stabilization mechanisms

Emulsions are thermodynamically unstable mixtures, characterized by the presence of at least two immiscible phases and an emulsifier that, when provided with enough mixing energy, are able to maintain stability over time [98]. The role of the emulsifier is essential to assure stable long-term properties. In general, emulsifiers are active surface substances, enabling their positioning at the oil–water interface, reducing the interfacial tension, hindering (or delaying) aggregation phenomena [99]. Typically, the hydrophilic part of the emulsifier is located in the aqueous phase, while the hydrophobic tail remains enclosed in the oil phase [82, 100]. During emulsion formation, the surfactant molecules require time to move to the interface, forming a layer to reach the interfacial tension equilibrium, a phenomenon related with their adsorption kinetics [82]. This pattern is dependent on emulsifiers' nature, taking from minutes (e.g., some saponins) to hours (e.g., some proteins), besides being dependent on environmental conditions (e.g., pH, temperature) [82]. To note that, even emulsions are commonly stabilized by a monolayer structure around the droplets, multilayer structures can also be formed. The multilayer pattern favors the electrostatic and steric repulsion of the droplets, improving stability while providing additional protection to the internal phase [16].

The emulsion stabilization mechanism can differ depending on the nature of the used surfactant. In this context four principal stabilization mechanisms are known, namely electrostatic repulsion, steric repulsion, Marangoni-Gibbs effect, and thin film stabilization mechanisms [101]. The electrostatic repulsion is related to ionic emulsifiers and consists on the formation of an electrical double layer at the droplet's interface, hindering their approximation. Steric repulsion is characteristic of nonionic and/or polymeric emulsifiers, and droplet's distance is kept due to the adsorption of the hydrophobic segment by the oil phase [101]. The Marangoni-Gibbs effect preserve emulsions' structure through the deformation of adjacent droplet's surface, avoiding their outflow, whereas the thin film stabilization mechanism avail the stability of the emulsion by generating a rigid and viscoelastic film, preventing droplets from destabilization effects [101].

Other factors can condition emulsion's stabilization mechanism, including the emulsifier content, the oil to water ratio or the preparation conditions (pH or temperature). For example, some phospholipids can have no charge at neutral pH, turning into anionic at acidic media, promoting molecule's swelling at the interface [100]. Moreover, the surfactant concentration can have also impact, e.g., sunflower lecithin in O/W emulsions, at low contents, create a layer surrounding the oil droplets, while at higher concentrations, the stabilization mechanism changes, producing, concurrently, liposomes that might destabilize the emulsion [10]. Considering amphiphilic polymers, when they are used as emulsifiers, they become positioned at the interface, just like the small molecules, but their ability to create intermolecular interactions can provide additional stabilization effects. Their effect on viscosity can also provide a positive stabilization effect [102]. The high hydrophilicity of most polysaccharides can difficult their emulsifier role, if considering the importance of the emulsifiers' hydrophilic/lipophilic character to interact with both phases. This constraint can be overcome by either chemical or physical strategies [103]. Namely, the suitability of anchoring hydrophobic groups into the polysaccharide structure can equilibrate the hydrophilic/lipophilic balance, that is the hydrophobisation of emulsifier's surface. Otherwise, alternative approaches imply the mixture of the polysaccharides with other polymers (co-surfactants) to favor the hydrophilic/lipophilic equilibrium and stabilization role.

3.3 Production methods

Food emulsions can be produced using several methods, classified as low-energy and high-energy processes, as represented schematically in Figure 4. The selection of the most appropriate method and respective equipment is based on the volume to process, characteristics of the initial mixture, emulsion's physicochemical properties, droplet size, and process costs [104]. In Table 3 a survey of recent works dealing with emulsion production trough different methods and using natural emulsifiers in their pure form or compounded with synthetic emulsifiers is presented. Moreover, their potential to encapsulate bioactives for food industry applications is also described.

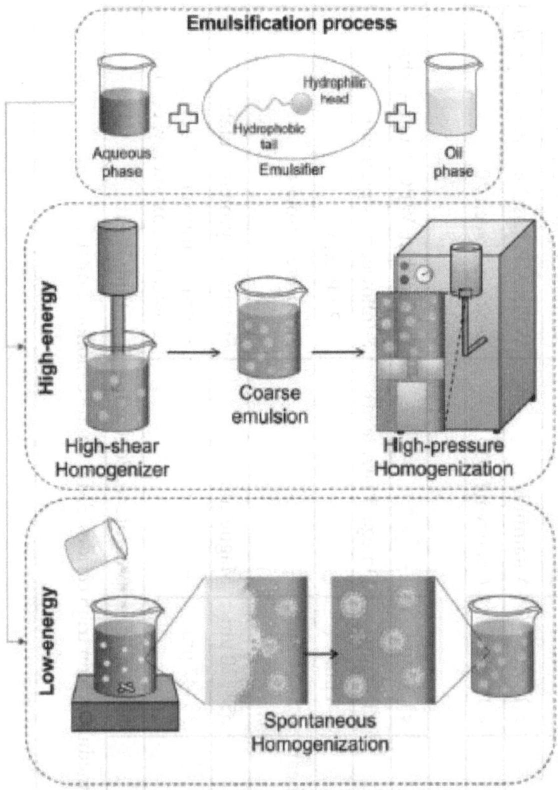

Figure 4. Schematic representation of the emulsification process through high- and low-energy methodologies.

Low-energy methods comprise, spontaneous emulsion, and emulsion phase inversion (e.g., phase inversion composition and phase inversion temperature), which occur due to environmental or composition changes namely temperature, pH, and ionic strength of the formulation [104]. Low-energy approaches are more cost effective than high-energy methods. However, they are limited to certain oils and emulsifiers, requiring also large amounts of surfactants, which is not desirable for many food applications [71]. In the work reported by Komaiko et al. [36], spontaneous emulsification lead to emulsions with large droplet size (>10 μm), comparatively with those produced by high-energy methods (<10 μm). The authors concluded that natural emulsifiers can be used in SE emulsions for applications where fine droplets are not essential (Table 3). By contrast, Mayer et al. [105] concluded that it was not possible to produce nanoemulsions using natural emulsifiers by the emulsion phase inversion method. These limitations imply that even natural-based emulsions can be prepared through low-energy methods, high-energy approaches are needed when natural emulsifiers are used.

Table 3. Studies applying different productive methods using natural emulsifiers or natural/synthetic blends to form emulsions and/or to encapsulate biocompounds for food industry applications.

Productive method	Emulsion type	Oil phase	Particle diameter	Bioactive compound	Emulsifiers	Ref.
Low-energy methods						
Spontaneous emulsification	O/W	MCT	> 10 μm	—	Sunflower phospholipids	[36]
Emulsion phase inversion	W/O	MCT and orange oil	> 10 μm	Vitamin E	WPI; SMP; Casein; *Quillaja saponin*	[105]
Phase inversion temperature	O/W	Peppermint oil	< 12 nm	Coenzyme Q10	Lecithin/Tween 20	[106]
High-energy methods						
High-pressure homogenization	O/W	Paprika oleoresin	<150 nm	—	Soy lecithin; WPC; Gum Arabic	[92]
Microfluidization	O/W	Fish oil	<150 nm	Omega-3 fatty acids	Sunflower phospholipids	[8]
Colloid mills	W/O	Rapeseed oil	4.8 μm	—	WPI; Pectin	[107]
Ultrasonic homogenizer	O/W	MCT; Palm oil; Soybean oil; Rapeseed oil	0.5–24.1 μm	—	Soy protein isolate	[6]
Membrane emulsification	W/O/W	Sunflower oil	35–320 μm	Magnesium	Starch; Pea protein isolate	[108]
Microchannel homogenizer	O/W	Soybean oil	35–47 μm	—	Sodium alginate; Pectin; Gum Arabic; Carboxymethyl cellulose	[109]
High-speed homogenization	O/W	Soybean oil	143.5 nm	β-carotene; Eugenol	Lecithin; WPI	[110]

WPI: Whey protein isolate; WPC: Whey protein concentrate; MCT: medium-chain triglycerides; SMP: Sucrose monopalmitate.

High-energy methods generate intensive forces promoting the water and oil phases disruption and their subsequent mixture. High-shear homogenizers are the most used equipment's for producing emulsions in the food industry. They consist on a rotor-stator or stirrer device able to mix the components at high speeds. Usually, large droplets are produced using this approach (1–10 μm) in comparison to alternative high-energy methods. High-pressure homogenization is also widely used in the food industry, being more effective to reduce the droplet size of a pre-emulsion. Generally, this coarse pre-emulsion is produced by high-shear homogenizers, then subjected to the high-pressure homogenization process. The equipment consists of a high-pressure pump (3–500 MPa) to pass the coarse emulsion through a narrow homogenizing valve, generating intensive disruptive forces (shear and cavitation), breaking down the droplets into smaller ones [80, 81].

Many studies have been conducted using two high-energy sequential methods (high-shear and -pressure homogenizers) to produce emulsions/nanoemulsions with natural emulsifiers [111, 112, 113]. Flores-Andrade and co-workers performed a study with soy lecithin, whey protein concentrate (WPC) and gum Arabic as natural emulsifiers, and paprika oleoresin as the oil phase. The coarse emulsion was produced by a high-speed homogenizer, then treated in a high-pressure homogenizer. O/W nanoemulsions were produced, being WPC more effective to form small droplets (d < 150 nm) than the other tested emulsifiers [92].

Microfluidization is the most effective method for producing emulsions with fine droplets (d < 100 nm). This approach is based on feeding the fluid into the homogenizer, which consists of a mixture chamber with two channels. Intensive disruptive forces are generated when these two fluid streams collide at high speed, breaking larger droplets and intermingling the fluids [3]. As the high-pressure homogenizers, microfluidizers were used after preparing a pre-emulsion by high-shear mixers [42, 114]. Ultrasound technique uses high-intense ultrasonic waves, generating intense shear and pressure gradients. The droplets are disrupted mainly by cavitation and turbulent effects [99, 115].

Currently, high-energy approaches are commonly used in the food industry due to their large-scale production capacity and the possibility to process a wide range of raw ingredients [71]. Although several high-energy emulsification devices are available, high-shear and pressure homogenizers, microfluidizers and ultrasound equipment's are the most used in the production of natural emulsifiers-based emulsions.

3.4 Emulsion stability

Emulsion stability is an important parameter indicating its ability to resist physicochemical changes over time [116]. For food emulsions, the required stability varies according to the intended final application. For example, short-term stability of minutes to hours, is enough for intermediate food emulsions such as cake batter and ice cream mixtures, while long-term stability is required for long shelf-life products, including mayonnaise and salad dressings [117]. For the latter ones, the development of effective strategies to retard emulsion destabilization implies the identification of the main mechanisms leading to this effect [73].

Emulsion instability can occur due to physical and/or chemical processes. The physical instability is responsible for modifying the emulsion droplets spatial distribution and structure, including gravitational separation (creaming/sedimentation), flocculation, coalescence, and Ostwald ripening phenomena (Figure 5). These effects depend on the emulsion composition and structure, besides the storage conditions, namely temperature variation and mechanical stirring [74, 116]. Moreover, the physical phenomena are interrelated and can influence each other during emulsion storage [77].

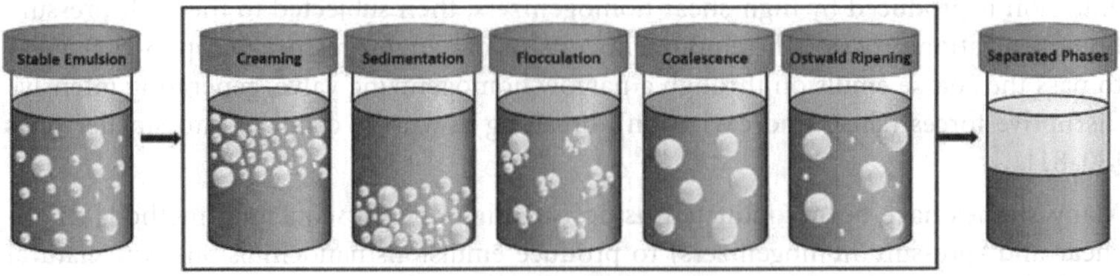

Figure 5. Common types of instability phenomena in emulsions.

Gravitational separation is driven by density differences between the droplets and the continuous phase. The droplets are subjected to gravitational forces tending to accumulate in the top (creaming) or in the bottom (sedimentation) of the system. Most edible oils present densities lower than water, favoring creaming in O/W emulsions, whereas sedimentation is usually observed in W/O emulsions [116]. Considering the impact of gravitational forces in the large droplets, the separation usually occurs for emulsions with droplet sizes higher than 100 nm or in a final stage of a sequence of instability phenomena [116]. By contrast, for lower droplet sizes, e.g., nanoemulsions, Brownian motion dominates over gravitational forces. Thus, reducing the droplet size is a suitable strategy to retard gravitational separation, with the emulsifier playing an important role to effectively reduce droplets' size [2, 74]. Furthermore, the emulsifier' layers tend to minimize the density difference between the emulsion phases, thus reducing the velocity of gravitational separation. Other strategies include modifying the rheology of the continuous phase or increasing the concentration of the droplets [74, 116].

Ostwald Ripening consists of the increase of the droplets size due to the diffusion of small droplets into larger ones, effect driven by their solubility in the continuous phase. This effect is promoted when the droplet's size decreases [73], being also influenced by the emulsifiers' properties. Namely, Ostwald Ripening can be retarded by decreasing the interfacial tension of the phases, favored when small-molecule surfactants are used or when using emulsifiers able to form rigid shell around the droplets. By contrast, emulsifiers prone to solubilize the oil and water phases through the formation of colloidal structures (e.g., micelles) accelerate the Ostwald Ripening [2].

Flocculation and coalescence mechanisms are related to droplets aggregation, effect leading to droplet size increase [74]. In flocculation the association of at least two droplets in an aggregate occurs, whereas in the coalescence, the droplets merge into a larger one [77]. Both phenomena are highly dependent on the selected emulsifier [77, 116], namely their nature and colloidal interactions' capacity [2].

4. Pickering Emulsions

Pickering emulsions are defined as systems stabilized by solid colloidal particles adsorbed at the oil–water interface in a practically irreversible process, creating a coating around the droplets, either in the form of a single or multiple layer, generating a strong steric barrier providing high stability [118]. In the context of Pickering emulsions, the search for natural-based particles is currently a hot topic to face market demands for novel clean label products (absent of emulsifiers) [119]. Pickering emulsions (Figure 6) are raising high interest in the recent years. They are characterized by a long-term stability and have green connotations due to the absence of conventional emulsifiers. These attributes comply with the recent trends of food industry towards the use of sustainable and healthy technologies [16]. The stability of Pickering emulsions is related with the intrinsic properties of the oil and water phases (e.g., type, oil/water ratio, pH, ionic strength) and of the particle stabilizers (e.g., wettability, particle morphology, size and concentration). Particles presenting a contact angle (θ) below 90° are generally suitable for preparing O/W emulsions, whereas θ values greater than 90° indicate good stabilizers for W/O emulsions. At 90°, the particle is immersed equally in both phases [120].

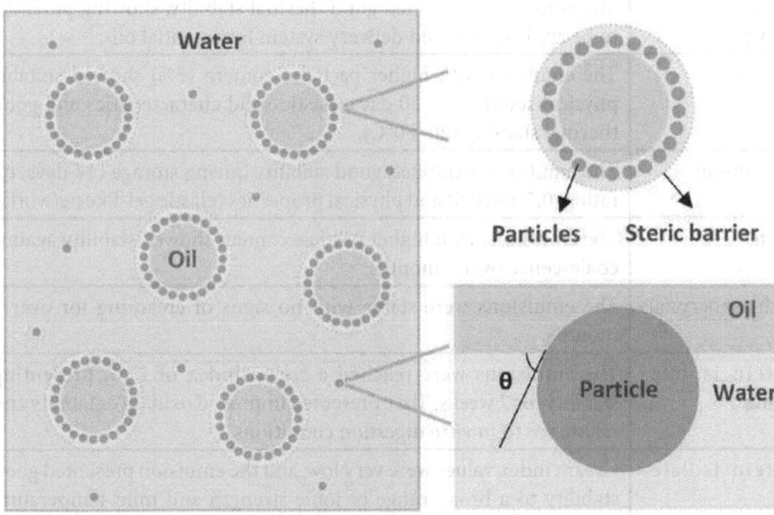

Figure 6. Schematic representation of a Pickering emulsion putting in evidence the particle stabilizers where θ represent the contact angle.

Regarding natural-based particles, three main typologies of stabilizers can be used, namely nanoparticles, microgels and fibrils. Examples include protein derived stabilizers, namely nanoparticles based on corn zein, and colloidal particles of kafirin and gliadin [118, 121, 122, 123]. Although many polysaccharides have high hydrophilic character, some can include hydrophobic side groups (e.g., beet pectin and modified starch) or even active proteins attached to the surface (e.g., gum Arabic) [120], offering potential to act as Pickering stabilizers. Other polysaccharides widely used to produce Pickering bionanoparticles include chitin, chitosan and cellulose. To overcome particle's limitations as Pickering stabilizers, the formation of complexes has been also proposed, namely complexes such as polysaccharide-polysaccharide, protein–protein, and polysaccharide-protein [124]. Examples include zein-xanthan [125], and tea water insoluble proteins/κ-carrageenan complexes [126].

In the context of the recent trends in Pickering emulsions, research aiming at finding new biological particles, the use of high internal phase emulsions (HIPPE), and the development of bio-based films from Pickering emulsions are becoming topics of high interest for the development of novel food applications. Table 4 presents an overview of recent works dealing with the preparation of Pickering emulsions based on novel biological particles together with the description of the main results envisaging potential food applications.

Table 4. Examples of bionanoparticles as Pickering stabilizers. All the systems are of O/W type.

Particle materials	Main results	Ref.
Apple pomace	Smaller particles led to emulsions with smaller droplet size, showing higher stability over time (30 days), in addition to improved physical properties (gel-like samples) and antioxidant activity.	[127]
Bamboo shoots dietary fiber	The emulsions were stable for 4 weeks avoiding coalescence against pH and ionic strength changes and pasteurization conditions.	[128]
Chitosan-sodium tripolyphosphate	The emulsions presented good thermal stability, showing potential to be applied as a food delivery system for essential oils.	[129]
Gliadin-pectin	The emulsion with higher particles content (2%) showed suitable physical stability for 30 days, elastic-solid characteristics and good thermal stability (20–80°C).	[130]
Hordein-chitosan	The emulsions exhibited good stability during storage (14 days, oil ratio = 0.5 and 0.6) and physical properties (elastic gel-like network).	[131]
Pea protein	The emulsions with higher particle content showed stability against coalescence over 3 months.	[30]
Sago starch nanocrystals	The emulsions were stable with no signs of creaming for over 2 months.	[132]
Soy protein isolate-anthocyanin	The emulsions were reached a cream index of 17%, presenting stability for 7 weeks. They presented improved oxidative stability and resistance to *in vitro* digestion conditions.	[133]
Soy protein isolate-chitosan	Cream index values were very low, and the emulsion presented good stability to a broad range of ionic strength and mild temperature conditions (4–60°C).	[134]
Tea protein	Emulsions with gel-like properties were produced, presenting no creaming over 50 days.	[135]
Zein-corn fiber gum	High oil concentrations (oil ratio = 0.5) led to higher stability and the formation of a gel-like structure.	[136]
Zein-gum arabic	The emulsion showed a high stability against coalescence and Ostwald ripening during 30 days of storage (oil ratio = 0.7).	[137]
Zein-pectin	The emulsions maintained excellent physical stability for 1 month. In addition, they demonstrated good performance as delivery systems of essential oils.	[138]
β-lactoglobulin-gum arabic	The particles provided stability against coalescence and Ostwald ripening for up to 12 weeks, in addition to improve chemical stability.	[139]

HIPPEs are characterized by having a high volume fraction of internal phase (generally higher than 74%), together with relatively low particles concentration resulting in an extremely compacted droplet's structure [140]. HIPPEs are becoming a novel approach of increasing interest in the food industry, since it combines diverse advantages, namely a semi-solid texture with the ability to encapsulate high amounts of bioactive compounds [141]. HIPPEs allow to control the droplet size distribution, manipulate the morphology and rheological properties, generally presenting enhanced stability against physical, chemical and microbiological stresses [142]. They are positioned as extremely promising substitutes for foods such as margarine, mayonnaise or ice creams [143, 144]. For example, Liu et al. studied wheat gluten as stabilizer in a HIPPE to develop a novel mayonnaise substitute [145]. They obtained excellent results concerning texture and sensory attributes when compared with commercial products.

Bio-based films made from hydrophilic particles added with hydrophobic compounds is another emerging approach in the scope of new applications developed from Pickering emulsions [146]. These strategies provide the ability to improve the stability of the base materials (hydrophilic), in addition to facilitate the combination with hydrophobic materials (e.g., waxes, fatty oils and oils) leading to systems with enhanced moisture barrier properties [147].

5. New Trends in Food Emulsion Systems

The wide variety of emulsion-based systems using natural emulsifiers makes their applicability attractive for various products, particularly in the food industry. The nature and function of emulsifiers, and the formed emulsion type (e.g., nano/micro-scale, simple or double character) can tailor appearance, sensorial characteristics, and attractiveness of foods. Among their diverse functions, the increasing use of emulsions as functionality carriers should be highlighted. In fact, recent works have demonstrated their potential and versatility for the encapsulation of flavors, and to protect and deliver specific bioactives in foods or beverages, helping to strengthen nutritional balances, and enabling the production of reduced-fat products. A summary of examples addressing new trends of emulsion-based products with potential in the food industry are included in Table 5, with some highlighs provided next.

Table 5. Applications of natural-based emulsifiers in food industry.

Emulsion type	Oil phase	Emulsifier	Highlights	Ref.
O/W	Vegetable oil	Orange pulp and peel powders	Base emulsions for food applications	[148]
O/W	Orange essential oil	Pea protein concentrate and soy protein isolate	Encapsulation of flavors for the food industry (powder form)	[149]
O/W	Sunflower, soybean, MCT, and orange oils	Crude saponins isolated from onion skin wastes	Food nanoemulsions (stable emulsions, except at acidic pH and high ionic strength)	[40]
O/W	Paprika oleoresin	Whey protein and gum Arabic, and soy lecithin	Food and beverage systems for the delivery of carotenoids	[92]
O/W	Hemp seed oil	*Aesculus hippocastanum* L. extract	Food nanoemulsions with enhanced nutritional properties	[150]

Emulsion type	Oil phase	Emulsifier	Highlights	Ref.
O/W	Almond, mustard, olive, and soyabean oils	Biosurfactants isolated from *Candida albicans* SC5314 and *Candida glabrata* CBS138 yeast strains	Food emulsions with improved antibacterial capacity	[151]
O/W	Corn and sunflower oils) /	Biosurfactant isolated from *Candida utilis*	Food emulsions with promising properties for salad dressings	[63]
O1/W/O2	O1 - Sunflower; O2 - palm oil	Primary emulsion - gelatin, xanthan gum; Secondary emulsion - solid fat crystals	Oil encapsulation systems for texturizing reduced-fat agents	[152]

Lopes Francisco et al. [149] reported an emulsifying system with encapsulation potential based on commercial pea and soy proteins. The work involved the encapsulation of an orange essential oil rich in d-limonene using a O/W emulsion followed by spray drying to obtain powder microparticles. It was demonstrated the ability of pea and soy proteins to act as emulsifiers in the encapsulation of orange essential oil, getting a slightly higher efficiency if using soy protein as the natural emulsifier. These promising results can help consolidate a platform aiming at developing new protective systems to encapsulate flavors for foods, complying with the increasing demand from this industrial sector for natural-based systems.

At the nanoscale, Flores-Andrade et al. [92] reported the preparation of O/W nanoemulsions by high-pressure homogenization, using amphiphilic biopolymers to stabilize paprika oleoresin, namely whey protein, gum Arabic, phospholipids, and soy lecithin. The results demonstrated the effective oil encapsulation, preserving carotenoids (e.g., lipophilic colorants) from chemical degradation, besides positioning this strategy as an attractive route to design new protective and delivery carriers for bioactive compounds aimed at food and/or beverage products.

The potential of double emulsions was also demonstrated by Cetinkaya et al. [152] that evidenced the reduction of the saturated fat content in O1/W/O2 emulsions prepared by fat crystallization according to a two-stage process. Firstly, the primary O1/W emulsion was prepared using sunflower oil and xanthan gum and gelatin as emulsifiers, which was then stabilized in a second oil phase (palm oil), resulting in a structured O1/W/O2 system. Microstructure examination revealed that the accumulation of fat crystals at the interface contributed to stabilize the internal water phase containing the encapsulated sunflower phase. These complex structures showed potential to directly encapsulate hydrophobic oils and act as texturizing reduced-fat agents, which might be of particular interest for the edible oils industry.

6. Conclusions

This chapter presents an up-to-date overview of current trends in natural emulsifiers and their application in emulsion technology directed to food applications. For this purpose, first, the evolution of food emulsifiers' scenario over the last 10 years was analyzed through the Bibliometrix-R package (RStudio) and VosViewer software. This analysis indicated a clear driving force towards using natural emulsifiers and the re-emerging importance of the Pickering emulsions. These facts are expected to impact the market growth following the prospectus of available market analysis. The six main identified families of natural emulsifiers were phospholipids, saponins, proteins, polysaccharides, emulsifiers from microbial sources and Pickering stabilizers. Some of them already find extensive use in practical food applications. However, others, mainly natural-based emulsifiers from microbial sources and Pickering stabilizers, despite their high potential, are still needing research investment and regulation clarification (e.g., related to the use of nanoparticles and the use of microbial strains classified as pathogenic in foods). From a technological perspective, the main concepts related to the typology, production methods, stabilization mechanisms, and instability phenomena were presented. Highlighting the increasing interest in Pickering emulsions, a summary of the most recent applications of these systems, including the so-called HIPPEs and their advantages in reduced-fat products development, was provided. To conclude, an analysis of current trends in food emulsion-based products was discussed, putting in evidence the emulsions increasing role as delivery systems of bioactives to support innovative fortified foods advances and the increasing interest in systems based on double emulsions, which provide the opportunity to combine bioactives of different nature. Overall, the field of natural-based emulsifiers combined with the new trends in emulsion technology can, hopefully, be the basis of a new generation of healthy and nutritious food products.

References

1. McClements DJ, Bai L, Chung C. Recent advances in the utilization of natural emulsifiers to form and stabilize emulsions. Annu Rev Food Sci Technol. 2017;205-236. DOI:10.1146/annurev-food-030216-030154
2. Ozturk B, McClements DJ. Progress in natural emulsifiers for utilization in food emulsions. Curr Opin Food Sci. 2016;7:1-6. DOI:10.1016/j.cofs.2015.07.008
3. McClements DJ. Food Emulsions: Principles, Practices and Techniques. Third Edit. Vol. 53, Journal of Chemical Information and Modeling. Boca Raton: CRC press; 2016
4. F. Csáki K, Sebestyén É. Who will carry out the tests that would be necessary for proper safety evaluation of food emulsifiers? Food Sci Hum Wellness. 2019;8:126-35. DOI:10.1016/j.fshw.2019.04.001
5. Csáki KF. Synthetic surfactant food additives can cause intestinal barrier dysfunction. 2011;76:676-681. DOI:10.1016/j.mehy.2011.01.030
6. Taha A, Hu T, Zhang Z, Bakry AM, Khalifa I, Pan S, et al. Effect of different oils and ultrasound emulsification conditions on the physicochemical properties of emulsions stabilized by soy protein isolate. Ultrason Sonochem. 2018;49:283-293. DOI: 10.1016/j.ultsonch.2018.08.020
7. Bai L, Huan S, Li Z, McClements DJ. Comparison of emulsifying properties of food-grade polysaccharides in oil-in-water emulsions: Gum arabic, beet pectin, and corn fiber gum. Food Hydrocoll. 2017;66:144-153. DOI:10.1016/j.foodhyd.2016.12.019
8. Komaiko J, Sastrosubroto A, McClements DJ. Encapsulation of ω-3 fatty acids in nanoemulsion-based delivery systems fabricated from natural emulsifiers: Sunflower phospholipids. Food Chem. 2016;203:331-339. DOI:10.1016/j.foodchem.2016.02.080

9. Shu G, Khalid N, Chen Z, Neves MA, Barrow CJ, Nakajima M. Formulation and characterization of astaxanthin-enriched nanoemulsions stabilized using ginseng saponins as natural emulsifiers. Food Chem. 2018;255:67-74. DOI:10.1016/j.foodchem.2018.02.062

10. McClements DJ, Gumus CE. Natural emulsifiers - biosurfactants, phospholipids, biopolymers, and colloidal particles: Molecular and physicochemical basis of functional performance. Adv Colloid Interface Sci. 2016;234:3-26. DOI:10.1016/j.cis.2016.03.002

11. Boukid F, Rosell CM, Castellari M. Pea protein ingredients : A mainstream ingredient to (re) formulate innovative foods and beverages . Trends Food Sci Technol. 2021;110:729-742. DOI: 10.1016/j.tifs.2021.02.040

12. Day L. Proteins from land plants e Potential resources for human nutrition and food security. Trends Food Sci Technol. 2013;32:25-42. DOI: 10.1016/j.tifs.2013.05.005

13. Hinderink EBA, Schröder A, Sagis L, Schroën K, Berton-Carabin CC. Physical and oxidative stability of food emulsions prepared with pea protein fractions. Lwt. 2021;146:111424. DOI: 10.1016/j.lwt.2021.111424

14. Jiang S, Yildiz G, Ding J, Andrade J, Rababahb TM, Almajwalc A, et al. Pea protein nanoemulsion and nanocomplex as carriers for protection of cholecalciferol (Vitamin D3). Food Bioprocess Technol. 2019;12:1031-1040. DOI: 10.1007/s11947-019-02276-0

15. Shao Y, Tang CH. Characteristics and oxidative stability of soy protein-stabilized oil-in-water emulsions: Influence of ionic strength and heat pretreatment. Food Hydrocoll. 2014;37:149-158. DOI:10.1016/j.foodhyd.2013.10.030

16. Tan C, McClements DJ. Application of advanced emulsion technology in the food industry: A review and critical evaluation. Foods. 2021;10:812. DOI: 10.3390/foods10040812

17. Xiao J, Li C, Huang Q. Kafirin nanoparticle-stabilized pickering emulsions as oral delivery vehicles: Physicochemical stability and in vitro digestion profile. 2015;63:10263-10270. DOI: DOI: 10.1021/acs.jafc.5b04385

18. Lu X, Li C, Huang Q. Combining in vitro digestion model with cell culture model: Assessment of encapsulation and delivery of curcumin in milled starch particle stabilized Pickering emulsions. Int J Biol Macromol. 2019;139:917-924. DOI:10.1016/j.ijbiomac.2019.08.078

19. Lv P, Wang D, Chen Y, Zhu S, Zhang J, Mao L, et al. Pickering emulsion gels stabilized by novel complex particles of high-pressure-induced WPI gel and chitosan : Fabrication , characterization and encapsulation. Food Hydrocoll. 2020;108:105992. DOI:10.1016/j.foodhyd.2020.105992

20. Hoffmann M, Mück D, Grossmann L, Greiner L, Klausmann P, Henkel M, et al. Surfactin from Bacillus subtilis displays promising characteristics as O/W-emulsifier for food formulations. Colloids Surfaces B Biointerfaces. 2021;203:111749. DOI: 10.1016/j.colsurfb.2021.111749

21. Campos JM, Stamford TLM, Rufino RD, Luna JM, Stamford TCM, Sarubbo LA. Formulation of mayonnaise with the addition of a bioemulsifier isolated from Candida utilis. Toxicol Reports. 2015;2:1164-1170. DOI: 10.1016/j.toxrep.2015.08.009

22. Research and Markets. Food Emulsifiers Market Research Report by Function, by Source, by Application - Global Forecast to 2025 - Cumulative Impact of COVID-19. Report. 2021

23. MarketResearch.com. Global Food Emulsifiers Market - 2020-2027. Report. 2021

24. Aria M, Cuccurullo C. bibliometrix: An R-tool for comprehensive science mapping analysis. J Informetr. 2017;11:959-975. DOI:10.1016/j.joi.2017.08.007

25. van Eck NJ, Waltman L. Software survey : VOSviewer , a computer program for bibliometric mapping. Scientometrics. 2010;84:523-538. DOI: 10.1007/s11192-009-0146-3

26. Wang L, Hu Y, Yin S, Yang X, Lai F, Wang S. Fabrication and Characterization of Antioxidant Pickering Emulsions Stabilized by Zein/Chitosan Complex Particles (ZCPs). J. Agricul. Food Chem. 2015; 63:2514-2524. DOI:10.1021/jf505227a

27. de Melo ANF, de Souza EL, da Silva Araujo VB, Magnani M. Stability, nutritional and sensory characteristics of French salad dressing made with mannoprotein from spent brewer's yeast. LWT - Food Sci Technol. 2015;62:771-774. DOI:10.1016/j.lwt.2014.06.050

28. Bai L, Huan S, Gu J, Julian D. Fabrication of oil-in-water nanoemulsions by dual-channel micro fl uidization using natural emulsi fi ers : Saponins , phospholipids , proteins , and polysaccharides. Food Hydrocoll. 2016;61:703-711. DOI:10.1016/j.foodhyd.2016.06.035

29. Fu D, Deng S, Julian D, Zhou L, Zou L, Yi J, et al. Encapsulation of β -carotene in wheat gluten nanoparticle-xanthan gum- stabilized Pickering emulsions : Enhancement of carotenoid stability and bioaccessibility. Food Hydrocoll. 2019;89:80-89. DOI:10.1016/j.foodhyd.2018.10.032

30. Zhang S, Holmes M, Ettelaie R, Sarkar A. Pea protein microgel particles as Pickering stabilisers of oil-in-water emulsions: Responsiveness to pH and ionic strength. Food Hydrocoll. 2020;102:105583. DOI:10.1016/j.foodhyd.2019.105583

31. Takahashi M, Morita T, Wada K, Hirose N, Fukuoka T, Imura T, et al. Production of Sophorolipid Glycolipid Biosurfactants from Sugarcane Molasses Using Starmerella bombicola NBRC 10243.J. Oleo Sci. 2011;273:267-273. DOI:10.5650/jos.60.267

32. Pratap A, Wadekar S, Kale S. Non-traditional Oils as newer feedstock for Rhamnolipids production by Pseudomonas aeruginosa (ATCC 10145). J. Amer. Oil Chemists' Soc. 2011;88:1935-1943. DOI:10.1007/s11746-011-1875-z

33. Toribio J, Escalante AE, Sobero G. Rhamnolipids : Production in bacteria other than Pseudomonas aeruginosa. Eur. J. Lipid Sci. Tech. 2010;112:1082-1087. DOI:10.1002/ejlt.200900256

34. Shaligram NS, Singhal RS. Surfactin – A Review on Biosynthesis , Fermentation , Purification and Applications. Food Technol. Biotechnol. 2010;48:119-134

35. Ribeiro BG, Sarubbo LA. Biosurfactants : Production and Application Prospects in the Food Industry. Biotechnol Prog. 2020;36:3030. DOI: 10.1002/btpr.3030

36. Komaiko J, Sastrosubroto A, McClements DJ. Formation of oil-in-water emulsions from natural emulsifiers using spontaneous emulsification: Sunflower phospholipids. J Agric Food Chem. 2015;63:10078-10088. DOI:10.1021/acs.jafc.5b03824

37. Ali AH, Zou X, Abed SM, Korma SA, Jin Q, Wang X. Natural phospholipids: Occurrence, biosynthesis, separation, identification, and beneficial health aspects. Crit Rev Food Sci Nutr.2019;59:253-275. DOI:10.1080/10408398.2017.1363714

38. Perazzo A, Gallier S, Liuzzi R, Guido S, Caserta S. Quantitative methods to detect phospholipids at the oil-water interface. Adv Colloid Interface Sci. 2021;290:102392. DOI:10.1016/j.cis.2021.102392

39. Delacharlerie S, Petrut R, Deckers S, Flöter E, Blecker C, Danthine S. Structuring effects of lecithins on model fat systems: A comparison between native and hydrolyzed forms. LWT - Food Sci Technol. 2016;72:552-558. DOI:10.1016/j.lwt.2016.04.043

40. Dahlawi SM, Nazir W, Iqbal R, Asghar W, Khalid N. Formulation and characterization of oil-in-water nanoemulsions stabilized by crude saponins isolated from onion skin waste. RSC Adv. 2020;10:39700-39707. DOI:10.1039/D0RA07756A

41. Jarzębski M, Siejak P, Smułek W, Fathordoobady F, Guo Y, Pawlicz J, et al. Plant extracts containing saponins affects the stability and biological activity of hempseed oil emulsion system. Molecules. 2020;25:1-16. DOI:10.3390/molecules25112696

42. Li Z, Dai L, Wang D, Mao L, Gao Y. Stabilization and rheology of concentrated emulsions using the natural emulsifiers quillaja saponins and rhamnolipids. J Agric Food Chem. 2018;66:3922-3929. DOI: 10.1021/acs.jafc.7b05291

43. Burakova Y, Madera R, Wang L, Buist S, Lleellish K, Schlup JR, et al. Food-grade saponin extract as an emulsifier and immunostimulant in emulsion-based subunit vaccine for pigs. J Immunol Res. 2018;8979838:1-8. DOI:10.1155/2018/8979838

44. Schreiner TB, Colucci G, Santamaria-Echart A, Fernandes IP, Dias MM, Pinho SP, et al. Evaluation of saponin-rich extracts as natural alternative emulsifiers: A comparative study with pure Quillaja Bark saponin. Colloids Surfaces A Physicochem Eng Asp. 2021. DOI:10.1016/j.jhazmat.2020.124562

45. Reichert CL, Salminen H, Weiss J. Quillaja saponin characteristics and functional properties. Annu Rev Food Sci Technol. 2019;10:43-73. DOI: 10.1146/annurev-food-032818-122010

46. Hoffmann H, Reger M. Emulsions with unique properties from proteins as emulsifiers. Adv Colloid Interface Sci. 2014;205:94-104. DOI:10.1016/j.cis.2013.08.007

47. Dapueto N, Troncoso E, Mella C, Zúñiga RN. The effect of denaturation degree of protein on the microstructure, rheology and physical stability of oil-in-water (O/W) emulsions stabilized by whey protein isolate. J Food Eng. 2019;263:253-261. DOI:10.1016/j.jfoodeng.2019.07.005

48. Burgos-Díaz C, Wandersleben T, Marqués AM, Rubilar M. Multilayer emulsions stabilized by vegetable proteins and polysaccharides. Curr Opin Colloid Interface Sci. 2016;25:51-57. DOI:10.1016/j.cocis.2016.06.014

49. Weigel F, Weiss J, Decker EA, McClements DJ. Lutein-enriched emulsion-based delivery systems: Influence of emulsifiers and antioxidants on physical and chemical stability. Food Chem. 2018;242:395-403. DOI:10.1016/j.foodchem.2017.09.060

50. Gumus CE, Decker EA, McClements DJ. Formation and stability of ω-3 oil emulsion-based delivery systems using plant proteins as emulsifiers: Lentil, pea, and faba bean proteins. Food Biophys. 2017;12:186-197. DOI 10.1007/s11483-017-9475-6

51. Chen X, McClements DJ, Zhu Y, Chen Y, Zou L, Liu W, et al. Enhancement of the solubility, stability and bioaccessibility of quercetin using protein-based excipient emulsions. Food Res Int. 2018;114:30-37. DOI:10.1016/j.foodres.2018.07.062

52. Padial-Domínguez M, Espejo-Carpio FJ, Pérez-Gálvez R, Guadix A, Guadix EM. Optimization of the emulsifying properties of food protein hydrolysates for the production of fish oil-in-water emulsions. Foods. 2020;9:636. DOI:10.3390/foods9050636

53. Qiu C, Zhao M, Decker EA, McClements DJ. Influence of protein type on oxidation and digestibility of fish oil-in-water emulsions: Gliadin, caseinate, and whey protein. Food Chem. 2015;175:249-257. DOI:10.1016/j.foodchem.2014.11.112

54. Liu F, Wang D, Sun C, McClements DJ, Gao Y. Utilization of interfacial engineering to improve physicochemical stability of β-carotene emulsions: Multilayer coatings formed using protein and protein-polyphenol conjugates. Food Chem. 2016;205:129-139. DOI:10.1016/j.foodchem.2016.02.155

55. Costa C, Medronho B, Filipe A, Mira I, Lindman B, Edlund H, et al. Emulsion formation and stabilization by biomolecules: The leading role of cellulose. Polymers. 2019;11:1-18. DOI:10.3390/polym11101570

56. Silva EK, Rosa MTM, Meireles MAA. Ultrasound-assisted formation of emulsions stabilized by biopolymers. Curr Opin Food Sci. 2015;5:50-59. DOI:10.1016/j.cofs.2015.08.007

57. European Commission. Commission Regulation (EU) No 1130/2011. Off J Eur Union. 2011;2011:178-204

58. Shekhar S, Sundaramanickam A, Balasubramanian T. Biosurfactant Producing Microbes and their Potential Applications: A Review. Crit Rev Environ Sci Technol. 2015;45:1522-1554. DOI:10.1080/10643389.2014.955631

59. Jahan R, Bodratti AM, Tsianou M, Alexandridis P. Biosurfactants, natural alternatives to synthetic surfactants: Physicochemical properties and applications. Adv Colloid Interface Sci. 2020;275:102061. DOI:10.1016/j.cis.2019.102061

60. Uzoigwe C, Burgess JG, Ennis CJ, Rahman PKSM. Bioemulsifiers are not biosurfactants and require different screening approaches. Front Microbiol. 2015;6:1-6. DOI: 10.3389/fmicb.2015.00245

61. Cortés-Sánchez A de J. Surfactants of microbial origin and its application in foods. Sci Res Essays. 2020;15:11-7. DOI: 10.5897/sre2019.6656

62. Mnif I, Ghribi D. Glycolipid biosurfactants: main properties and potential applications in agriculture and food industry. J Sci Food Agric. 2016;96(13):4310-4320. DOI:10.1002/jsfa.7759

63. Campos JM, Stamford TLM, Sarubbo LA. Characterization and application of a biosurfactant isolated from Candida utilis in salad dressings. Biodegradation. 2019;30:313-324. DOI:10.1007/s10532-019-09877-8

64. Ribeiro BG, Guerra JMC, Sarubbo LA. Potential Food Application of a Biosurfactant Produced by Saccharomyces cerevisiae URM 6670. Front Bioeng Biotechnol. 2020;8:1-13. DOI: 10.3389/fbioe.2020.00434

65. Zouari R, Besbes S, Ellouze-Chaabouni S, Ghribi-Aydi D. Cookies from composite wheat-sesame peels flours: Dough quality and effect of Bacillus subtilis SPB1 biosurfactant addition. Food Chem. 2016;194:758-769. DOI:10.1016/j.foodchem.2015.08.064

66. Kiran GS, Priyadharsini S, Sajayan A, Priyadharsini GB, Poulose N, Selvin J. Production of lipopeptide biosurfactant by a marine Nesterenkonia sp. and its application in food industry. Front Microbiol. 2017;8:1-11. DOI: 10.3389/fmicb.2017.01138

67. Kavitake D, Kalahasti KK, Devi PB, Ravi R, Shetty PH. Galactan exopolysaccharide based flavour emulsions and their application in improving the texture and sensorial properties of muffin. Bioact Carbohydrates Diet Fibre. 2020;24:100248. DOI:10.1016/j.bcdf.2020.100248

68. Moutinho LF, Moura FR, Silvestre RC, Romão-Dumaresq AS. Microbial biosurfactants: A broad analysis of properties, applications, biosynthesis, and techno-economical assessment of rhamnolipid production. Biotechnol Prog. 2020;37:1-14. DOI: 10.1002/btpr.3093

69. Kalam S, Abu-Khamsin SA, Kamal MS, Patil S. A review on surfactant retention on rocks: mechanisms, measurements, and influencing factors. Fuel. 2021;293:120459. DOI:10.1016/j.fuel.2021.120459

70. Aguirre-Ramírez M, Silva-Jiménez H, Banat IM, Díaz De Rienzo MA. Surfactants: physicochemical interactions with biological macromolecules. Biotechnol Lett. 2021;43:523-535. DOI:10.1007/s10529-020-03054-1

71. McClements DJ, Rao J. Food-Grade Nanoemulsions: Formulation, Fabrication, Properties, Performance, Biological Fate, and Potential Toxicity. Crit Rev Food Sci Nutr. 2011;51:285-330. DOI:10.1080/10408398.2011.559558

72. Wu J, Yan F, Jia Q, Wang Q. QSPR for predicting the hydrophile-lipophile balance (HLB) of non-ionic surfactants. Colloids Surfaces A Physicochem Eng Asp. 2021;611:125812. DOI:10.1016/j.colsurfa.2020.125812

73. Mcclements DJ. Critical Review of Techniques and Methodologies for Characterization of Emulsion Stability. Crit Rev Food Sci Nutr. 2007;47:611-649. DOI:10.1080/10408390701289292

74. Mao L, Miao S. Structuring Food Emulsions to Improve Nutrient Delivery During Digestion. Food Eng Rev. 2015;7:439-451. DOI:10.1007/s12393-015-9108-0

75. Chung C, McClements DJ. Structure–function relationships in food emulsions: Improving food quality and sensory perception. Food Struct. 2014;1:106-126. DOI:10.1016/j.foostr.2013.11.002

76. Muschiolik G, Dickinson E. Double Emulsions Relevant to Food Systems: Preparation, Stability, and Applications. Compr Rev Food Sci Food Saf. 2017;16:532-555. DOI:10.1111/1541-4337.12261

77. Ravera F, Dziza K, Santini E, Cristofolini L, Liggieri L. Emulsification and emulsion stability: The role of the interfacial properties. Adv Colloid Interface Sci. 2021;288:102344. DOI:10.1016/j.cis.2020.102344

78. McClements DJ. Nanoemulsions versus microemulsions: terminology, differences, and similarities. Soft Matter. 2012;8:1719-1729. DOI:10.1039/C2SM06903B

79. Rao J, McClements DJ. Food-grade microemulsions, nanoemulsions and emulsions: Fabrication from sucrose monopalmitate & lemon oil. Food Hydrocoll. 2011;25:1413-1423. DOI:10.1016/j.foodhyd.2011.02.004

80. Dickinson E. Double Emulsions Stabilized by Food Biopolymers. Food Biophys. 2011;6:1-11. DOI:10.1007/s11483-010-9188-6

81. Silva-Avellaneda E, Bauer-Estrada K, Prieto-Correa RE, Quintanilla-Carvajal MX. The effect of composition, microfluidization and process parameters on formation of oleogels for ice cream applications. Sci Rep. 2021;11:7161. DOI:10.1038/s41598-021-86233-y

82. Chung C, Sher A, Rousset P, Decker EA, McClements DJ. Formulation of food emulsions using natural emulsifiers: Utilization of quillaja saponin and soy lecithin to fabricate liquid coffee whiteners. J Food Eng. 2017;209:1-11. DOI:10.1016/j.jfoodeng.2017.04.011

83. Paximada P, Howarth M, Dubey BN. Double emulsions fortified with plant and milk proteins as fat replacers in cheese. J Food Eng. 2021;288:110229. DOI:10.1016/j.jfoodeng.2020.110229

84. Jiang Z-Q, Wang J, Stoddard F, Salovaara H, Sontag-Strohm T. Preparation and Characterization of Emulsion Gels from Whole Faba Bean Flour. Foods. 2020;9:755. DOI: doi:10.3390/foods9060755

85. Du Le H, Loveday SM, Nowak E, Niu Z, Singh H. Pectin emulsions for colon-targeted release of propionic acid. Food Hydrocoll. 2020;103:105623. DOI:10.1016/j.foodhyd.2019.105623

86. Abbasi H, Karimi S, Gharibzahedi SMT. Rhamnolipid as a unique emulsifier to stabilize sesame oil-in-water beverage emulsions formed by ultrasound-induced cavitation: Optimizing the formulation and physical properties. J Food Process Preserv. 2020;44:1-12. DOI:10.1111/jfpp.14810

87. Ge A, Iqbal S, Kirk T V., Chen XD. Modulating the rheological properties of oil-in-water emulsions using controlled WPI-polysaccharide aggregation in aqueous phases. J Food Eng. 2021;297:110488. DOI: 10.1016/j.foodhyd.2019.05.028

88. Hu X, Karthik P, Chen J. Enhanced oral oil release and mouthfeel perception of starch emulsion gels. Food Res Int. 2021;144:110356. doi: 10.1016/j.foodres.2021.110356

89. Donsì F, Annunziata M, Sessa M, Ferrari G. Nanoencapsulation of essential oils to enhance their antimicrobial activity in foods. LWT - Food Sci Technol. 2011;44:1908-1914. DOI: 10.1016/j.lwt.2011.03.003

90. Nejatian M, Abbasi S. Formation of concentrated triglyceride nanoemulsions and nanogels: natural emulsifiers and high power ultrasound. RSC Adv. 2019;9:28330-28344. DOI: 10.1039/c9ra04761a

91. Ozturk B, Argin S, Ozilgen M, McClements DJ. Formation and stabilization of nanoemulsion-based vitamin e delivery systems using natural surfactants: Quillaja saponin and lecithin. J Food Eng. 2014;142:57-63. DOI: 10.1016/j.jfoodeng.2014.06.015

92. Flores-Andrade E, Allende-Baltazar Z, Sandoval-González PE, Jiménez-Fernández M, Beristain CI, Pascual-Pineda LA. Carotenoid nanoemulsions stabilized by natural emulsifiers: Whey protein, gum Arabic, and soy lecithin. J Food Eng. 2021;290:110208. DOI: 10.1016/j.jfoodeng.2020.110208

93. Wanyi W, Lu L, Zehan H, Xinan X. Comparison of emulsifying characteristics of different macromolecule emulsifiers and their effects on the physical properties of lycopene nanoemulsions. J Dispers Sci Technol. 2020;41:618-627. DOI: 10.1080/01932691.2019.1610421

94. Abbasi S, Radi M. Food grade microemulsion systems: Canola oil/lecithin:n-propanol/water. Food Chem. 2016;194:972-979. DOI: 10.1016/j.foodchem.2015.08.078

95. Chatzidaki MD, Mitsou E, Yaghmur A, Xenakis A, Papadimitriou V. Formulation and characterization of food-grade microemulsions as carriers of natural phenolic antioxidants. Colloids Surfaces A Physicochem Eng Asp. 2015;483:130-136. DOI: 10.1016/j.colsurfa.2015.03.060

96. Jalali-Jivan M, Abbasi S. Novel approach for lutein extraction: Food grade microemulsion containing soy lecithin and sunflower oil. Innov Food Sci Emerg Technol. 2020;66:102505. DOI: 10.1016/j.ifset.2020.102505

97. Amiri-Rigi A, Abbasi S. Stability assessment of lycopene microemulsion prepared using tomato industrial waste against various processing conditions. J Sci Food Agric. 2017;97:4922-4928. DOI: 10.1002/jsfa.8368

98. Zhang H, Yang H, Wang F, Zhao H, Li X, Zhou B, et al. Study on the stabilization of emulsion formed by Two different inclusion Complexes. Colloids Surfaces A Physicochem Eng Asp. 2020;594:124651. DOI:10.1016/j.colsurfa.2020.124651

99. Taha A, Ahmed E, Ismaiel A, Ashokkumar M, Xu X, Pan S, et al. Ultrasonic emulsification: An overview on the preparation of different emulsifiers-stabilized emulsions. Trends Food Sci Technol. 2020;105:363-377. DOI:10.1016/j.tifs.2020.09.024

100. Van Nieuwenhuyzen W. Production and utilization of natural phospholipids. 1906 AH Li. Consulting L, editor. Polar Lipids: Biology, Chemistry, and Technology. The Netherlands: AOCS Press; 2015. 245-276 p

101. Sweeta Akbari, Abdurahman Hamid Nour. Emulsion types, stability mechanisms and rheology: A review. Int J Innov Res Sci Stud. 2018;1:14-21. https://ssrn.com/abstract=3324905

102. Zhao H, Kang W, Yang H, Huang Z, Zhou B, Sarsenbekuly B. Emulsification and stabilization mechanism of crude oil emulsion by surfactant synergistic amphiphilic polymer system. Colloids Surfaces A Physicochem Eng Asp. 2021;609:125726. DOI:10.1016/j.colsurfa.2020.125726

103. Zhang R, Belwal T, Li L, Lin X, Xu Y, Luo Z. Recent advances in polysaccharides stabilized emulsions for encapsulation and delivery of bioactive food ingredients: A review. Carbohydr Polym. 2020;242:116388. DOI: 10.1016/j.carbpol.2020.116388

104. Grumezescu AM. Emulsions. Nanotechnology in the Agri-Food Industry. Grumezesc AM, editor. Elsevier; 2016

105. Mayer S, Weiss J, McClements DJ. Vitamin E-enriched nanoemulsions formed by emulsion phase inversion: Factors influencing droplet size and stability. J Colloid Interface Sci. 2013;402:122-130. DOI: 10.1016/j.jcis.2013.04.016

106. Chen H, Guan Y, Zhong Q. Microemulsions based on a sunflower lecithin-tween 20 blend have high capacity for dissolving peppermint oil and stabilizing coenzyme Q10. J Agric Food Chem. 2015;63:983-989. DOI: 10.1021/jf504146t

107. Koch L, Hummel L, Schuchmann HP, Emin MA. Structural changes and functional properties of highly concentrated whey protein isolate-citrus pectin blends after defined, high temperature treatments. LWT - Food Sci Technol. 2017;84:634-642. DOI: 10.1016/j.lwt.2017.06.026

108. Pu X, Wolf B, Dragosavac M. Generation of magnesium enriched water-in-oil-in-water food emulsions by stirred cell membrane emulsification. J Food Eng. 2019;247:178-187. DOI: 10.1016/j.jfoodeng.2018.11.022

109. Tan TB, Nakajima M, Tan CP. Effect of polysaccharide emulsifiers on the fabrication of monodisperse oil-in-water emulsions using the microchannel emulsification method. J Food Eng. 2018;238:188-194. DOI: 10.1016/j.jfoodeng.2018.06.026

110. Guan Y, Wu J, Zhong Q. Eugenol improves physical and chemical stabilities of nanoemulsions loaded with β-carotene. Food Chem. 2016;194:787-796. DOI: 10.1016/j.foodchem.2015.08.097

111. Bouhoute M, Taarji N, Vodo S, Kobayashi I, Zahar M, Isoda H, et al. Formation and stability of emulsions using crude extracts as natural emulsifiers from Argan shells. Colloids Surfaces A Physicochem Eng Asp. 2020;591:124536. DOI: 10.1016/j.colsurfa.2020.124536

112. Chung C, Sher A, Rousset P, McClements DJ. Use of natural emulsifiers in model coffee creamers: Physical properties of quillaja saponin-stabilized emulsions. Food Hydrocoll. 2017;67:111-119. DOI: 10.1016/j.foodhyd.2017.01.008

113. Cha Y, Shi X, Wu F, Zou H, Chang C, Guo Y, et al. Improving the stability of oil-in-water emulsions by using mussel myofibrillar proteins and lecithin as emulsifiers and high-pressure homogenization. J Food Eng. 2019;258:1-8. DOI: 10.1016/j.jfoodeng.2019.04.009

114. Chung C, Sher A, Rousset P, McClements DJ. Influence of homogenization on physical properties of model coffee creamers stabilized by quillaja saponin. Food Res Int. 2017;99:770-777. DOI: 10.1016/j.foodres.2017.06.060

115. Modarres-Gheisari SMM, Gavagsaz-Ghoachani R, Malaki M, Safarpour P, Zandi M. Ultrasonic nano-emulsification – A review. Ultrason Sonochem. 2019;52:88-105. DOI: 10.1016/j.ultsonch.2018.11.005

116. McClements DJ. Food Emulsions: Principles, Practices, and Techniques. 3rd ed. CRC Press; 2015. 714 p

117. Das KP, Kinsella JE. Stability Of Food Emulsions: Physicochemical Role Of Protein And Nonprotein Emulsifiers. In: Advances in Food and Nutrition Research. Elsevier Inc.; 1990. p. 81-201

118. Yan X, Ma C, Cui F, McClements DJ, Liu X, Liu F. Protein-stabilized Pickering emulsions: Formation, stability, properties, and applications in foods. Trends Food Sci Technol. 2020;103:293-303. DOI: 10.1016/j.tifs.2020.07.005

119. Sharkawy A, Barreiro MF, Rodrigues AE. Chitosan-based Pickering emulsions and their applications: A review. Carbohydr Polym. 2020;250:116885. DOI: 10.1016/j.carbpol.2020.116885

120. Cui F, Zhao S, Guan X, McClements DJ, Liu X, Liu F, et al. Polysaccharide-based Pickering Emulsions: Formation, Stabilization and Applications. Food Hydrocoll. 2021;106812. DOI: 10.1016/j.foodhyd.2021.106812

121. Ribeiro EF, Morell P, Nicoletti VR, Quiles A, Hernando I. Protein- and polysaccharide-based particles used for Pickering emulsion stabilisation. Food Hydrocoll. 2021;119:106839. DOI: 10.1016/j.foodhyd.2021.106839

122. Niroula A, Gamot TD, Ooi CW, Dhital S. Biomolecule-based pickering food emulsions: Intrinsic components of food matrix, recent trends and prospects. Food Hydrocoll. 2021;112:106303. DOI: 10.1016/j.foodhyd.2020.106303

123. Sarkar A, Dickinson E. Sustainable food-grade Pickering emulsions stabilized by plant-based particles. Curr Opin Colloid Interface Sci. 2020;49:69-81. DOI: 10.1016/j.cocis.2020.04.004

124. Xia T, Xue C, Wei Z. Physicochemical characteristics, applications and research trends of edible Pickering emulsions. Trends Food Sci Technol. 2021;107:1-15. DOI: 10.1016/j.tifs.2020.11.019

125. Santos J, Alcaide-González MA, Trujillo-Cayado LA, Carrillo F, Alfaro-Rodríguez MC. Development of food-grade Pickering emulsions stabilized by a biological macromolecule (xanthan gum) and zein. Int J Biol Macromol. 2020;153:747-754. DOI: 10.1016/j.ijbiomac.2020.03.078

126. Ren Z, Li Z, Chen Z, Zhang Y, Lin X, Weng W, et al. Characteristics and application of fish oil-in-water pickering emulsions structured with tea water-insoluble proteins/κ-carrageenan complexes. Food Hydrocoll. 2021;114:106562. DOI: 10.1016/j.foodhyd.2020.106562

127. Lu Z, Ye F, Zhou G, Gao R, Qin D, Zhao G. Micronized apple pomace as a novel emulsifier for food O/W Pickering emulsion. Food Chem. 2020;330:127325. DOI: 10.1016/j.foodchem.2020.127325

128. He K, Li Q, Li Y, Li B, Liu S. Water-insoluble dietary fibers from bamboo shoot used as plant food particles for the stabilization of O/W Pickering emulsion. Food Chem. 2020;310:125925. DOI: 10.1016/j.foodchem.2019.125925

129. Alehosseini E, Jafari SM, Tabarestani HS. Production of D-limonene-loaded Pickering emulsions stabilized by chitosan nanoparticles. Food Chem. 2021;354:129591. DOI: 10.1016/j.foodchem.2021.129591

130. Jiang Y, Zhu Y, Li F, Li D, Huang Q. Gliadin/amidated pectin core–shell nanoparticles for stabilisation of Pickering emulsion. Int J Food Sci Technol. 2020;55:3278-3288. DOI: 10.1111/ijfs.14590

131. Li F, Li X, Huang K, Luo Y, Mei X. Preparation and characterization of pickering emulsion stabilized by hordein-chitosan complex particles. J Food Eng. 2021;292:110275. DOI: 10.1016/j.jfoodeng.2020.110275

132. Azfaralariff A, Fazial FF, Sontanosamy RS, Nazar MF, Lazim AM. Food-grade particle stabilized pickering emulsion using modified sago (Metroxylon sagu) starch nanocrystal. J Food Eng. 2020;280:109974. DOI: 10.1016/j.jfoodeng.2020.109974

133. Ju M, Zhu G, Huang G, Shen X, Zhang Y, Jiang L, et al. A novel pickering emulsion produced using soy protein-anthocyanin complex nanoparticles. Food Hydrocoll. 2020;99:105329. DOI: 10.1016/j.foodhyd.2019.105329

134. Yang H, Su Z, Meng X, Zhang X, Kennedy JF, Liu B. Fabrication and characterization of Pickering emulsion stabilized by soy protein isolate-chitosan nanoparticles. Carbohydr Polym. 2020;247:116712. DOI: 10.1016/j.carbpol.2020.116712

135. Ren Z, Chen Z, Zhang Y, Lin X, Li B. Characteristics and rheological behavior of Pickering emulsions stabilized by tea water-insoluble protein nanoparticles via high-pressure homogenization. Int J Biol Macromol. 2020;151:247-256. DOI: 10.1016/j.ijbiomac.2020.02.090

136. Zhu Q, Lu H, Zhu J, Zhang M, Yin L. Development and characterization of pickering emulsion stabilized by zein/corn fiber gum (CFG) complex colloidal particles. Food Hydrocoll. 2019;91:204-213. DOI: 10.1016/j.foodhyd.2019.01.029

137. Dai L, Sun C, Wei Y, Mao L, Gao Y. Characterization of Pickering emulsion gels stabilized by zein/gum arabic complex colloidal nanoparticles. Food Hydrocoll. 2018;74:239-248. DOI: 10.1016/j.foodhyd.2017.07.040

138. Jiang Y, Wang D, Li F, Li D, Huang Q. Cinnamon essential oil Pickering emulsion stabilized by zein-pectin composite nanoparticles: Characterization, antimicrobial effect and advantages in storage application. Int J Biol Macromol. 2020;148:1280-1289. DOI: 10.1016/j.ijbiomac.2019.10.103

139. Su J, Guo Q, Chen Y, Dong W, Mao L, Gao Y, et al. Characterization and formation mechanism of lutein pickering emulsion gels stabilized by β-lactoglobulin-gum arabic composite colloidal nanoparticles. Food Hydrocoll. 2020;98:105276. DOI: 10.1016/j.foodhyd.2019.105276

140. Bizmark N, Du X, Ioannidis MA. High Internal Phase Pickering Emulsions as Templates for a Cellulosic Functional Porous Material. ACS Sustain Chem Eng. 2020;8:3664-3672. DOI: 10.1021/acssuschemeng.9b06577

141. Gao H, Ma L, Cheng C, Liu J, Liang R, Zou L, et al. Review of recent advances in the preparation, properties, and applications of high internal phase emulsions. Trends Food Sci Technol. 2021;112:36-49. DOI: 10.1016/j.tifs.2021.03.041

142. Abdullah, Weiss J, Ahmad T, Zhang C, Zhang H. A review of recent progress on high internal-phase Pickering emulsions in food science. Trends Food Sci Technol. 2020;106:91-103. DOI: 10.1016/j.tifs.2020.10.016

143. Zhao Q, Zaaboul F, Liu Y, Li J. Recent advances on protein-based Pickering high internal phase emulsions (Pickering HIPEs): Fabrication, characterization, and applications. Compr Rev Food Sci Food Saf. 2020;19:1934-1968. DOI: 10.1111/1541-4337.12570

144. Shi A, Feng X, Wang Q, Adhikari B. Pickering and high internal phase Pickering emulsions stabilized by protein-based particles: A review of synthesis, application and prospective. Food Hydrocoll. 2020;109:106117. DOI: 10.1016/j.foodhyd.2020.106117

145. Liu X, Guo J, Wan ZL, Liu YY, Ruan QJ, Yang XQ. Wheat gluten-stabilized high internal phase emulsions as mayonnaise replacers. Food Hydrocoll. 2018;77:168-175. DOI: 10.1016/j.foodhyd.2017.09.032

146. Niro CM, Medeiros JA, Freitas JAM, Azeredo HMC. Advantages and challenges of Pickering emulsions applied to bio-based films: a mini-review. J Sci Food Agric. 2021;101:3535-3540. DOI: 10.1002/jsfa.11029

147. Li Q, Ma Q, Wu Y, Li Y, Li B, Luo X, et al. Oleogel Films through the Pickering Effect of Bacterial Cellulose Nanofibrils Featuring Interfacial Network Stabilization. J Agric Food Chem. 2020;68:9150-9157. DOI: 10.1021/acs.jafc.0c03214

148. Huang L, Liu J, Addy M, Ding B, Cheng Y, Peng P, et al. Physicochemical and emulsifying properties of orange fibers stabilized oil-in-water emulsions. LWT - Food Sci Technol. 2020;133:110054. DOI:10.1016/j.lwt.2020.110054

149. Francisco CRL, de Oliveira Júnior FD, Marin G, Alvim ID, Hubinger MD. Plant proteins at low concentrations as natural emulsifiers for an effective orange essential oil microencapsulation by spray drying. Colloids Surfaces A Physicochem Eng Asp. 2020;607:125470. DOI:10.1016/j.colsurfa.2020.125470

150. Jarzębski M, Smułek W, Siejak P, Rezler R, Pawlicz J, Trzeciak T, et al. Aesculus hippocastanum L. as a stabilizer in hemp seed oil nanoemulsions for potential biomedical and food applications. Int J Mol Sci. 2021;22:1-18. DOI: 10.3390/ijms22020887

151. Gaur VK, Regar RK, Dhiman N, Gautam K, Srivastava JK, Patnaik S, et al. Biosynthesis and characterization of sophorolipid biosurfactant by Candida spp.: Application as food emulsifier and antibacterial agent. Bioresour TechnoL. 2019;285:121314. DOI:10.1016/j.biortech.2019.121314

152. Cetinkaya T, Altay F, Ceylan Z. A new application with characterized oil-in-water-in-oil double emulsions: Gelatin-xanthan gum complexes for the edible oil industry. Lwt. 2021;138:110773. DOI:10.1016/j.lwt.2020.110773

Chapter-7

Role of Natural Additives on Quality and Shelf Life Extension of Fish and Fishery Products

1. Introduction

Fish and fishery products have become increasingly popular due to their high demand and nutritional value. According to Food and Agriculture Organization (FAO), worldwide production of finfish, mollusks (mainly bivalves), and crustaceans are 54.3, 17.7, and 9.4 million tons respectively [1]. Human consumption of fish is around 88% of total production and among them, 44% consist of live, fresh, or chilled products, and 35% consist of frozen products [2]. But, among animal-derived products, fish is considered as the most perishable commodity as it contains a high amount of water, high post mortem pH (greater than 6), non-protein-nitrogen content, free amino acids, lower content of connective tissues, and presence of an osmoregulant, trimethylamine oxide (TMAO) [3]. Spoilage or the deterioration process refers to any change in the condition of food in terms of taste, smell, appearance, or texture and becomes undesirable or unacceptable for human consumption. Generally, the process involves 3 stages; rigor mortis, autolysis, and putrefaction. Rigor mortis or the muscle stiffening will last for hours (time may vary with temperature) after their catch. Subsequently softening occurs due to enzymatic or oxidative self-digestion, and completed by microbiological processes (putrefaction) [4].

Every year, chemical and microbial deterioration alone contributes 25% of gross primary product loss (agricultural and fishery products). Besides this, there are several other factors such as harvesting season, type of species, capturing method, handling, the time lag from catch to processing, method of processing, storage temperature, etc. that also influence the rate of spoilage. During spoilage, the breakdown of various components and formation of new compounds responsible for the off- flavor, off odor, discoloration, and texture damage of the fish meat takes place [5]. Therefore, certain food additives have been added to maintain the quality, and the shelf-life of fish and fishery products. The main aim is to combat microbial contamination as well as oxidation for the extension of product's shelf life. Generally, lipid oxidation leads to quality deterioration, and some of them can be detected by organoleptic evaluation, but microbial contamination especially pathogenic microorganisms mostly do not produce sensory deterioration, which act as a challenge for food safety. It emphasizes the importance of the application of antimicrobials in the preservation techniques [6]. In the case of fish and fishery products, preservation techniques draw more scientific attraction, since they represent internationally traded products. Even though many strategies have been developed to prevent chemical and microbial spoilage by chemical preservation, there is still a need for the use of natural preservatives, considering consumer safety. Thus, various researches and efforts have been made to invent more natural alternative solutions in the field of food preservation.

2. Quality Changes in Fish and Fishery Products

Fish and fishery products deteriorate rapidly as a consequence of various biochemical breakdowns and microbial activities on the chemical composition of meat. The spoilage involves autolysis or self-digestion of these compounds by digestive enzymes or by free radicals [7]. The major spoilage process is lipid degradation, which mainly occurs through oxidation or hydrolysis. The oxidation could be various types such as photo-oxidation, thermal oxidation, enzymatic oxidation, and auto-oxidation. It can also be accelerated by prooxidants within the body such as hemoglobin, myoglobin, cytochrome c, etc. The process involves the reaction of unsaturated fatty acids of triglycerides with atmospheric oxygen to form unstable primary products like free fatty acids (FFAs), dienes, and peroxides and secondary products like aldehydes, ketones, alcohols, hydrocarbons, volatile organic acids, trienes, epoxy compounds and carbonyls [8]. The process of lipid hydrolysis or lipolysis is breaking down of triglycerides into FFAs by the action of enzyme lipases. Accumulation of these FFAs stimulates protein denaturation, texture damage, and drip loss by the formation of protein-lipid cross-linkages [9]. Generally, protein denaturation in fish occurs mainly by the action of proteolytic enzymes in the muscle (cathepsins) and the intestinal tract (trypsins), which results in muscle solubilization and leads to undesirable texture damage. End products like amino acids, peptides, amines, H_2S, ammonia, indole, etc. will be formed and will all act as a medium for microbial growth. Microbial breakdown of amino acids will lead to bitterness, souring, bad odor, sliminess, etc. of the flesh [10].

For fish and fishery products, gram-negative bacteria like *Shewanella, Photobacterium, Pseudomonas, Moraxella, Acinetobacter, Flavobacterium*, Aeromonadaceae, and Vibrionaceae and the gram-positive bacteria such as *Bacillus, Micrococcus, Lactobacillus, Clostridium*, and *Corynebacterium* are considered as the major spoilers [3, 10]. The spoilage resulting in off-flavors is due to the formation of specific alcohols, aldehydes, acids, ketones, and sulfur and nitrogen compounds. One of the other major non-nitrogen compounds formed is trimethylamine (TMA) by the action of several spoilage bacteria on TMAO, an osmoregulant, present in fish (mostly marine and some freshwater fish), and cause a high (positive) redox potential (Eh) in the flesh. Under anoxic conditions, many of the spoilage bacteria utilize TMAO as a terminal hydrogen acceptor, thus allowing them to grow, and resulting in the formation of TMA. TMA reacts with lipids in the muscle to produce the off odor of low-quality fish. This could be a reason for rapid spoilage occurring in seafood than other muscle foods [11]. Thus, microbial spoilage can be determined by TMA level in the product. In the case of shrimps, at above 10°C, indole-positive organisms such as Aeromonas cause subsequent conversion of tryptophan to indole, which is associated with the off-odor of decomposition of shrimp. Thus high levels of indole in the flesh is an indicator of high temperature in the chilled storage process [12].

Clams and oysters undergo fermentative type spoilage also [13]. Generally, the microbial contamination in the fish mainly occurs through microbes associated with the habitat, invasion during processing, handling, and long-term storage. Growth of spoilers differs by habitats like freshwater or marine, temperate or tropical water, and storage or processing conditions. The microbial and chemical stability of food during processing and storage will be determined by the available water for microbial growth, called water activity (a_w). Yeast requires a_w of minimum 0.7 for their growth, and except *Staphylococcus aureus,* most bacteria require at least 0.9 a_w to grow [14]. Thus, it can be said that microbiologically stable fish product is with an a_w less than 0.6 [15]. Thus, the water content in the product should be minimum to prevent microbial spoilage. Moreover, pathogenic microbes of public health concern are also taken into consideration as they can produce hazardous toxins. Some of these are; toxin produced by *Clostridium botulinum* (botulinum toxin) in processed food, Scombrotoxin as a result of the microbial conversion of histidine to histamine. Bacteria involved in this process include *Morganella morganii, Klebsiella pneumoniae, Hafnia alvei, Pseudomonas putrefaciens,* and *Clostridium perfringens.* Shellfishes can accumulate various algal toxins like brevetoxins, okadaic acids, domoic acids, saxitoxins, etc., and cause serious illness to humans [16].

Another important spoilage mechanism is post-mortem nucleotide catabolism, resulting in ATP depletion and subsequent formation of hypoxanthine (Hx) (Figure 1). The breakdown products do not affect the safety but sensory quality undergoes some changes [17, 18]. Based on these compounds formed, the freshness can be expressed. The ratio of inosine (Ino) and Hx to total nucleotides and their catabolic derivatives will give the K value, an indicator of loss of freshness [19]. The ratio of Hx to the total of Inosine monophosphate (IMP), Ino, and Hx will give the H value, an indicator of Hx accumulation (bitterness), and its limit for human consumption has been suggested as 60% [20]. Another quality indicator is the F value. It is the ratio of IMP to the total of IMP, Ino, and Hx, and fish with F-value of 10% and higher is considered unacceptable [21]. Thus there is a huge need for the use of additives in the food industry. Application of food additives and low-temperature preservation leads to diminution of most of the spoilage process to a greater extent.

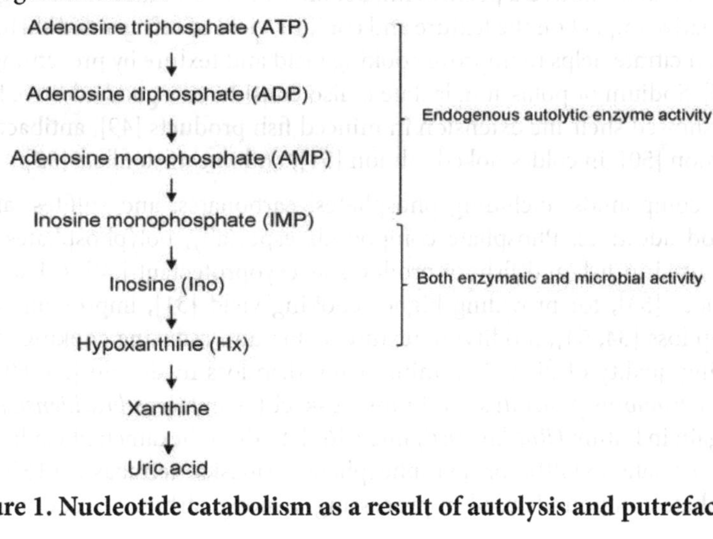

Figure 1. Nucleotide catabolism as a result of autolysis and putrefaction.

3. Role of Chemical Additives and Natural Alternative Solutions

By definition, additives are the substances that are added to maintain or improve the safety, freshness, taste, texture, or appearance of food. Generally additives can occur in fish and fishery products during production, processing, storage, packaging, and transportation. Additives can be of two types; Synthetic or chemical, and natural additives. Some of them are listed in Table 1.

3.1 Chemical additives

Among chemical additives, the most common and widely used chemical is sodium chloride (NaCl). Salt drying and brining is the most traditional as well as an effective method of food preservation, and several studies have been made to explore all the preservation properties of NaCl. In Nile tilapia fillets, NaCl improved the weight and minimized drip loss [34], showed weight gain in white shrimp (*Litopenaeus vannamei*) [35], and had anti-melanotic activity along with the shelf-life extension in shrimp (*Xyphopenaeus kroyeri*) [36]. Like salt, sugar is also easily available and is a widely used additive for seafood products. Sugar treatment can significantly reduce pH value and decrease volatile bases like total volatile base nitrogen (TVB-N) [37]. It also showed a cryoprotectant action in frozen surimi (wet protein concentrate) and other products [38, 39], protection of myofibrillar protein [40], decrease the accumulation of biogenic amines in sausages and dry-cured grass carp [41, 42], and prevention of protein denaturation in minced fish meat [43]. The combination of both sugar and salt could also delay spoilage and improve many sensory qualities [37]. The product 'gravad' traditionally manufactured in Nordic countries is prepared by such a combination of sugar and salt [44]. Additives such as table salt and organic acids like acetic acid or citric acid in the Marination technique not only prevent microbial growth but also improve organoleptic properties of fish and fishery products [45]. In seafood, the addition of organic acids provides great preservative action as an antimicrobial agent. Acetic acid and lactic acid, either single-use or combination had a growth-inhibiting effect against pathogens like *Listeria monocytogenes* and *Escherichia coli* [46]. The inhibitory effect of these acids against *L. monocytogenes* was also reported from mussels [47]. Generally, the addition of citric acid showed a positive impact on TVB-N accumulation, toughness, and pH, but a negative impact on the texture and cooking yield of refrigerated shrimp. In such a case, sodium citrate helps to improve cooking yield and texture by preventing excessive pH drop [48]. Sodium or potassium lactate is also considered a good additive for seafood products. It showed shelf-life extension in minced fish products [49], antibacterial effect in sliced salmon [50], in cold-smoked salmon [51], and in catfish fillets [52].

Many other compounds, including phosphates, carbonates, and sulfites, are used as major seafood additives. Phosphate compounds especially, polyphosphates (PP) have been widely used in fish and fishery products as cryoprotectant [38], gel strength, and flavor enhancer [53], for providing higher cooking yield [31], improving weight, and reducing drip loss [34, 54], modifying texture, color, and reducing cooking loss [55, 56, 57], improving quality of fillet [58], minimizing drip loss in shrimp [59, 60], drip loss in sea robin (*Prionotus punctatus*) and pink cuskeel (*Genypterus brasiliensis*) fillet [31], and weight gain in kutum (*Rutilus frisii*) fillets [61]. Sodium hexametaphosphate (SHMP) or tripolyphosphate (STPP), or pyrophosphate- tribasic/ tetrabasic (TSPP) are the major phosphate compounds used in processing. Among carbonates, sodium carbonate

(Na_2CO_3) sodium bicarbonate ($NaHCO_3$), and magnesium carbonate ($MgCO_3$) have been widely used. Weight gain is observed when white shrimp are treated with Na_2CO_3 and $NaHCO_3$ [35]. The addition of $NaHCO_3$ also provides the highest expansion volume for yellow pike conger crackers [62]. Sulfites have been widely used as additives due to their desirable technical properties like preventing melanosis or discoloration. The most predominant sulfiting agent is sodium sulfite used to prevent melanosis in crustaceans like shrimp, lobster, crab, crayfish, etc. [32]. Nitrite is another chemical commonly used as an antimicrobial agent, and effective against *C. botulinum* and its toxin production [63]. A combination of nitrite and sorbic acid would also give the best result as it can inhibit most yeasts. A combination of sorbic acid with benzoic acid could preserve brined shrimp [64]. Moreover, additives such as flavor enhancers, sweeteners, colorants, etc. are used to enhance the appeal of the food. Monosodium glutamate (MSG), calcium chloride ($CaCl_2$), and Disodium guanylate/ inosinate are the major flavor enhancers. Commonly used sweetening agents are saccharin, sucralose, glycerol, acesulfame potassium, aspartame, sodium cyclamate, neotam, and neohesperidine. Widely used colorants include carmine, carmosine, caramel, paprika, annatto dye, iron oxides and hydroxides, ponceau, cochineals, titanium dioxide (TiO_2), FD&C Yellow, and astaxanthin (Table 1). Butylated hydroxyl toluene (BHT), butylated hydroxyanisole (BHA), tertbutylhydroquinone (TBHQ), propyl gallate (PG), and sodium acetate are widely used synthetic antioxidants to prevent lipid oxidation through free radicals scavenging, breaking chain reactions, peroxide decomposition, and decreasing oxygen concentrations and thereby increasing the shelf life [50]. But preservatives include sulfites, nitrates, benzoates, sorbates, formaldehyde, and others that may possess carcinogenic side effects. Thus nowadays the use of chemical preservatives in food industries steadily decreases and consumers are turning to the use of natural additives.

3.2 Natural additives

3.2.1 Plant-derived products

The use of plant-derived natural compounds such as essential oils, plant extracts, hydrocolloids, phenolic compounds, etc. is very popular in seafood preservation. Their strong antimicrobial and antioxidant activities present great potential for use in the food industry [64, 65, 66].

Plant extracts and essential oils can be derived from plant petals, leaves, fruits, peels, stems, roots, and xylems and their antioxidant effects are due to volatile organic compounds, terpenoid, and phenolic components in the plant. The inhibitory effects of essential oil on gram-negative bacteria are less than that of gram-positive bacteria as their lipopolysaccharide cell wall of gram-negative bacteria blocked the invasion of hydrophobic oils into the cell membrane [67]. Using essential oils (EOs) and plant extracts to extend shelf-life and maintain the quality of fish and fishery products has been reported frequently. Some of the recent studies of their application in fish and fishery products are represented in Table 2. However strong odor and taste, high volatility, complex chemical composition, low bioavailability and stability, and factors affecting chemical compositions like plant genetic variability, extraction techniques, etc. are some limitations for the application of these phytogenic additives [99]. Like other plant-derived products, seaweed and algal extracts are emerging as a rich source of natural antioxidants, along with many nutritional values. The three important widely used hydrocolloids are; agar-agar, align, and Carrageenan. As thickening agent agar-agar is used mainly in fish paste products. Carrageenan is used to enhance the gelling property of fish mince [100, 101, 102], and organoleptic properties of mussels and squids [103]. The sodium salt of alginic acid is widely used as a stabilizer and thickener in coating films. Sodium alginate coating with rosemary extract reduced the accumulation of biogenic amines and bacterial count in Abalone (*Haliotis discus hannai*) [104]. Coating with gingerol delayed lipid oxidation, protein degradation, nucleotide breakdown, and inhibited microbial growth in Seabream (*Pagrosomus major*) [105]. Coating with tea polyphenols had significantly lowered the levels of TVB-N, lipid oxidation, and protein decomposition in Japanese Sea Bass (*Lateolabrax japonicas*) fillets [106]. The use of alginate-calcium film coating with *Citrus wilsonii* extract delays the deterioration and results in a higher sensory score for *L. vannamei* [107]. Significant reduction in the TVB-N, TMA, and thiobarbituric acid reactive substance (TBARS) has been detected during chilled storage with the presence of *Gracilaria gracilis* extract in shrimp [108], *G. verrucosa* extract in Indian mackerel [109], and extracts of *Hypnea musciformis* and *A. muscoides* in black tiger shrimp [110]. Similarly, seaweeds like *Sargassum kjellmanianurn* [111], and *Grateloupia filicina* [112] exhibits a good antioxidant activity and prevent lipid oxidation in fish oil. Extracts of seaweeds such as *Fucus vesiculosus* inhibited the hemoglobin-mediated lipid oxidation in washed cod muscle and cod protein isolates [113], and extracts of *Durvillaea antarctica* (cochayuyo/ulte), *Pyropia columbina* (red luche), and *Ulva lactuca* (sea lettuce) improved the lipid and sensory qualities in canned salmon [114]. Some phenolic compounds like flavonoids, phenolic acids, hydroxycinnamic acid, and lignans are also used as plant derived natural additives [115]. In surimi-derived products, several hydrocolloids like konjac enhance the gelling property [116]. Other products like starch [101, 102], gums such as garrofin, guar, xanthan [117], etc. also provide a gelling effect and assure elasticity of the product. Iota carrageenan and xanthan had a cryoprotective effect too [118]. Other plant-derived products such as soybean protein, wheat gluten, and starch are also used as additives for fish-paste products [119].

Table 1. Lists of additives used in fish and fishery products.

Additive function	Categories	Examples	Application	Ref.
To maintain palatability and wholesomeness (preservatives)	Antimicrobial agents	Benzoates, Sorbates, NaCl NO_3^-, NO_2^- Organic acids, EOs	Surimi/minced fish products dried, salted or cured fish Fish fillets	[22, 23]
	Antioxidants	BHA, BHT, TBHQ, PG, Ca/Na propionate, vit E, Ascorbate, Citric acid, Erythorbic acid EOs	Fish oil Fish fillets	[24, 25, 26]
To enhance the appeal of foods	Flavor enhancers	MSG, $CaCl_2$, Citric acid, Disodium guanylate/ inosinate	Ready to cook or ready to eat products	[25]
	Sweeteners	Sucrose, lactose, glucose, fructose, glycerol, sorbitol. Acesulfame K, Aspartame, Sodium cyclamate, Saccarin, Sucralose, Neotam, Neohesperidine	Crab meat Fish fillets Marinades	[27, 28]
	Colorants	Carmine, carmosine, caramel, paprika, annatto dye Iron oxides and hydroxides, Ponceau, Cochineals, Oleoresin of turmeric, TiO_2, FD&C Yellow, Astaxanthin	Paste products, pre-cooked crustaceans, salmon substitutes, surimi, fish roe and smoked fish.	[28, 29]
	Texturizing agents; (Emulsifiers/Stabilizers/ Water-binding agents)	Polysorbates, DATEM, Agar, Alginates, Carrageenan, gum, Calcium stearate, Lecithin, Yeast, Ammonium alum,	Fish and shrimp paste/mince products, Surimi	[25, 30]
To aid in the processing	Moisture control	CaO, Calcium stearate, STPP	Processed products	[25]
	Reduce thaw drip	Na_2CO_3, SHMP, STPP, TSPP	Frozen: clams, crab, fillets, lobster, shrimp and minced fish	[31]
	Prevent cracking of glaze	Na_3PO_4, Na_2HPO_4	Frozen products	[31]
	Anticaking agents	$Ca_3(PO4)_2$, Na_2SiO_3, Ca_2SiO_4, $MgCO_3$, $Na_4P_2O_7$	Paste/minced products	[25]
	Prevent discolouration/blackspot/browning	Na_2SO_3 4-Hexylresorcinol, EDTA	Crustaceans	[32]
	Packaging gases	O_2, CO_2, N_2, Ar, O_3, N_2O, He, SO_2, Cl	Fish and fishery products	[33]
	Cryoprotectants	PP	Paste products, fillets, Frozen crustaceans	[28]

Table 2. Summary of some of the recent studies of application of essential oils, and plant extracts in fish and fishery products.

Plant derived compounds	Preservative properties	Product	Ref.
Essential oils			
Rosemary	Retarded microbial growth, delay chemical deterioration, sensory qualities, and extend the shelf life for 14 days during storage.	Bonito Fish Patties	[68]
Rosemary and basil	Inhibited the formation of TVB-N and lipid oxidation products during storage	Atlantic Mackerel	[69]
Rosemary, laurel, thyme, and sage	Antimicrobial, antioxidant properties, and also enhanced the organoleptic quality	Rainbow trout fillet	[70]
Rosemary, cinnamon fennel and cardamom	The counts of *S. aureus, E. coli* and *Bacillus cereus* were reduced	Carp fingers	[71]
Green pepper	Inhibited the growth of wild type strain of *P. aeruginosa* and attenuated its virulence properties	Fish-based products	[72]
Orange leaf	Enriched with Gelatin film showed shelf-life extension of 10 days	Shrimps	[73]
Carvacrol, bergamot and grapefruit	Improved the quality of fresh fish and extended the shelf-life up to 4 days.	Seabream	[74]
Oregano, thyme, and star anise	Inhibited microbial growth and delaying lipid oxidation	Grass carp	[75]
Ginger	Significant reduction in the TVBN and lipid oxidation	Cobia steaks	[76]
Clove, cumin, and spearmint	Retarded sensory deterioration and formation of biogenic amines.	Red drum fillets	[77]
Black cumin	Higher sensory quality, and extended shelf life	Fresh fish fillets	[78]
limonene	Maintained spoilage bacteria at a lower level and extended the shelf-life of 15 days	Gilthead sea bream fillets	[79]
Allyl Isothiocyanate	Extended the shelf-life by maintain specific spoilage organisms at a lower level	Gilthead sea bream	[80]
Oregano	Activity against *Listeria* spp.	Salmon	[81]
Fennel	Coating with chitosan nanoparticles reduced the PV, TVBN, TBARS and microbial count	*Huso huso* fish fillets	[82]
Satureja thymbra leaves	Reduced peroxide value and eliminates secondary oxidation products	Gilthead seabream fillets	[83]
Lemon	Lowered accumulation of histamine, and improved sensory characteristics	Salted sardines	[84]
cinnamon bark	Inhibited *Aeromonas* and *Shewanella*, and reduced the accumulation of TVBN, putrescine, cadaverine	Grass carp	[85]

Plant derived compounds	Preservative properties	Product	Ref.
Plant extracts			
Cumin seed and Wild mint leaf	Retarded microbial growth, chemical deterioration, and improved sensory characteristics	Rainbow trout fillets	[86]
Shallot fruit and ajwain seed	Delayed lipid oxidation and microbiological spoilage	Semi-fried coated rainbow trout fillets	[87]
Allium paradoxum and *Eryngium caucasicum trauve.*	Significantly delayed oxidative deterioration and maintain lower bacterial growth	Silver carp fillets	[88]
Punica granatum peels and *Hibiscus sabdariffa* calyxes	Colorant, preservative, and antimicrobial action	Burger and surimi	[89]
Nothopanax sutellarium leaf	Potential to preserve the fresh fish during transportation	Nile tilapia	[90]
Mint leaf and citrus peel	Retarded the quality changes and extended the shelf life	Indian mackerel	[91]
Syzygium australe and *S. luehmannii* fruit and leaf	Inhibition to *Shewanella* spp.	Fresh and cold storage fish	[92]
Kakadu plum	Inhibited bacterial growth for 15 days at 4°C	Fish fillets	[93]
Clove, Sage and kiwifruit peel	Antioxidant and antimicrobial potential, and extend the shelf life	Fish fingers	[94]
Pomegranate, rosemary, and olive	Delayed the lipid oxidation, and microbial count	Fish patties	[95]
Plectranthus amboinicus leaf	Improved the color, rehydration and water activity	Fish oil fortified soup powder	[96]
Green Tea Leaves and Fenugreek Seeds	Decreased TVBN, TBARS, total bacterial count, and pH	Shrimp	[97]
Basil leaf	Antimicrobial effect and longest shelf life	Mullet fillet	[98]

3.2.2 Animal-derived products

Nowadays, animal-derived products like chitosan, gelatin, and whey proteins are widely used as food additives. Chitosan is a natural polymer obtained from chitin, a component of the exoskeleton of shellfish and fungal cell walls. Gelatin is a protein derived from the raw collagen of animal body parts. Whey protein is one of the two proteins, other than casein, found in milk. The bioactive coating of food products with these compounds provides antioxidant and antimicrobial properties and can thereby increase the shelf life of the product. Direct addition of compounds into the packaging materials also provided more potent preservative action [120]. Some of the recent studies on the application of chitosan, gelatin, and whey protein as edible coatings in fish and fish product preservation are represented in Table 3. But in moist environments, edible films and coating showed relatively low stiffness and strength, thus limited their use in specific conditions. Another animal-derived product, bioactive peptide (specific protein fragments) showed antimicrobial [137], and antioxidant activities [138]. In fish paste products, the products like plasma hydrolysate, plasma protein, ovomucoid, egg albumin, egg white, etc. were added as additives for improving strength [119]. The binding effect of egg whites and hydrolyzed beef plasma proteins in surimi gels [139], gel enhancing effect of bovine plasma powder, and egg white powder in arrow tooth surimi [140] were also reported.

Table 3. Some of the recent studies on the application of chitosan, gelatin, and whey protein as edible coatings in fish and fish product preservation.

Fish product	Preservative action	Ref.
Chitosan-based		
Grass carp	Inhibited cathepsin activities and thereby retarded the proteolysis	[121]
Olive flounder	Coating combined with clove oil improved the quality and shelf life was extended by 6 days	[122]
White shrimp	Coating combined with pomegranate peel extract reduced the melanosis, TVBN and microbial count	[123]
Mackerel	Coating combined with gallic acid decreased microbial growth, protein decomposition, biogenic amine formation, lipid oxidation and nucleotide breakdown and shelf life was extended	[124]
Fish burgers	The chitosan film containing lactoperoxidase system suppress the increase of *Pseudomonas* spp. and *Shewanella* spp. and TBARS value	[125]
Mori Fish	Spoilage reduction when coated with rosemary extract on storage	[126]
Gelatin-based		
Rainbow trout fillets	Coating incorporating oregano essential oil minimized the formation of volatile bases, oxidation products, and the growth of total and psychrotrophic bacteria	[127]
Shrimp	Fish gelatin reduced lipid oxidation and extend shelflife	[128]
Shrimp	Coating enriched with orange leaf essential oil extend shelf life about 10 days	[73]
Tilapia fillets	Combined with grape seed extract reduced the formation of undesirable metabolites like TMA and histidine significantly	[129]
Chitosan-gelatin-based		
Golden-pomfret fillets	Inhibits myofibril degradation	[130]

Fish product	Preservative action	Ref.
White shrimp	Decreased the total and psychrotrophic bacteria and increased the shelf-life	[131]
Salmon fillet	Incorporated with garlic and lime extracts extended the shelf life by antibacterial and antioxidant activity	[132]
Minced trout fillet	Incorporated with grape seed extract *Ziziphora clinopodioides* essential oil reduced the spoilage	[133]
Whey protein-based		
Pike-Perch	Coating with lactoperoxidase system and α-tocopherol extended the shelflife significantly	[134]
Rainbow trout fillet	Reduced microbial growth, and TVB-N and TBA	[135]
Rainbow trout fillet	Inhibited *Shewanella and P. fluorescens* and shelf life was extended	[136]

3.2.3 Microbial-derived products

Bacteriocin, a major bacterial-derived bio preservative (mostly from *Lactobacillus*) has potent antimicrobial properties. The mode of action is interfering cell wall synthesis of bacteria by pore formation and squeezing out of the inner material thereby restricting their growth [141]. Along with this antimicrobial action, other properties like nontoxicity, active in a wide range of pH and temperature, etc. making them generally recognized as safe (GRAS) additive [142]. The most common bacteriocins produced by *Lactobacillus* are Nicin, lacticin, pediocin, etc. Many bacteriocins are known to be more effective against endospore-forming bacteria. Bacteriocins were used to reduce the counts of *Salmonella* and *Vibrio* spp. in marine fishes and loligo [143], *Listeria inaqua,* and *Pseudomonas* spp. in fish homogenates [144], and aerobic and anaerobic bacteria in cold smoked salmon [145]. A novel bacteriocin BCC7293 from *Weissella hellenica* showed activity against *L. monocytogenes, S. aureus, P. aeruginosa, A. hydrophila, E. coli,* and *S. Typhimurium* in Pangasius fillets [146]. Bacteriocin FGC-12 and DY4–2 produced by *Lactobacillus plantarum* showed some inhibitory effect on *Vibrio parahaemolyticus* in shrimp [147], and *Pseudomonas fluorescens* in turbot fillet [148] respectively. Bacteriocin LJR1produced by *Pediococcus pentosaceus* showed activity against *L. monocytogenes* in white shrimp [149]. Bacteriocin GP1 produced by *Lactobacillus rhamnosus* active against Coliforms, *Aeromonas*, and *Vibrio* spp. in fish fillets [150]. The combination of bacteriocins with other preservation techniques usually results in better action. Microencapsulated *Ziziphora clinopodioides* essential oil and Nisin showed the strongest effect on preserving the sensorial quality of fish burgers [151]. However the use of bacteriocin is limited due to its high cost. Another microbial-derived product kojic acid, a natural product of many fungi like *Aspergillus* and *Penicillium*, has certain anti-enzymatic browning and antibacterial effects, especially against gram-negative bacteria [152]. A combination of kojic acid and tea polyphenols showed an antibacterial effect against spoilage bacteria in refrigerated seabass (*Lateolabrax japonicas*) [153]. ε-Polylysine is another microbial-derived product with excellent preservative properties. It was isolated originally from bacteria *Streptomyces albulus.* Treatment with ε-Polylysine lowered the TVB-N, putrescine, cadaverine, and hypoxanthine and extended the shelf-life of shrimp [154]. The addition of ε-polylysine chitosan and carrageenan showed shelf life extension of Chinese shrimp (*Fenneropenaeus chinensis*) [155], and chitosan-based coatings combined with ε-polylysine and rosmarinic acid contributed to the reduction of TVB-N, TMA, and ATP-related compounds in Half-smooth tongue sole fillets [156]. A combination of plant, animal, and microbial-derived products showed the strongest preservative action than the independent use.

4. Conclusions

As a perishable food commodity, most of the world's supply of fish and fishery products are lost through chemical and microbial spoilage than other reasons like improper storage, handling and processing damage. Thus, the increasing demand for good quality fish products has intensified the search for applications of additives in preservation strategies. It is well known that none of the additives offer complete protection against spoilage, but can improve the quality of fish as well as shelf life to a greater extent. By considering the potential health hazards associated with chemicals as well as consumer preference, application of natural products from cheap and underutilized resources enabling food safety holds promise.

References

1. FAO yearbook: Fishery and aquaculture statistics 2017. Rome: Food and Agriculture Organization of the United Nations. 2019:50-52
2. FAO. The state of world fisheries and aquaculture 2020. Rome: Food and Agriculture Organization of the United Nations. 2020:60-61. https://doi.org/10.4060/ca9229en
3. Gram L, and Huss HH. Microbiological spoilage of fish and fish products. International journal of food microbiology. 1996;33:121-137
4. Sriket C. Proteases in fish and shellfish: Role on muscle softening and prevention. International Food Research Journal. 2014; 21:433-445
5. Baird-Parker TC. The production of microbiologically safe and stable foods. The microbiological safety and quality of food. 2000;1:3-18
6. Mahmud A, Abraha B, Samuel M, Abraham W, Mahmud E. Fish preservation: A multi-dimensional approach. MOJ Food Processing and Technology. 2018;6:303-310
7. FAO. Post-harvest changes in fish. In: FAO Fisheries and Aquaculture Department, Food and Agriculture Organization, Rome, Italy. 2005
8. Ghaly AE, Dave D, Budge S, Brooks MS. Fish spoilage mechanisms and preservation techniques. American journal of applied sciences. 2010;7:859
9. Mackie IM. The effects of freezing on flesh proteins. Food Reviews International. 1993;9:575-610
10. Gram L, Dalgaard P. Fish spoilage bacteria — Problems and solutions. Current Opinion in Biotechnology. 2002;13:262-266
11. Gram L, Trolle G, Huss HH. Detection of specific spoilage bacteria from fish stored at low (0 C) and high (20 C) temperatures. International journal of food microbiology. 1987;4:65-72
12. Miget RJ. Microbiology of crustacean processing: shrimp, crawfish, and prawns. In Microbiology of marine food products. Springer, Boston, MA.1991:65-87

13. Fieger EA, Novak AF. Microbiology of shellfish deterioration. Fish as food. 1961:561-611

14. Ratkowsky DA, Olley J, McMeekin TA, Ball A. Temperature and bacteria. Journal of Bacteriology. 1982;149:1-5

15. Abbas KA, Saleh AM, Mohamed A, Lasekan O. The relationship between water activity and fish spoilage during cold storage: A review. Journal of Food Agriculture and Environment. 2009:86-90

16. Özogul F, Hamed I. Marine-based toxins and their health risk. InFood Quality: Balancing Health and Disease. Academic Press. 2018:109-144

17. Masniyom P. Deterioration and shelf-life extension of fish and fishery products by modified atmosphere packaging. Songklanakarin Journal of Science & Technology. 2011;33

18. Visciano P, Schirone M, Tofalo R, Suzzi G. Biogenic amines in raw and processed seafood. Frontiers in microbiology. 2012;3:188

19. Howgate P. A review of the kinetics of degradation of inosine monophosphate in some species of fish during chilled storage. International journal of food science & technology. 2006;41:341-353

20. Luong JHT, Male KB, Masson C, Nguyen AL. Hypoxanthine ratio determination in fish extract using capillary electrophoresis and immobilized enzymes. Journal of Food Science. 1992;57:77-81. doi: 10.1111/j.1365-2621.1992.tb05429

21. Kuley E, Özogul F, Özogul Y. Effects of aluminium foil and cling film on biogenic amines and nucleotide degradation products in gutted sea bream stored at 2±1 C. European Food Research and Technology. 2005;221:582-591

22. Syamdidi M. The Use of Chemical Additives for Fisheries Product Preservation. Squalen Bulletin of Marine and Fisheries Postharvest and Biotechnology. 2013;7:79-87

23. Desai MA, Soni KA, Nannapaneni R, Schilling MW, Silva JL. Reduction of Listeria monocytogenes in raw catfish fillets by essential oils and phenolic constituent carvacrol. Journal of Food Science 2012;77:16-22

24. Kaitaranta JK. Control of lipid oxidation in fish oil with various antioxidative compounds. Journal of the American Oil Chemists Society. 1992;69:810-813

25. Taylor SL, Nordlee JA. Chemical additives in seafood products. Clinical reviews in allergy. 1993;11:261-291

26. Veeck AP, Klein B, Ferreira LF, Becker AG, Heldwein CG, Heinzmann BM, Baldisserotto B, Emanuelli T. Lipid stability during the frozen storage of fillets from silver catfish exposed in vivo to the essential oil of Lippia alba (Mill.) NE Brown. Journal of the Science of Food and Agriculture. 2013;93:955-960

27. Patel S. Plant essential oils and allied volatile fractions as multifunctional additives in meat and fish-based food products: a review. Food Additives & Contaminants: Part A. 2015;32:1049-1064

28. Fishery Products Additives Regulations Legal Notice No 65/2003

29. Jeyakumari A. Surimi and surimi based products. Central Institute of Fisheries Technology, Cochin. 2014

30. Lee CM. Ingredient and formulation technology for surimi-based products. Surimi Technology. 1992:273-302

31. Gonçalves AA, Ribeiro JL. Optimization of the freezing process of red shrimp (Pleoticus muelleri) previously treated with phosphates. International Journal of refrigeration. 2008;31:1134-1144

32. Gonçalves AA, de Oliveira AR. Melanosis in crustaceans: A review. LWT-Food Science and Technology. 2016;65:791-799

33. DeWitt CA, Oliveira A. Modified atmosphere systems and shelf life extension of fish and fishery products. Foods. 2016;5:48

34. Gonçalves AA, Souza MA, Regis RC. Effects of different levels of food additives on weight gain, cook-related yield loss, physicochemical and sensorial quality of Nile tilapia fillets (Oreochromis niloticus). International Food Research Journal. 2018;25

35. Chantarasuwan C, Benjakul S, Visessanguan W. Effects of sodium carbonate and sodium bicarbonate on yield and characteristics of Pacific white shrimp (Litopenaeus vannamei). Food science and technology international. 2011;17:403-414

36. Fossati AA, Bergmann GP, Ribeiro LA, Júnior DP, Schneider TM, Kindlein L. Effects of different additives on colorimetry and melanosis prevention of Atlantic seabob shrimp (Xyphopenaeus kroyeri) stored under refrigeration. International Journal of Fisheries and Aquaculture. 2016;8:74-80

37. Fan H, Luo Y, Yin X, Bao Y, Feng L. Biogenic amine and quality changes in lightly salt-and sugar-salted black carp (Mylopharyngodon piceus) fillets stored at 4 C. Food Chemistry. 2014;159:20-28

38. Sych J, Lacroix C, Adambounou LT, Castaigne F. The effect of lowor non-sweet additives on the stability of protein functional properties of frozen cod surimi. Journal of Food Science and Technology. 1991;26: 185-197

39. Sultanbawa Y, Li-Chan EC. Cryoprotective effects of sugar and polyol blends in ling cod surimi during frozen storage. Food research international. 1998;31:87-98

40. Lee CM. Surimi process technology. Food technology (Chicago). 1984;38:69-80

41. Bover-Cid S, Izquierdo-Pulido M, Vidal-Carou MC. Changes in biogenic amine and polyamine contents in slightly fermented sausages manufactured with and without sugar. Meat Science. 2001;57:215-221

42. Zhang J, Liu Z, Hu Y, Fang Z, Chen J, Wu D, Ye X. Effect of sucrose on the generation of free amino acids and biogenic amines in Chinese traditional dry-cured fish during processing and storage. Journal of food science and technology. 2011;48:69-75

43. Toyoda K. The surimi manufacturing process. Surimi technology. 1992:79-112

44. Lyhs U, Lahtinen J, Fredriksson-Ahomaa M, Hyytiä-Trees E, Elfing K, Korkeala H. Microbiological quality and shelf-life of vacuum-packaged 'gravad'rainbow trout stored at 3 and 8 C. International Journal of Food Microbiology. 2001;70:221-230

45. Kilinc B, Cakli S. Chemical, enzymatical and textural changes during marination and storage period of sardine (Sardina pilchardus) marinades. European Food Research and Technology. 2005;221:821-827

46. Young KM, Foegeding PM. Acetic, lactic and citric acids and pH inhibition of Listeria monocytogenes Scott A and the effect on intracellular pH. The Journal of applied bacteriology. 1993;74:515-520

47. Conner DE, Scott VN, Bernard DT. Growth, inhibition, and survival of Listeria monocytogenes as affected by acidic conditions. Journal of Food Protection. 1990;53:652-655

48. Agrafioti PT, Katsanidis E. Effects of additives on the selected quality attributes and cooking yield of squid: Modelling and optimization. International Journal of food properties. 2012;15:579-589

49. Birkeland S, Rotabakk BT. Effects of Additives and Packaging Method on Quality and Microbiological Characteristics in Mild Thermal Processed Fish Mince. Journal of Aquatic Food Product Technology. 2014;23:368-384

50. Sallam KI. Antimicrobial and antioxidant effects of sodium acetate, sodium lactate, and sodium citrate in refrigerated sliced salmon. Food control. 2007;18:566-575

51. Vogel BF, Ng YY, Hyldig G, Mohr M, Gram L. Potassium lactate combined with sodium diacetate can inhibit growth of Listeria monocytogenes in vacuum-packed cold-smoked salmon and has no adverse sensory effects. Journal of food protection. 2006;69:2134-2142

52. Williams SK, Rodrick GE, West RL. Sodium lactate affects shelf life and consumer acceptance of fresh catfish (Ictalurus nebulosus, marmoratus) fillets under simulated retail conditions. Journal of Food Science. 1995;60:636-639

53. Hui YH. Handbook of food science, technology, and engineering. CRC press. 2006

54. Wangtueai S, Tongsiri S, Maneerote J, Supaviriyakorn W. Effect of phosphate on frozen Nile tilapia fillets. Food and Applied Bioscience Journal. 2014;2:203-215

55. Lampila LE. Functions and uses of phosphates in the seafood industry. Journal of Aquatic Food Product Technology. 1993;1:29-41

56. Chang CC, Regenstein JM. Water uptake, protein solubility, and protein changes of cod mince stored on ice as affected by polyphosphates. Journal of Food Science. 1997;62:305-309

57. Gonçalves AA. Phosphates for seafood processing. Phosphates: Sources, Properties, and Applications. Hauppauge. 2012:83-112

58. Kilinc B, Cakli S, Cadun A, Sen B. Effects of phosphate dip treatments on chemical, microbiological, color, textural, and sensory changes of rainbow trout (Onchorhyncus mykiss) fillets during refrigerated storage. Journal of Aquatic Food Product Technology. 2009;18:108-119

59. Moawad RK, Ashour MM, Mohamed GF, El-Hamzy EM. Effect of food grade trisodium phosphate or water dip treatments on some quality attributes of decapitated white marine shrimp (Penaeus spp.) during frozen storage. Journal of Applied Sciences Research. 2013;9:3723-3734

60. Carneiro CD, Mársico ET, Ribeiro RD, Conte Junior CA, Álvares TS, De Jesus EF. Quality Attributes in Shrimp Treated with Polyphosphate after Thawing and Cooking: A Study Using Physicochemical Analytical Methods and Low-Field 1 H NMR. Journal of Food Process Engineering. 2013;36:492-499

61. Etemadian Y, Shabanpour B, Mahoonak AS, Shabani A. Combination effect of phosphate and vacuum packaging on quality parameters of Rutilus frisii kutum fillets in ice. Food Research International. 2012;45:9-16

62. Peranginangin R, Fawzya YN, Sugiyono S, Muljanah I. Food additives and effect of thickness on fish crackers quality.1997:106-114

63. Tompkin RB, Branen AL, Davidson PM. Antimicrobials in food. Nitrite. Davidson, PM. 2005:169-236

64. Einarsson H, Lauzon HL. Biopreservation of brined shrimp (Pandalus borealis) by bacteriocins from lactic acid bacteria. Applied and Environmental Microbiology. 1995;61:669-676

65. Benkeblia, N. Antimicrobial activity of essential oil extracts of various onions (Allium cepa) and garlic (Allium sativum). LWT-Food Science and Technology. 2004;37:263-268. https://doi.org/10.1016/j.lwt.2003.09.001

66. Chouliara E, Karatapanis A, Savvaidis I, Kontominas M. Combined effect of oregano essential oil and modified atmosphere packaging on shelf-life extension of fresh chicken breast meat, stored at 4 °C. Food Microbiology. 2007;24:607-617

67. Bajpai VK, Rahman A, Kang SC. Chemical composition and inhibitory parameters of essential oil and extracts of Nandina domestica Thunb. to control food-borne pathogenic and spoilage bacteria. International Journal of Food Microbiology. 2008;125:117-122

68. Guran HS, Oksuztepe G, Coban OE, Incili GK. Influence of different essential oils on refrigerated fish patties produced from bonito fish (Sarda sarda Bloch, 1793). Czech Journal of Food Sciences. 2015;33:37-44

69. Karoui R, Hassoun A. Efficiency of rosemary and basil essential oils on the shelf-life extension of Atlantic mackerel (Scomber scombrus) fillets stored at 2 C. Journal of AOAC International. 2017;100:335-344

70. Ozogul Y, Yuvka İ, Ucar Y, Durmus M, Kösker AR, Öz M, Ozogul F. Evaluation of effects of nanoemulsion based on herb essential oils (rosemary, laurel, thyme and sage) on sensory, chemical and microbiological quality of rainbow trout (Oncorhynchus mykiss) fillets during ice storage. Lebensmittel-Wissenschaft & Technologie. 2017;75:677-684

71. Abdeldaiem MH, Ali HG, Ramadan MF. Impact of different essential oils on the characteristics of refrigerated carp (Cyprinus carpio) fish fingers. Journal of Food Measurement and Characterization. 2017;11:1412-1420

72. Myszka K, Olejnik A, Majcher M, Sobieszczańska N, Grygier A, Powierska-Czarny J, Rudzińska M. Green pepper essential oil as a biopreservative agent for fish-based products: Antimicrobial and antivirulence activities against Pseudomonas aeruginosa KM01. Lebensmittel-Wissenschaft & Technologie. 2019;108:6-13

73. Alparslan Y, Yapıcı HH, Metin C, Baygar T, Günlü A, Baygar T. Quality assessment of shrimps preserved with orange leaf essential oil incorporated gelatin. LWT-Food Science and Technology. 2016;72:457-466

74. Navarro-Segura L, Ros-Chumillas M, López-Cánovas AE, García-Ayala A, López-Gómez A. Nanoencapsulated essential oils embedded in ice improve the quality and shelf life of fresh whole seabream stored on ice. Heliyon. 2019;5:1804

75. Huang Z, Liu X, Jia S, Zhang L, Luo Y. The effect of essential oils on microbial composition and quality of grass carp (Ctenopharyngodon idellus) fillets during chilled storage. International journal of food microbiology. 2018;266:52-59

76. Remya S, Mohan CO, Venkateshwarlu G, Sivaraman GK, Ravishankar CN. Combined effect of O2 scavenger and antimicrobial film on shelf life of fresh cobia (Rachycentron canadum) fish steaks stored at 2 C. Food Control. 2017;71:71-78

77. Cai L, Cao A, Li Y, Song Z, Leng L, Li J. The effects of essential oil treatment on the biogenic amines inhibition and quality preservation of red drum (Sciaenops ocellatus) fillets. Food Control. 2015;56:1-8

78. Ozpolat E, Duman M. Effect of black cumin oil (Nigella sativa L.) on fresh fish (Barbus grypus) fillets during storage at 2±1 C. Food Science and Technology. 2017;37:148-152

79. Giarratana F, Muscolino D, Beninati C, Ziino G, Giuffrida A, Panebianco A. Activity of R (+) limonene on the maximum growth rate of fish spoilage organisms and related effects on shelf-life prolongation of fresh gilthead sea bream fillets. International journal of food microbiology. 2016;237:109-113

80. MuSColino D, GiArrAtAnA F, BEninAti C, Ziino G, GiuFFriDA A, PAnEBiAnCo A. Effects of allyl isothiocyanate on the shelf-life of gilthead sea bream (Sparus aurata) fillets. Czech Journal of Food Sciences. 2016;34:160-165

81. Pedrós-Garrido S, Clemente I, Calanche JB, Condón-Abanto S, Beltrán JA, Lyng JG, Brunton N, Bolton D, Whyte P. Antimicrobial activity of natural compounds against Listeria spp. and their effects on sensory attributes in salmon (Salmo salar) and cod (Gadus morhua). Food Control. 2020;107:106768

82. Maghami M, Motalebi AA, Anvar SA. Influence of chitosan nanoparticles and fennel essential oils (Foeniculum vulgare) on the shelf life of Huso huso fish fillets during the storage. Food science & nutrition. 2019;7:3030-3041

83. Choulitoudi E, Bravou K, Bimpilas A, Tsironi T, Tsimogiannis D, Taoukis P, Oreopoulou V. Antimicrobial and antioxidant activity of Satureja thymbra in gilthead seabream fillets edible coating. Food and bioproducts processing. 2016;100:570-577

84. Alfonzo A, Martorana A, Guarrasi V, Barbera M, Gaglio R, Santulli A, Settanni L, Galati A, Moschetti G, Francesca N. Effect of the lemon essential oils on the safety and sensory quality of salted sardines (Sardina pilchardus Walbaum 1792). Food Control. 2017;73:1265-1274

85. Huang Z, Liu X, Jia S, Luo Y. Antimicrobial effects of cinnamon bark oil on microbial composition and quality of grass carp (Ctenopharyngodon idellus) fillets during chilled storage. Food Control. 2017;82:316-324

86. Raeisi S, Quek SY, Ojagh SM, Alishahi AR. Effects of Cumin (Cuminum cyminum L.) Seed and Wild Mint (Mentha Longifolia L.) Leaf Extracts on the Shelf Life and Quality of Rainbow Trout (Oncorhynchus Mykiss) Fillets Stored at 4C±1. Journal of Food Safety. 2016;36:271-281

87. Raeisi S, Sharifi-Rad M, Quek SY, Shabanpour B, Sharifi-Rad J. Evaluation of antioxidant and antimicrobial effects of shallot (Allium ascalonicum L.) fruit and ajwain (Trachyspermum ammi (L.) Sprague) seed extracts in semi-fried coated rainbow trout (Oncorhynchus mykiss) fillets for shelf-life extension. LWT-Food Science and Technology. 2016;65:112-121

88. Raeisi S, Ojagh SM, Sharifi-Rad M, Sharifi-Rad J, Quek SY. Evaluation of Allium paradoxum (MB) G. Don. and Eryngium caucasicum trauve. Extracts on the shelf-life and quality of silver carp (Hypophthalmichthys molitrix) fillets during refrigerated storage. Journal of Food Safety. 2017;37:12321

89. Tayel AA, Almabady NA, Sorour NM, Diab AM. Application of natural plant extracts as colorants, preservatives, and anti-listerial agents in processed fish products. Journal of Food Safety. 2018;38:12435

90. Safrida S, Hardania DI, Khairil K, Muhammad N, Nur YI. Evaluation natural preservatives of Nothopanax scutellarium merr. leaf extract in physical characteristics of nile tilapia (Orechromis niloticus). InIOP Conference Series: Earth and Environmental Science. IOP Publishing 2021;667:12060

91. Viji P, Binsi PK, Visnuvinayagam S, Bindu J, Ravishankar CN, Gopal TK. Efficacy of mint (Mentha arvensis) leaf and citrus (Citrus aurantium) peel extracts as natural preservatives for shelf life extension of chill stored Indian mackerel. Journal of food science and technology. 2015;52:6278-6289

92. Murhekar S, Wright MH, Greene AC, Brownlie JC, Cock IE. Inhibition of Shewanella spp. growth by Syzygium australe and Syzygium luehmannii extracts: natural methods for the prevention of fish spoilage. Journal of food science and technology. 2017;54:3314-3326

93. Wright MH, Shalom J, Matthews B, Greene AC, Cock IE. Terminalia ferdinandiana Exell: Extracts inhibit Shewanella spp. growth and prevent fish spoilage. Food microbiology. 2019;78:114-122

94. Abdel-Wahab M, El-Sohaimy SA, Ibrahim HA, El-Makarem HS. Evaluation the efficacy of clove, sage and kiwifruit peels extracts as natural preservatives for fish fingers. Annals of Agricultural Sciences. 2020;65:98-106

95. Martínez L, Castillo J, Ros G, Nieto G. Antioxidant and antimicrobial activity of rosemary, pomegranate and olive extracts in fish patties. Antioxidants. 2019;8:86

96. Lekshmi RG, Jayathilakan K, Sarika K, Priya ER, Greeshma SS, Sultana K, Tejpal CS, Mathew S. Effect of Plectranthus amboinicus leaf extract on the quality attributes of microencapsulated fish oil fortified soup powder. Krishi. 2019

97. Hatab S, Lin K, Miao W, Chen M, Lin J, Deng S. Potential Utilization of Green Tea Leaves and Fenugreek Seeds Extracts as Natural Preservatives for Pacific White Shrimp During Refrigerated Storage. Foodborne pathogens and disease. 2018;15:498-505

98. Shofiani S, Rostini I, Afrianto E. The Effect of Concentrations of Basil Leaves Extract as Natural Preservatives in Mullet Fillet on Bacterial Growth in Low-Temperature Storage. Asian Journal of Fisheries and Aquatic Research. 2020:30-41

99. Stevanović ZD, Bošnjak-Neumüller J, Pajić-Lijaković I, Raj J, Vasiljević M. Essential oils as feed additives—future perspectives. Molecules. 2018;23:1717

100. Borderías J, Montero P, Marti de Castro MA. Gelificación de serrín de merluza (Merluccius australis)/Gelling of hake (Merluccius australis) sawdust. Food Science and Technology International. 1996;2:293-299

101. Gómez-Guillén C, Borderías AJ, Montera P. Thermal gelation properties of two different composition sardine (Sardina pilchardus) muscles with addition of non-muscle proteins and hydrocolloids. Food Chemistry. 1997;58:81-87

102. Park JW. Ingredient technology and formulation development. Food Science and Technology-New York-Marcel Dekker-. 2000:343-392

103. Guldas M, Hecer C. Influences of the selected additives on the weight loss and organoleptic properties of marinated mussels and squids. Acta Veterinaria Brno. 2013;81:263-267

104. Hao R, Liu Y, Sun L, Xia L, Jia H, Li Q , Pan J. Sodium alginate coating with plant extract affected microbial communities, biogenic amine formation and quality properties of abalone (Haliotis discus hannai Ino) during chill storage. LWT-Food Science and Technology. 2017;81:1-9

105. Cai L, Wang Y, Cao A, Lv Y, Li J. Effect of alginate coating enriched with 6-gingerol on the shelf life and quality changes of refrigerated red seabream (Pagrosomus major) fillets. RSC Advances. 2015;5:36882-36889

106. Nie X, Wang L, Wang Q, Lei J, Hong W, Huang B, Zhang C. Effect of a sodium alginate coating infused with tea polyphenols on the quality of fresh japanese sea bass (Lateolabrax japonicas) fillets. Journal of food science. 2018;83:1695-1700

107. Liu X, Jia Y, Hu Y, Xia X, Li Y, Zhou J, Liu Y. Effect of Citrus wilsonii Tanaka extract combined with alginate-calcium coating on quality maintenance of white shrimps (Litopenaeus vannamei Boone). Food control. 2016;68:83-91

108. Balti R, Mansour MB, Zayoud N, Le Balc'h R, Brodu N, Arhaliass A, Massé A. Active exopolysaccharides based edible coatings enriched with red seaweed (Gracilaria gracilis) extract to improve shrimp preservation during refrigerated storage. Food Bioscience. 2020;34:100522

109. Arulkumar A, Paramasivam S, Miranda JM. Combined effect of icing medium and red alga Gracilaria verrucosa on shelf life extension of Indian Mackerel (Rastrelliger kanagurta). Food and Bioprocess Technology. 2018;11:1911-1922

110. Arulkumar A, Satheeshkumar K, Paramasivam S, Rameshthangam P, Miranda JM. Chemical Biopreservative Effects of Red Seaweed on the Shelf Life of Black Tiger Shrimp (Penaeus monodon). Foods. 2020;9:634

111. Yan XJ. Quantitative determination of phlorotannins from some Chinese common brown seaweeds. Studia Marina Sinica. 1996;37:61-65

112. Athukorala Y, Lee KW, Song C, Ahn CB, Shin TS, Cha YJ, Shahidi F, Jeon YJ. Potential antioxidant activity of marine red alga Grateloupia filicina extracts. Journal of Food Lipids. 2003;10:251-265

113. Wang T, Jónsdóttir R, Kristinsson HG, Thorkelsson G, Jacobsen C, Hamaguchi PY, Ólafsdóttir G. Inhibition of haemoglobin-mediated lipid oxidation in washed cod muscle and cod protein isolates by Fucus vesiculosus extract and fractions. Food Chemistry. 2010;123:321-330

114. Ortiz J, Vivanco JP, Aubourg SP. Lipid and sensory quality of canned Atlantic salmon (Salmo salar): Effect of the use of different seaweed extracts as covering liquids. European Journal of Lipid Science and Technology. 2014;116:596-605

115. Maqsood S, Benjakul S, Shahidi F. Emerging role of phenolic compounds as natural food additives in fish and fish products. Critical reviews in food science and nutrition. 2013;53:162-179

116. Park JW. Temperature-tolerant fish protein gels using konjac flour. Journal of Muscle Foods. 1996;7:165-174

117. Montero P, Hurtado JL, Pérez-Mateos M. Microstructural behaviour and gelling characteristics of myosystem protein gels interacting with hydrocolloids. Food Hydrocolloids. 2000;14:455-461

118. Da Ponte DJ, Roozen JP, Pilnik W. Effects of iota carrageenan, carboxymethyl cellulose and xanthan gum on the stability of formulated minced fish products. International Journal of Food Science & Technology. 1987;22:123-133

119. Bashir KM, Kim JS, An JH, Sohn JH, Choi JS. Natural food additives and preservatives for fish-paste products: a review of the past, present, and future states of research. Journal of Food Quality. 2017

120. Ahmed J, Mulla M, Arfat YA. Mechanical, thermal, structural and barrier properties of crab shell chitosan/graphene oxide composite films. Food Hydrocolloids. 2017;71:141-148

121. Yu D, Regenstein JM, Zang J, Xia W, Xu Y, Jiang Q, Yang F. Inhibitory effects of chitosan-based coatings on endogenous enzyme activities, proteolytic degradation and texture softening of grass carp (Ctenopharyngodon idellus) fillets stored at 4 C. Food chemistry. 2018;262:1-6

122. Li XP, Zhou MY, Liu JF, Xu YX, Mi HB, Yi SM, Li JR, Lin H. Shelf-life extension of chilled olive flounder (Paralichthys olivaceus) using chitosan coatings containing clove oil. Journal of food processing and preservation. 2017;41:13204

123. Yuan G, Lv H, Tang W, Zhang X, Sun H. Effect of chitosan coating combined with pomegranate peel extract on the quality of Pacific white shrimp during iced storage. Food Control. 2016;59:818-823

124. Wu C, Li Y, Wang L, Hu Y, Chen J, Liu D, Ye X. Efficacy of chitosan-gallic acid coating on shelf life extension of refrigerated Pacific mackerel fillets. Food and Bioprocess Technology. 2016;9:675-685

125. Ehsani A, Hashemi M, Afshari A, Aminzare M, Raeisi M, Zeinali T. Effect of different types of active biodegradable films containing lactoperoxidase system or sage essential oil on the shelf life of fish burger during refrigerated storage. LWT. 2020;117:108633

126. Nawaz T, Fatima M, Shah SZ, Afzal M. Coating effect of rosemary extract combined with chitosan on storage quality of mori (Cirrhinus mrigala). Journal of Food Processing and Preservation. 2020;44:14833

127. Hosseini SF, Rezaei M, Zandi M, Ghavi FF. Effect of fish gelatin coating enriched with oregano essential oil on the quality of refrigerated rainbow trout fillet. Journal of Aquatic Food Product Technology. 2016;25:835-842

128. Mirzapour-Kouhdasht A, Moosavi-Nasab M. Shelf-life extension of whole shrimp using an active coating containing fish skin gelatin hydrolysates produced by a natural protease. Food science & nutrition. 2020;8:214-223

129. Zhao X, Wu JE, Chen L, Yang H. Effect of vacuum impregnated fish gelatin and grape seed extract on metabolite profiles of tilapia (Oreochromis niloticus) fillets during storage. Food chemistry. 2019;293:418-428

130. Feng X, Bansal N, Yang H. Fish gelatin combined with chitosan coating inhibits myofibril degradation of golden pomfret (Trachinotus blochii) fillet during cold storage. Food Chemistry. 2016;200:283-292

131. Farajzadeh F, Motamedzadegan A, Shahidi SA, Hamzeh S. The effect of chitosan-gelatin coating on the quality of shrimp (Litopenaeus vannamei) under refrigerated condition. Food Control. 2016;67:163-170

132. Thaker M, Hanjabam MD, Gudipati V, Kannuchamy N. Protective Effect of Fish Gelatin-Based Natural Antimicrobial Coatings on Quality of Indian Salmon Fillets during Refrigerated Storage. Journal of Food Process Engineering. 2017;40:12270

133. Kakaei S, Shahbazi Y. Effect of chitosan-gelatin film incorporated with ethanolic red grape seed extract and Ziziphora clinopodioides essential oil on survival of Listeria monocytogenes and chemical, microbial and sensory properties of minced trout fillet. LWT-Food Science and Technology. 2016;72:432-438

134. Shokri S, Ehsani A. Efficacy of whey protein coating incorporated with lactoperoxidase and α-tocopherol in shelf life extension of Pike-Perch fillets during refrigeration. LWT-Food Science and Technology. 2017;85:225-231

135. Erbay EA, Dağtekin BB, Türe M, Yeşilsu AF, Torres-Giner S. Quality improvement of rainbow trout fillets by whey protein isolate coatings containing electrospun poly (ε-caprolactone) nanofibers with Urtica dioica L. extract during storage. Lwt. 2017;78:340-351

136. Shokri S, Ehsani A, Jasour MS. Efficacy of lactoperoxidase system-whey protein coating on shelf-life extension of rainbow trout fillets during cold storage (4 C). Food and Bioprocess Technology. 2015;8:54-62

137. Mosquera M, Gimenez B, Ramos S, Lopez-Caballero ME, Gomez-Guill ′ en, MdC, Montero P. Antioxidant, ACE-Inhibitory, and antimicrobial activities of peptide fractions obtained from dried giant squid tunics. Journal of Aquatic Food Product Technology. 2016;25:444-455. https://doi.org/10.1080/10498850.2013.819543

138. Irshad I, Kanekanian A, Peters A, Masud T. Antioxidant activity of bioactive peptides derived from bovine casein hydrolysate fractions. Journal of Food Science and Technology. 2015;52:231-239

139. Park JW. Functional protein additives in surimi gels. Journal of Food Science. 1994;59:525-527

140. Wasson DH, Reppond KD, Babbitt JK, French JS. Effects of additives of proteolytic and functional properties of arrowtooth flounder surimi. Journal of Aquatic Food Product Technology. 1993;1:147-165

141. Sharma G, Dang S, Gupta S, Gabrani R. Antibacterial activity, cytotoxicity, and the mechanism of action of bacteriocin from Bacillus subtilis GAS101. Medical Principles and Practice. 2018;27:186-192

142. Woraprayote W, Malila Y, Sorapukdee S, Swetwiwathana A, Benjakul S, Visessanguan W. Bacteriocins from lactic acid bacteria and their applications in meat and meat products. Meat Science. 2016;120:118-132

143. Ashwitha A, Thamizharasan K, Vithya V, Karthik R. Effectiveness of bacteriocin from Bacillus subtilis (KY808492) and its application in biopreservation. Journal of Fisheries Sciences com. 2017;11:36-42

144. Schelegueda LI, Gliemmo MF, Campos CA. Antimicrobial synergic effect of chitosan with sodium lactate, nisin or potassium sorbate against the bacterial flora of fish. Journal of Food Research. 2012;1:272-281

145. Ye M, Neetoo H, Chen H. Effectiveness of chitosan-coated plastic films incorporating antimicrobials in inhibition of Listeria monocytogenes on cold-smoked salmon. International Journal of Food Microbiology. 2008;127:235-240

146. Woraprayote W, Pumpuang L, Tosukhowong A, Zendo T, Sonomoto K, Benjakul S, Visessanguan W. Antimicrobial biodegradable food packaging impregnated with Bacteriocin 7293 for control of pathogenic bacteria in pangasius fish fillets. Lebensmittel-Wissenschaft & Technologie. 2018;89:427-433

147. Lv X, Du J, Jie Y, Zhang B, Bai F, Zhao H, Li J. Purification and antibacterial mechanism of fish-borne bacteriocin and its application in shrimp (Penaeus vannamei) for inhibiting Vibrio parahaemolyticus. World Journal of Microbiology and Biotechnology. 2017;33:1-2

148. Lv X, Ma H, Sun M, Lin Y, Bai F, Li J, Zhang B. A novel bacteriocin DY4-2 produced by Lactobacillus plantarum from cutlassfish and its application as bio-preservative for the control of Pseudomonas fluorescens in fresh turbot (Scophthalmus maximus) fillets. Food Control. 2018;89:22-31

149. Ladha G, Jeevaratnam K. Characterization of purified antimicrobial peptide produced by Pediococcus pentosaceus LJR1, and its application in preservation of white leg shrimp. World Journal of Microbiology and Biotechnology. 2020;36:1-2

150. Sarika AR, Lipton AP, Aishwarya MS. Biopreservative efficacy of bacteriocin GP1 of Lactobacillus rhamnosus GP1 on stored fish filets. Frontiers in nutrition. 2019;6:29

151. Shahinfar R, Khanzadi S, Hashami M, Azizzadeh M, Bostan A. The effect of Ziziphora clinopodioides essential oil and nisin on chemical and microbial characteristics of fish burger during refrigerated storage. Iranian Journal of Chemistry and Chemical Engineering (IJCCE). 2017;36:65-75

152. Wu Y, Shi YG, Zeng LY, Pan Y, Huang XY, Bian LQ, Zhu YJ, Zhang RR, Zhang J. Evaluation of antibacterial and anti-biofilm properties of kojic acid against five food-related bacteria and related subcellular mechanisms of bacterial inactivation. Food Science and Technology International. 2019;25:3-15

153. Wang R, Hu X, Agyekumwaa AK, Li X, Xiao X, Yu Y. Synergistic effect of kojic acid and tea polyphenols on bacterial inhibition and quality maintenance of refrigerated sea bass (Lateolabrax japonicus) fillets. Lebensmittel-Wissenschaft & Technologie. 2021;137:110452

154. Jia S, Liu Y, Zhuang S, Sun X, Li Y, Hong H, Lv Y, Luo Y. Effect of ε-polylysine and ice storage on microbiota composition and quality of Pacific white shrimp (Litopenaeus vannamei) stored at 0° C. Food microbiology. 2019;83:27-35

155. Zhang Z, Xia G, Yang Q, Fan X, Lyu S. Effects of chitosan-based coatings on storage quality of Chinese shrimp. Food science & nutrition. 2019;7:4085-4094

156. Li NA, Liu W, Shen Y, Mei J, Xie J. Coating effects of ε-polylysine and rosmarinic acid combined with chitosan on the storage quality of fresh half-smooth tongue sole (Cynoglossus semilaevis Günther) fillets. Coatings. 2019;9:273

Chapter-8

Active and Intelligent Packaging of Cheese: Developments and Future Scope

1. Introduction

Packaging industry stands at third position globally, next to food and petroleum industries contributing nearly 2% of Gross National Product in developed nations [1]. Approximately 51% of all packaging applications are dedicated to food sector [2]. Consumer inclination towards safe and healthy food have led to the development of state-of-the-art and unique approaches in food processing and packaging. One such development is the introduction of smart packaging technologies. Smart packaging although interchangeably used for intelligent packaging at times, refers to combination of active and intelligent packaging [3]. The Framework Regulation on Food Contact Materials (1935/2004) defines "*active materials and articles*" as materials intended to extended the shelf-life or to maintain or improve the condition of packaged food; they are designed to deliberately incorporate components that would release or absorb substances into or from the packaged food or the environment surrounding the food. Similarly, according to Framework Regulation (EC) No. 1935/2004 materials and articles which monitors the condition of packaged food or the environment surrounding the food are defined as "*intelligent materials and articles*" [2]. Most of the smart packaging interventions in food sector are limited to fruits and vegetables, fish products, meat and seafood [4] indicating huge scope and potential to be explored for dairy products.

Active and intelligent packaging market was estimated at 17.50 billion US $ in 2019 and expected to reach at 25.16 billion US $ by 2025 witnessing a CAGR of 6.78%. Asia Pacific region was identified as the fastest growing market including China, Japan, India and South Korea and North America as the largest market with WestRock®, Honeywell®, BASF® and Amcor Ltd. as the major market players. Oxygen and moisture scavengers are the utmost commercialized forms of active packaging. Gas scavengers for food was the most marketed active packaging technique in USA during 2018–2019 [5]. During past ten years, the research interestedness in active and intelligent packaging has increased steadily as indicated by the trend of peer-reviewed publications in Figure 1 during 2010–2019. As per a survey conducted by O'Callaghan and Kerry (2016) [6] for applicability of smart packaging to cheese, the future is highly optimistic with consumers willing to pay more on receiving the information provided by these advanced technologies. However, to the best of our knowledge, not a single article has reviewed the application and future research directions of smart packaging technologies in cheese. Therefore, the present review offers insight to active and intelligent packaging systems for cheese and future research aspects.

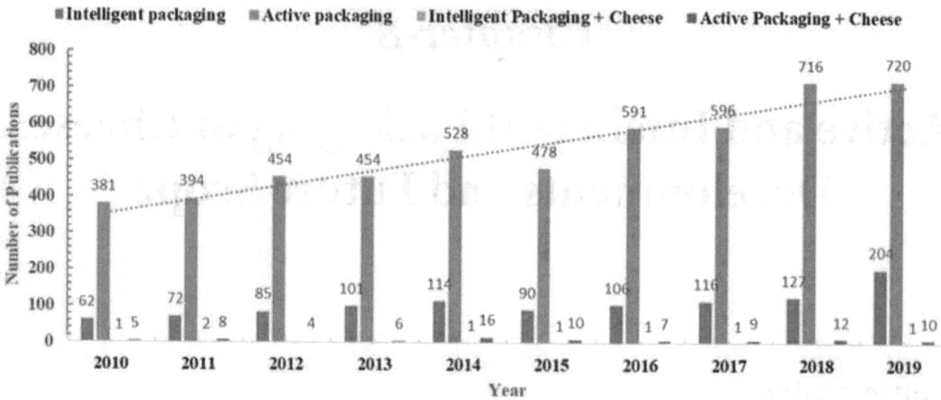

Figure 1. Graph illustrating the number of publications on active packaging, intelligent packaging and cheese during the year 2010–2019 (Source: compiled from SCOPUS using document search with title, abstract, keywords).

2. Status of Cheese Market

World cheese production has shown significant increase from 5.43 million tonnes in 1961, 14.58 million tonnes in 1995 to 22.65 million tonnes in 2015 [7]. About 3000 varieties of cheeses are produced throughout the world and the annual total cheese consumption during 2015–2028 is expected to grow at a CAGR of 1.4% [8]. EU 28 (European Union consisting of 28 countries) stood at first position in cheese export by exporting 841.8 thousand tonnes of cheese. The USA accounted for almost 20% of the world's cheese production and exported 348.5 thousand tonnes of cheese contributing 13.8% of the total export share during 2018 while Japan and Russia were the top export destination [8]. Approximately 40% of world's milk is converted to cheese with France, USA, Iceland, Finland and other developed nations being the major players in cheese production and consumption [7]. The total cheese production in USA was 5,908 million kg, with an import of 176 million kg [8]. Mozzarella is the highest produced cheese variety in USA and several other major cheese producing nations [9]. Additionally, the retail prices of cheese in almost all the countries had shown an upsurge during last ten years [8]. The detailed information about cheese production, consumption, import, export quantity of several countries and retail price of selected cheeses are presented in Table 1. The total whole cow milk cheese in India was 2250 tonnes in 2014 [7]. It is true that India is not a traditionally structured 'cheese nation' but it is gaining pace with increased domestic consumption and exports. India offers only 40 varieties of cheese of which about 60 per cent of the market is dominated by processed cheeses, 30 per cent by cheese spreads and the remaining 10 per cent by flavored and Mozzarella cheese [10].

Table 1. Cheese production, consumption, imports, exports (in '000 tonnes) and retail price during 2018–2019.

Country	Production	Consumption	Imports	Exports	Retail Price		
					Cheese type	Currency	Price/kg
EU28	9376	9652	59 (H)	842 (H)			
Germany	2339	2002	32	130	Gouda	EUR	5.98
France	1725 (A)	1721	—	117	Emmental	EUR	8.43
Italy	1101 (A)	1320	10	100	Mozzarella	EUR	4.46
Netherlands	880 (A)	420	—	140	Gouda	EUR	10.98
Poland	825	723	—	53	Gouda	PLN	20.69
Denmark	452	166	—	73			
United Kingdom	426	795	—		Cheddar	GBP	7.28
Ireland	224	31	—	49	NS	EUR	9.60
Austria	200	200	—	—			
Spain	179 (A)	416	—	—	NS	EUR	8.60
Czech Republic	135	201	—	—	Edam	CZK	144.73
Belgium	109	164	—	—	NS	EUR	9.65
Lithuania	102	58		—	Tilsit	EUR	7.34
Finland	87	142	—	—	Edam	EUR	9.08
Hungary	84	129	—	—	Trappist	HUF	1700.00
Sweden	82	201	—	—	Herrgardsost	SEK	90
Latvia	47	39	—	—	Hard cheese	EUR	7.89
Estonia	45	32	—	—	Gouda	EUR	8.24
Slovakia	38 (A)	74	—	—	Edam	EUR	6.55
Cyprus	3 (A)	22	—	—		—	
Luxemburg	3	16	—	—	—	—	—
Other EU	—	—	17	179	—	—	—
North and Central America							
USA	5908	5668	176	348	Cheddar	USD	11.87
Canada	443	538	31	—	NS	CAD	14.70
Mexico	419	539	123	—	—	—	—
El Salvador	—	—	39	—	—	—	—
Nicaragua	—	—	—	41	—	—	—
South America							
Brazil	755	781	—	—	Mozzarella	BRL	30.49
Argentina	579	574	—	49	Quartirolo-type	ARS	184.24
Chile	101 (B)	198	—	—	Gouda	CLP	6396.00
Colombia	97	100	—	—	—	—	—
Uruguay	45	33	—	—	NS	UYU	143.22
Other Europe							
Russia	473	811	263	—	NS	RUB	412.60

Country	Production	Consumption	Imports	Exports	Retail Price		
					Cheese type	Currency	Price/kg
Belarus	332	128	—	210	—	—	—
Switzerland	190 (A)	186	62	68	NS	CHF	13.32
Ukraine	168	198	—	—	Russian (50% fat)	UAH	172.00
Norway	82 (C)	101	—	—	—	—	—
Iceland	11	9	—	—	—	—	—
Asia							
Turkey	753 (D)	714	—	—	—	—	—
Israel	146 (A)	160	—	51	Edam	ILS	41.30
India	48 (E)	—	—	—	Mozzarella	INR	380.00
Japan	45 (F)	321	297	—	Processed	JPY	1890.00
China	41 (G)	149	124	—	—	—	—
Kazakhstan	28	47	—	—	—	—	—
Republic of Korea	4	156	124	—	NS	KRW	16,225.0
Saudi Arabia	—	—	172	—	—	—	—
Indonesia	—	—	30	—	—	—	—
Philippines	—	—	38	—	—	—	—
Oceania							
New Zealand	385 (G)	48		323	Cheddar	NZD	8.84
Australia	344	350	98	176	Cheddar	AUD	13.25
Africa							
Egypt	395	482	—	61	NS	EGP	59.41
South Africa	108	109	—	—	NS	ZAR	117.19
Zimbabwe	3	9	—	—	NS	USD	4.00
Total selected countries	21,277						
Rest of world	—	—	865	381	—	—	—
World	—	—	2550	2550	—	—	—

(A) Cow's milk cheese only; (B) Based on production of big dairies; (C) 2018: Cow's milk cheese- 72,600 tonnes; (D) 2018: Cow's milk cheese- 658,500 tonnes; (E) Refers to co-operative dairies only; (F) Natural cheese production; (G) Including processed cheese; (H) Excluding Intra-EU trade; NS- Not specified (*Source: compiled from Bulletin of the International Dairy Federation 501/2019*).

3. Presently Used Cheese Packaging Systems

In order to simplify the cheese packaging requirements, its mandatory to classify them in several categories depending on their moisture content (hard, semi-hard, soft, very-soft), shapes (wheels or half-wheel cheese, cheese slabs also known as portioned cheese, sliced cheese, cheese squares, soft and creamy cheese, grated, diced and processed cheese) and preservation techniques (cheese preserved in brine, wax coated, modified atmosphere or vacuum packaged). The very hard, extra hard, hard to semi-hard category of cheese possess moisture content in the range of 36–52% and includes Edam, Gouda, Swiss,

Parmesan, Cheshire and Romano [11]. Rindless types of cheese are ripened in their packaging material alike to cheeses having their surface covered with molds, bacteria or yeasts producing enzymes responsible for ripening [12]. The important factors for selecting packaging materials of very hard to hard varieties of cheese are ripening time, temperature, cheese surface area to volume ratio, gas production (if any), cheese product form (sliced, grated, portions) and permeability of packaging materials [13]. The packaging systems for rindless cheeses includes laminates of polyethylene terephthalate-low density polyethylene (PET-LDPE) (300/50 μm thickness), cover film of oriented (O)PET-LDPE (23/75 μm thickness), tubular bags of oriented polyamide (OPA)-LDPE (15/40 μm thickness) and trough film of PET-HMLDPE (high molecular weight LDPE) (200/25/25 μm thickness). Wax coatings (mineral, paraffin and microcrystalline wax) are used to prevent mold growth, moisture evaporation and high gas barrier properties [11]. Modified atmosphere packaging (MAP) with high barrier materials (PA/EVOH (ethylene vinyl alcohol), LLDPE/EVA (ethylene vinyl acetate)/Ionomers) is generally used for portioned or sliced hard cheese owing to their large surface area exposure to light and oxygen. Vacuum packaging is not preferred for cheese with eyes (Swiss, Gouda, Edam) as it rupture the eyes structure [14].

The semi-soft and soft varieties of cheese contain 52–80% moisture and can be further categorized broadly in three groups (i) ripened by bacteria e.g. Brick, Munster; (ii) ripened by surface mold e.g. Limburger, Brie, Camembert and (iii) internally mold ripened e.g. Gorgonzola, Roquefort, Stilton [15]. Packaging requirements of bacteria ripened cheeses is affected by presence of light, humidity, pH and gases. Internally mold ripened cheese should be packed in O_2, CO_2 and water permeable packages e.g. polystyrene, polyvinyl chloride or thermoformed packages etc., for optimum mold growth [3]. For externally ripened cheese, packaging should not take place until mold had grown to certain extent and packaging material with certain permeability to O_2 and H_2O are prerequisite to avoid growth of anaerobic proteolytic bacteria and moisture condensation inside cheese pack, respectively. *Penicillium camemberti* converts lactate to CO_2 and H_2O, hence perforated OPP (oriented polypropylene) is the suitable material for gas and water passage [16].

Fresh or unripened cheeses (e.g. cottage, quark, cream etc.) have moisture content greater than 80% and are exposed to lactic acid fermentation. Such cheeses have very high chances of dehydration or whey expulsion owing to their high-water activity. Some of the suitable packaging material for fresh cheeses are injection molded HDPE or PP packages with side slits for whey drainage, paraffin or PVDC (polyvinylidene chloride) coated paper and LDPE or PP laminated aluminum (Al) foil (7–20 μm) [14]. Processed cheese is hot filled into pouches, polymer coated or lacquered Al foils (12–15 μm). Processed cheese slices are packed in laminates of PET-HDPE, PET-PVDC and OPP-EVOH-LDPE and processed cheese spreads in tubes of LDPE/EVOH/PET or metal tubes, PP or PET-LDPE cups heat sealed with Al foil, tin plate or enameled Al cans and glass cups closed with Al foil plastic laminate or lidded with an easy opening tin plate [17]. A comprehensive list of permitted additives and their recommended usage level is presented in Table 2, which could be utilized for the development of legally permitted smart packaging materials. Also, a few commercially available smart packaging systems used for cheese are listed in Figure 2.

Table 2. Additives permitted in different varieties of cheese as per FSSAI (Food Safety and Standards Authority of India).

Name of the additive (&INS No.)	Recommended maximum levels			
	Unripened cheese	#Ripened cheese	Plain processed cheese/ processed cheese, processed cheese spread	Note
Aspartame (951)	1000 mg/kg	—	—	If used in combination with aspartame-acesulfame salt (INS 962), combined maximum use level, expressed as aspartame, should not exceed this level.
Carotenoids	100 mg/kg	—	100 mg/kg	
Chlorophylls and Chlorophyllin, copper complexes	50 mg/kg	—	100 mg/kg (Chlorophyll- INS No.-140)	
Canthaxanthin (161 g)	15 mg/kg	15 mg/kg	—	For use in flavored products only
Caramel III - ammonia caramel (150c)	15000 mg/kg	—	—	
Caramel IV-sulfite ammonia caramel (150d)	50000 mg/kg	—	—	
Indigotine (Indigo carmine) (132)	200 mg/kg	—	—	For use in surface treatment only
*Lauric arginate ethyl ester (243)	200 mg/kg	—	—	Equivalent to 2 mg/dm2 surface application to a maximum depth of 5 mm, For use in surface treatment only
Natamycin (Pimaricin) (235)	40 mg/kg	40 mg/kg	40 mg/kg	
Phosphates	4400 mg/kg	—	9000 mg/kg	As phosphorus
Polysorbates	80 mg/kg	—	—	On the creaming mixture basis
Ponceau 4R (124)	100 mg/kg	—	—	For use in surface treatment only
Riboflavins	300 mg/kg	300 mg/kg	300 mg/kg	
*Sorbates	2000 mg/kg	3000 mg/kg	3000 mg/kg	As sorbic acid, For Chhana and paneer only)

Name of the additive (&INS No.)	Recommended maximum levels			
	Unripened cheese	#Ripened cheese	Plain processed cheese/ processed cheese, processed cheese spread	Note
Nisin (234)	12.5 mg/kg	12 mg/kg	12.5 mg/kg	For Chhana and paneer only
Propionic acid, sodium propionate, calcium propionate (singly or in combination, expressed as propionic acid) (280, 281, 282, 283)	3000 mg/kg	3000 mg/kg	—	
Glucono delta lactone (575)	GMP	—	—	
Sunset yellow FCF (110)	100 mg/kg	—	100 mg/kg	For use in surface treatment only
Calcium chloride (509)	200 mg/kg	200 mg/kg		Except cream cheese
Beta-carotenes, vegetable (160a(ii))	600 mg/kg	100 mg/kg	1000 mg/kg	Except Coulommiers
Carrageenan (407)	5000 mg/kg	—		For cream cheese only
Alginate of sodium/potassium/ calcium (410, 402, 404)	5000 mg/kg	—	—	For cream cheese only
Propylene glycol alginate (405)	5000 mg/kg	—	—	
Paprika extract (160c)	GMP	GMP	—	
Curcumin (100)	GMP	100 mg/kg	100 mg/kg	
Annatto (160b (i) and (ii))	GMP	$100 mg/kg @50 mg/kg	50 mg/kg	$(Norbixin based) @(Bixin based)
Lysozyme (1105)	—	GMP	—	
Sodium salts of mono/di/poly phosphoric acid (339, 450 (i, ii, iii), 451 (i), 452 (i))	—	9000 mg/kg	—	Total salt content should not exceed 9000 mg/kg calculated as phosphorous/carbonates /citrate/ chloride
Potassium salts of mono/di/poly phosphoric acid (340, 450 (iv, v), 451 (ii), 452 (ii))	—	9000 mg/kg	—	
Allura red AC (129)	—	—	100 mg/kg	

Name of the additive ([&]INS No.)	Recommended maximum levels			
	Unripened cheese	[#]Ripened cheese	Plain processed cheese/ processed cheese, processed cheese spread	Note
Diacetyltartaric and fatty acid esters of glycerol (472e)	—	—	10000 mg/kg	
Hydroxybenzoates, para	—	—	300 mg/kg	As para-hydroxybenzoic acid
Iron oxides	—	—	50 mg/kg	
Sodium aluminum phosphates	—	—	1600 mg/kg	For use in processed cheese only As aluminum
Pimaricin (Natamicin) (235)	—	2 mg/dm^2 surface.	—	For surface/rind treatment only Not present in depth below 5 mm

* Ingredients permitted in whey cheese includes Lauric arginate ethyl ester (INS No.-243) - 200 mg/kg and Sorbates (1000 mg/kg).[&]INS- International Numbering System for food additives.

[$]Indicates the amount of annatto if it is norbixin based.

[@]It indicates the amount of annatto if it is bixin based.

[#]Ripened cheese- Cheddar, Danbo, Edam, Gouda, Havarti, Tilisiter, Camembert, Brie, Saint Paulin, Samsoe, Emmentaler, Provolone, extra hard grating/sliced/cut/shredded cheese.

Source: Compiled from Manual of Food Safety and Standards Authority of India

Figure 2. Commercially available active and intelligent packaging systems for cheese (A) biodegradable active antifungal film Antipack™ AF, Handary, Brussels, Belgium (B) antimicrobial films with natamycin, VGP SL®, Barcelona, Spain (C) edible plastic films developed from casein by Lactips, France (D) pull timer™, time temperature indicator for indicating temperature abuse developed by Macfarlane labels and insignia technologies, Scotland. (Source: compiled from internet).

4. Active Packaging of Cheese: Concepts and Applications

"Active packaging" term was coined by food scientist Dr. Theodore Labuza [3], which includes oxygen absorbers, carbon dioxide absorbers/emitters, moisture absorbers, self-heating and self-cooling containers, antimicrobial packaging, ethanol emitters, flavor absorbers/releasers and microwave assisted containers [18]. The following section discusses different active packaging systems applicable to cheese and brief studies on active packaging materials for cheese and its products are also presented in Table 3.

Table 3. Types of active packaging materials/systems explored for cheese and cheese-based products.

Type of active packaging	Variety of cheese	Description
Antimicrobial packaging	Cottage cheese [19]	Sachets of allyl isothiocyanate were effective against yeast and mold
	Mozzarella cheese [13]	Lysozyme and ethylenediaminetetraacetic disodium salt (Na_2-EDTA) inhibited the growth of coliform and *Pseudomonadaceae* without affecting the lactic acid bacteria
	Kashar Cheese [20]	Zein and zein-wax coating with lysozyme, catechin and gallic acid. Lysozyme based film prevented the growth of *Listeria monocytogens*
	Mozzarella cheese [21]	Packages containing calcium lactate and lactic acid-based brine enhanced the shelf-life by 50%
	Surface ripened cheese [22]	Polyethylene films coated with polyvinyldichloride and containing natamycin/nisin possessed inhibitory effect against *Penicillium expansum* in surface ripened cheese i.e. *Blatacke zlato* and *Olomoucke tvaruzky*
	Zamorano sheep cheese [23]	Poly propylene and polyethylene terephthalate films with *Origanum vulgare* and ethyl lauroyl arginate essential oils inhibited *E.coli* O157:H7
	Saloio cheese [24]	Whey protein isolate coating containing natamycin reduced water loss, color changes and microbial growth throughout the storage period of 60 days
Oxygen absorbers	Low fat cheese (5% fat in dry matter) [25]	Microbial oxygen absorber; Contains microorganisms which utilizes oxygen e.g. *Lactococcus lactis* strain; Flavor and odor improved
	High fat cheese (60% fat in dry matter) [25]	Microbial oxygen absorber containing *Lactococcus lactis* strain; Flavor improved and no explicit difference on odor
	Cheddar cheese [26]	Microbial oxygen absorber containing *Lactococcus lactis* strain; Positive influence on shelf-life
	Delite 5% sliced cheese [26]	Microbial oxygen absorber containing *Lactococcus lactis* strain; Positive influence on product characteristics
Moisture absorbers	Saloio cheese [27]	*Humidipak*®, Moisture controlling sachets with sodium propionate impregnated over it to control mold growth. Extended the shelf significantly by decreasing the water loss
	Camembert cheese [28]	3-layered film with absorber/desorber film. 10% concentration of water absorbent, maintained attractive white appearance of cheese while 25% caused damage of the varnish layer due to swelling.
UV light absorbers	Cheese puffs [29]	Tricalcium phosphate-based UV light inhibitor could be incorporated directly into dry mix flavor powder of cheese puffs cooked in hot oil to prevent light induced rancidity and spoilage.

4.1 Moisture absorbers

The presence of moisture not only affects the package appearance but also leads to poor texture and quality of cheese both microbiologically and chemically. Moisture control in the cheese package reduces the water activity thus preventing microbial growth and leaching of soluble nutrients [17]. Moisture scavengers include desiccants like silica gel, molecular sieves, natural clays like calcium oxide, calcium chloride and modified starch in the form of pads, sheets, sachets and blankets [4]. Moisture control in cheese packages could also be attained by incorporating humectant between different layers of packaging material, while keeping the inside layer water permeable. A two layered packaging material for moisture sensitive products like soft cheese was developed by [30] Marbler & Parmentier, (1999). The packaging material consisted of first functional layer (coated paper) for storing and releasing moisture and second layer (plastic laminate) for controlling gas permeability as a function of moisture content. These types of packaging material find their utility for cheese matured inside the package. Pantaleao, Pintado, & Pocas (2007) [27] successfully demonstrated humidity controller (Humidipak®) with Saloio cheese for shelf-life extension. A dual compartment vacuum packaging system (Tenderpac®) developed by SEALPAC® (Germany) for neatly collecting the drip loss from meat products, could be optimized for fresh unripened cheeses like mozzarella, quarg and cottage [31].

4.2 Oxygen scavengers

Oxygen scavengers market size was 1.80 billion USD in 2016 which is estimated to reach 2.41 billion USD in 2022 at a compound annual growth rate (CAGR) of 5.1%. North America (USA, Canada and Mexico) is the leading market while Asia Pacific region (China, India, Japan and South Korea) is the fastest growing market [5]. Oxygen is majorly responsible for cheese spoilage as its presence facilitates the growth of aerobic microorganisms, oxidation of cheese components, nutritional value decline, off-flavors generation, unacceptable color changes, shelf-life reduction and decrease in food safety [32]. Therefore, control of oxygen content inside cheese package is of prime importance. Modified atmosphere packaging (MAP), vacuum packaging and oxygen absorbers are the alternatives available to reduce or completely remove oxygen from the package [25]. However, MAP and vacuum packaging require costly equipment for packing cheese and still do not remove the oxygen completely (residual oxygen could be up to 1% in the headspace). Vacuum packaging can affect the appearance and structure of soft cheeses adversely and oxygen can also permeate through the packaging film during later stages of storage or distribution [33]. Oxygen scavengers provide the best alternative to remove the oxygen permeating through the packaging film and also to overcome the challenges of MAP and vacuum packaging [34].

The shelf-life of cheese tarts increased to 48 days when packaged with an iron-oxide based oxygen scavenger as compared to 7 days for control samples [35]. An oxygen scavenging film containing a blend of ethylene, methyl acrylate and cyclohexene methyl acrylate copolymer as oxygen scavenger resin was developed to overcome the oxidative rancidity in cheeses, dried milk and meat products [36]. A study on the effectiveness of various packaging methods for Gouda cheese revealed that oxygen scavengers (ATCO FT 210) were as effective as vacuum packaging and MAP (40% CO_2 and 60% N_2) in prolonging its shelf-life [34]. Microbiological oxygen scavenging material consisting of *Lactococcus lactis* strain was reported to consume oxygen in cheese packs with limited production of acetoin and diacetyl [25]. Graviera cheese when packed using a combination of oxygen scavenger and ethanol emitter showed lower microbial growth as compared to 100% nitrogen modified atmosphere packages. An increase in sensory shelf-life for oxygen scavenger and ethanol emitter combined packages was also observed [33]. Negamold®, an ethanol vapor sachet was developed by Nippon Kayalan firm (Japan) for meat products. Later, Freund corporation (Japan) combined oxygen absorber with Negamold® and used it for cheese packaging [37]. Ozdemir & Sadikoglu (1998) [38] had also suggested the replacement of ethanol emitters in cheese packages with UV-excimer-laser-treatment of polymer films to generate bactericidal properties. BIOPACK is a polylactic acid-based packaging system consisting of oxygen scavengers and preservatives encapsulated in cyclodextrin with an objective to extend the best before date of cheeses from 2 to 3 months to 9 months with minimum effect on cost of package [39].

4.3 Free radical scavenger or antioxidant incorporated films

Cheeses like Cheddar, Swiss, Blue, Colby etc. are highly prone to lipid oxidation owing to their high fat content. Antioxidants are extensively used to prevent oxidation by scavenging free radical but due to augmented customer trend for additives free food products, incorporation into packaging material is the best option [40]. Antioxidants incorporation into packaging material not only prevents quality deterioration of the product but also stabilizes the polymer [41]. Synthetic antioxidants like butylated hydroxytoluene (BHT) and butylated hydroxy anisole (BHA) are conventionally used in cheese packing. As per Code of Federal Regulation (CFR 21/172.115), the maximum rate of BHT addition to cheese is 200 mg/kg of fat and specific migration limit of BHA is 30 mg/kg of food product as per EU 10/2011 regulations. Asadero cheese was vacuum packed in LDPE co-extruded film containing 8 and 14 mg/g of BHT. Cheese packed in LDPE film incorporated with 8 mg/g of BHT had oxidized flavor while film with 14 mg/g of BHT surpassed the legal limit of BHT addition [42]. Therefore, similar to natural counterparts of other additives the recent focus is on natural antioxidants. Pomegranate peel extract (PPE) incorporated into zein films for packaging of Himalayan Kalari cheese retarded the oxidation of fat and protein due to the presence of polyphenols in PPE [43]. Sliced cheese packed in red algae films incorporated with 1% grape fruit seed extract (GFSE) showed decreased peroxide and thiobarbituric acid value indicating the antioxidant capability of GFSE [44]. Gelatin-chitosan edible film with Boldo herb extract possessed antioxidant and antimicrobial activity and had preservative effect on sliced Prato cheese by preventing psychrotrophs [41]. Similarly, other natural antioxidants like green tea extract [45], catechins [46] and rosemary extract [40] had been explored for their antioxidant potential in cheese packaging but the major challenge with antioxidant incorporated films in cheese packaging is synchronization of antioxidant diffusion rate according to cheese requirement. Also, for natural antioxidant incorporation in continuous film production by extrusion, their stability or thermal degradation is the major concern [46].

4.4 Carbon dioxide absorbers

Cheeses packed with higher CO_2 may suffer from sensory related issues as its dissolution leads to formation of carbonic acid [14]. Taleggio cheese produced excessive 2.5 mmol kg-1 day-1 CO_2 when stored in nitrogen flushed packages at 6°C causing quality degradation [47]. However, carbon dioxide production is essential in some cheeses to achieve desired texture, eye formation in Emmental and Swiss cheese, and inhibition of microorganisms but excessive production could lead to puffed pouches or package burst [48]. When cheeses are preserved and sold at ambient temperature or when desired shelf life is high, the adverse effects of higher CO_2 concentration aggravates many folds [47]. In such circumstances, carbon dioxide absorbers could be used to remove the excess CO_2 and create a balanced internal cheese package atmosphere [2]. The only noticeable progress in segment of CO_2 absorbers for cheese is by Fellows (2009) [49], who developed a mechanism for CO_2 release from mold ripened cheese (e.g. Camembert) package using one-way valve while disallowing other gases to infiltrate. Crump (2012) [50] developed a CO_2 absorber pouch using polyethylene that contained 1.1 g of calcium hydroxide (200 mesh) and silica gel each in 2:1 mixture of water for shrink wrapped Swiss cheese (114 g) and reported that the product remained in good color with acceptable taste without any expansion due to CO_2 release during storage at 5°C for 4 months. The gas composition and volume of modified atmosphere packed semi-hard cheese (Kadett®, Arla Foods) packages were optimized using mathematical modeling based on gas solubility coefficients, initial carbon dioxide content in cheese and packaging material, thus avoiding consumer rejection due to volume changes [48].

4.5 Light stabilizers

Light, and principally UV light, may cause or accelerate various undesirable reactions like lipid oxidation in cheese. Also, riboflavin, an efficient photosensitizer, present in cheeses at levels of 0.30–0.60 mg/100 g, quickly captivates energy owing to its conjugated double bond and generates either free radicals or reactive oxygen species (ROS). These free radicals and ROS are the major causes of lipid oxidation, off-flavors, color bleaching and nutrient losses especially vitamin A in cheeses [51]. Light stabilizers are divided into five major categories namely: light absorbers, light screeners, excited-state quenchers, peroxide decomposers and free radical scavengers based on their mode of action [52]. Kristoffersen, Stussi, & Gould (1964) [53] reported reduced flavor deterioration in consumer packs of cheddar cheese using Uvinul D 49® as a UV light screening material. Uvinul® S-Pack is a novel FDA approved UV absorber for PET packaging films, which prevented the UV degradation of vitamins and β-carotene, thus highlighting its potential of preventing light degradation changes in cheeses kept in refrigerated illuminated cabinet of supermarkets [54]. Recently, flavonoids had been reported to facilitate the dissipation of photon energy to heat thus deterring photodegradation [22]. Thus, flavonoids incorporated packaging material as natural active element for UV light absorption may be explored for cheese.

4.6 Antimicrobial releasers

Antimicrobial packaging is the most researched forms of cheese active packaging. Antimicrobial agent at certain minimum concentration (known as minimum inhibitory concentration (MIC)) diminishes or impedes microbial growth [9]. Antimicrobial effect in cheeses is most commonly obtained by organic acids and its salt derivatives (sorbic acid, citric acid and their anhydrides), bacteriocins (nisin, lacticin and pediocin), fungicides (imazalil and natamycin), enzymes (lysozyme and lactoferrin), essential oils (basil leaf, thyme, oregano and cinnamon) and miscellaneous compounds like potassium metabisulphite, allyl isothiocyanate, EDTA (ethylenediaminetetraacetic acid) or a combination of these agents [22, 55, 56]. Antimicrobial agents which are sensitive to higher polymer processing temperature are usually applied as coatings. Gliadin based bioplastic films prepared by casting, and containing cinnamaldehyde as active ingredient inhibited fungal growth in cheese spreads [57]. Immobilization of antimicrobial agents like nisin on the surface of cheese packaging material is a convenient technique, however immobilization is appropriate for fluids because of direct contact between antimicrobial surface and entire liquid food [58]. Active polyethylene terephthalate film immobilized with silver nanoparticles extended the shelf-life of white fresh cheese up to 30 days [59]. Labels containing antimicrobial agents can also be used for enhancing cheese shelf life. Labels containing allyl isothiocyanate enhanced the shelf-life of Danish Danbo cheese to 28 weeks when used in combination with MAP as compared to 18 weeks with MAP alone [60].

Chitosan, a natural polysaccharide had been utilized for antimicrobial cheese packaging owing to its biodegradable, antimicrobial, filmogenic and metal complexation attributes [61]. Cellulose polymer based antimicrobial films incorporated with nisin and natamycin showed the potential for preservation of sliced Mozzarella cheese [62]. Electrospinning technique was utilized for incorporation of nisin (at the rate of 5 mg/mL) in polyethylene oxide nanofibers to inhibit *Listeria monocytogens* contamination in cheddar cheeses without affecting its sensory attributes [63]. A novel antimicrobial film based on hybrid organic–inorganic material commonly called as "anionic clays", consisting of layered double hydroxide intercalated with salicylate and carbonate anions increased the storage life of Mozzarella to three weeks at a storage temperature of 18°C [64]. DSM™ has developed Pack-Age® as a solution for ripening of cheese in a vapor pervious foil united with yeast and mold blockers [65].

4.7 Color and flavor releasers

Flavor emitters are mainly used to impart flavor to any packed product or scalp/downgrade any undesirable flavor due to harsher processing conditions, thereby improving sensorial attributes and chances of modifying product formulation [66]. It may be used for masking off-flavors but food processors may unfairly market their expired, unsafe or low-quality foods without letting the consumers know. ScentSational Technologies® is global leader in developing food packages with controlled release of legally permitted flavor into headspace of a pack at varying intervals and provision for adjustment of flavor intensity [31]. Recently, they have also ventured into developing customized and patented injection molded scented and/or flavored parts of any pack. Kraft foods had developed a system for controlled and prolonged release of volatile flavor upon opening and reopening of the package [67]. Such type of packaging innovation could also be used for cheese products like chiplets, slices, processed cheese etc. which are usually contained in multi-use packages.

Color releasing multilayered film is the novel technique for incorporating permitted food grade colors (Table 2) such as annatto over cheese surface. Such films generally find their application when low intensity shade of color is desired or color is adversely affected during any processing step, storage or distribution. Mohan, Ravishankar, & Gopal, (2010) [4] suggested the migration of edible food permitted red color from the wrapper of surimi to provide it a more desirable and acceptable color. Similarly, α, β-citral migrated from the cellulose acetate films and improved the yellowness of Coalho cheese without affecting its texture during 25 days of storage [68].

4.8 Miscellaneous active packaging systems for cheese

4.8.1 Self-cleaning rinds

Rindless cheeses are cooked or uncooked hard varieties of cheese that are ripened in plastic film which allows little or no gas or moisture movement e.g. Cheddar, Edam, Gouda and Swiss. Natural rind is the outer crust of cheese formed either during cheese making or storage under controlled humidity and temperature [3]. These rinds are highly susceptible to undesirable fungal growth and becomes slimy at times. Gerber, Koehler, Grass, & Stark (2012) [69] developed a three layered, self-cleaning and porous rind inoculated with *Penicillium roqueforti*. The base layer consisted of polyvinyl chloride (90 μm), living layer of agar (300 μm) with inoculum and porous cover layer of polycarbonate (10 μm) for diffusion of gases and nutrient supply. There is immense future potential for the development of antibacterial self-cleaning rinds using penicillin producing molds (*Penicillium jensenii*) for cheese varieties [69].

4.8.2 Microwave assisting films

Microwave susceptors are the substances which absorb microwave energy and convert it into heat energy. It consists of Al foil layer deposited on paperboard or polyester film for uniform heating treatment [18]. Emmi®, a USA based cheese manufacturing firm, provides different variants of fondue recipes (melted Swiss cheese) in microwaveable containers which are ready-to-(h)eat, convenient and recyclable [70]. These types of microwave assisted heating packs could be used for melted cheese recipes. The major concern with microwave assisted heating cheese containers is duration of microwave heating. Some pop-up sound mechanism could be attached with package which blows up and makes a noise on complete even heating of the package content [3].

4.8.3 Pesticide control agents

Pesticide control agents are generally used with secondary packaging systems to prevent insects, or for fungicidal control, during import and export of food products over distant horizons. Packaging material with pesticide control could also be used to prevent detrimental effects of pests and insects for cheeses like Cheddar, Parmesan etc. which require longer ripening period. The major concerns with these types of pesticide control agents containing packaging is their permissible limit and regulatory issues for use with cheeses. Natamycin is a GRAS status (as per FDA) fungicide which is produced during fermentation by *Streptomyces natalensis*. Romero et al. (2016) [71] showed positive effect of natamycin incorporated biodegradable triticale flour films on mold inhibition when used for wrapping soft cheese.

5. Intelligent Packaging of Cheese: Concepts and Applications

Intelligent packaging has not been researched extensively for cheese as reflected by very few publications in Figure 1. A few intelligent packaging systems investigated for cheese are presented in this section. However, large size of cheese market including import and export offers attractive opportunities. A list of different suppliers of commercially available smart packaging materials along with their head office, website and contact point are detailed in Table 4.

Table 4. Suppliers and Asian contact point of commercially available smart packaging systems.

Type of smart packaging	Company (Head Office)	Brand name	Website	Distributor/ Contact point in Asia
Oxygen scavenger	Clariant® Chemicals (Switzerland)	OXY-GUARD™, O-Buster®	www.clariant.com	Clariant Chemical, Vadodara
	Mitsubishi Gas Chemical (Japan)	Ageless	www.mgc.co.jp	Information & Advanced Materials Company, Oxygen Absorbers Division, Japan
	Toppan Printing (Japan)	Freshilizer	www.toppan.com	Max Speciality Films Limited, Punjab, India
	Multisorb Filtration Group® (New York, USA)	StabilOx®, Freshmax	www.multisorb.com	—
	Southcorp Packaging (Acquired by Visy®) (Australia)	Zero$_2$	www.visy.com.au	No facility in India. Available in Thailand.
	AGM Containers (USA)	ActiSorb®O		Clariant India, Maharashtra India
Time temperature indicator	Avery Dennison (California, USA)	TT Sensor™	www.averydennison.com	Bangalore, Karnataka
	IntroTech (Netherlands)	Monitor Mark®	www.introtech.eu	—
	Vitsab® (Limhamn, Sweden)	CheckPoint®	www.vitsab.com	—
	TempTime® Corporation (USA)	Fresh-Check®	www.temptimecorp.com	Lisaline Lifescience Technologies Pvt. Ltd., Thane, India

Type of smart packaging	Company (Head Office)	Brand name	Website	Distributor/ Contact point in Asia
Antimicrobial packaging	Life Materials Technology Limited (Hong Kong)	Agion®	www.life-materials.com	—
	Addmaster Limited (UK)	Biomaster®	www.addmaster.co.uk	Jebsen & Jessen, Indonesia (Contact point in Asia)
	VGP (Barcelona, Spain)	Natamycin	info@pimaricina.com	—
Ethylene scavenger	Evert-Fresh Corporation (USA)	Evert-Fresh	www.evertfresh.com	—
	Sekisui Jushi (Japan)	Neupalon	www.sjc-strapping.com	—
	Peakfresh Products Ltd. (Australia)	Peakfresh	www.peakfresh.com	—
Moisture absorbers	Sealed Air® Corporation (USA)	Dri-Loc®	www.sealedair.com	—
	SEALPAC® (Germany)	Tenderpac®	www.sealpacinternational.com	Synerchem Sdn. Bhd., Selangor, Malaysia (Contact point in Asia)
Integrity Indicator	Freshpoint Lab (Australia)	O_2 Sense	www.freshpoint.com	—
	Timestrip Ltd.	Timestrip	—	—
	Mitsubishi Gas Chemical (Japan)	Ageless Eye	www.mgc.co.jp	Information & Advanced Materials Company, Oxygen Absorbers Division, Japan
	Insignia Technologies Ltd. (Scotland)	Novas	www.insigniatechnologies.com	—
RFID	Temptrip LLC (USA)	Temptrip	www.temptrip.com	—
	Mondi Plc (Austria)	Intelligent Box	www.mondigroup.com	—

Type of smart packaging	Company (Head Office)	Brand name	Website	Distributor/ Contact point in Asia
Freshness Indicator	COX Technologies (USA)	Fresh Tag	www.cox-tec.com	—
	Timestrip (UK)	Timestrip®	www.timestrip.com	—
	Ripesense Ltd. (New Zealand)	ripeSense®	www.ripesense.co.nz	—
Microwave susceptors	Sirane Food Packaging Limited (UK)	Sira-Crisp™	www.sirane.com	Sirane East, Vostok, Russia
	VacPac Inc. (USA)	SmartPouch	www.vacpacinc.com	—
Source: compiled from internet using website of the companies.				

5.1 Gas indicators

Gas indicators or package integrity or leak indicators generally indicate the presence or absence of any gas (majorly oxygen) on the basis of certain chemical or enzymatic reactions. Cheeses are packed under modified atmospheres usually devoid of oxygen to enhance their shelf life. However, the gas composition of cheese package may change relying on the microbial growth inside the package, barrier properties of the packaging material, efficiency of packaging system, or physical damage, if any, that causes leakage [72]. So, knowing the level of oxygen is important to ensure cheese quality and safety in the entire supply chain and throughout its shelf-life. Redox dye-based oxygen indicators have been reported to indicate the package integrity and status of MAP in food non-destructively [73]. A schematic illustration of Mozzarella cheese package equipped with an oxygen indicator and oxygen scavenger with dye-based oxygen sensor is presented in Figures 3 and 4, respectively.

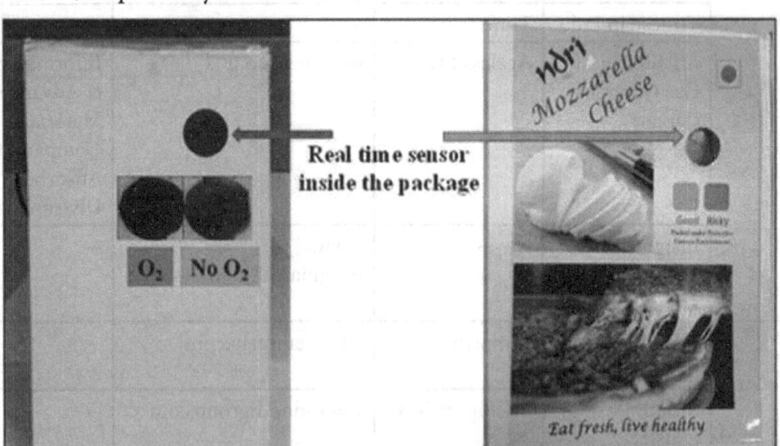

Figure 3. A schematic illustration of intelligent packaging system using an oxygen indicator applied to mozzarella cheese package (Source: [15]).

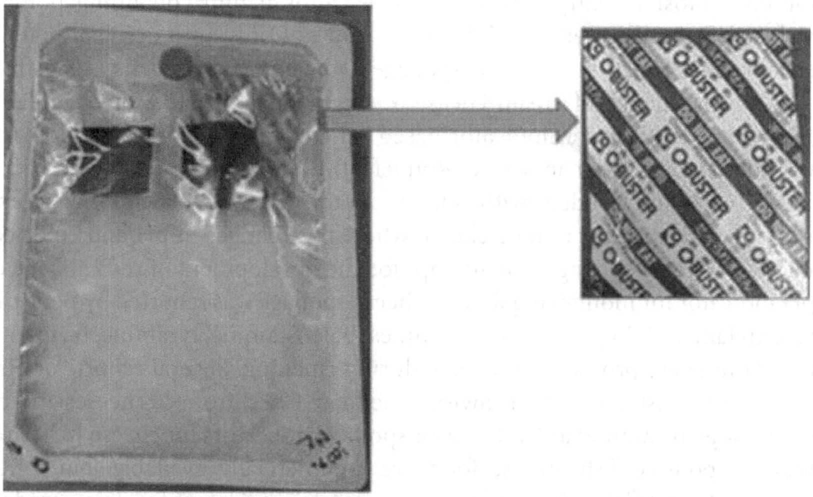

Figure 4. A schematic illustration of smart (active + intelligent) packaging system for mozzarella cheese package with oxygen indicator (shown in pink color) and oxygen scavenger (O-buster® oxygen scavenger) (Source: [15]).

A single use fluorescent-based oxygen sensor prepared using platinum octaethylporphyrin-ketone (PtOEPK), a phosphorescent oxygen-sensitive dye, sensed oxygen concentration changes in MAP cheddar cheese over a period of 4 months. The sensor was reported to possess sensitivity in the range between 0.02% and 100% oxygen. Correlation between oxygen concentration and microbial growth presented an opportunity for assessment of cheese quality using colorimetric oxygen sensor [74]. Similarly, dye based ultraviolet light activated oxygen sensor was successfully developed and characterized for its oxygen sensitivity, oxygen dependent color change and mechanical properties by Deshwal et al. (2018) [75]. The developed indicator was integrated with MAP Mozzarella cheese as an integrity/oxygen indicator, which could be helpful for stakeholders in the entire supply chain [15]. Hempel, Gillanders, Papkovsky, & Kerry, (2012) [76] successfully exploited optical oxygen sensors for detecting integrity (ingress of oxygen) of vacuum packaged cheddar cheese samples during its storage.

5.2 Freshness indicators

Freshness indicators, mostly colorimetric in nature, determine the safety, quality or freshness of product based on microbial growth or chemical change. They trigger a visual indication mechanism by detecting the metabolites of microbial or chemical change [77]. Possibilities of freshness detection of packaged milk, cream and cottage cheese using polymer-based labels was proposed by Chen & Zall (1987) [78]. Major approach for characterizing the deterioration of any cheese is by identifying the volatile organic compounds liberated during its storage (or ripening) using solid phase microextraction-gas chromatography/mass spectroscopy (SPME-GC–MS). Octane, hexanal and 2-pentyl-furan were the indicators for light exposure as obtained during the volatile profile of processed cheese [79]. Fourier Transform Infrared Spectroscopy (FTIR) and near infrared spectroscopy (NIR) have also been used to rapidly identify the chemical groups involved in the Crescenza cheese spoilage for possible development of freshness indicator [80]. Most recently, a biodegradable chitosan film containing pomegranate peels/Melissa officinalis essential oil demonstrated not only antimicrobial potential but also anthocyanins functionality as a spoilage indicator changing its color from blue to red due to pH change of cream cheese during spoilage [77]. A diverse blue cheese classification or identification indicator based on chromogenic array pattern of several pH dyes differentiated five cheeses i.e. Roquefort, Blue Stilton, blue cheese with leaves, blue cheese spread and Cheddar with 100% accuracy [81]. Such type of indicators can be used as freshness indicators of blue cheese where the changes in pH and color could be correlated with cheese spoilage. An attempt for the development of red cabbage extract-based pH indicator for monitoring Ricotta cheese spoilage was reported by Bento, Pereira, Chaves, & Stefani, (2015) [82]. Biogenic amines like histamine, tyramine, tryptamine and phenylethylamine are produced in cheese during ripening. Several reports of histamine poisoning in the past for Gouda, Swiss, Cheddar, Cheshire etc. cheeses indicate the potential of biogenic amines as freshness or spoilage indicators for cheese [83]. Freshness indicators for poultry, fish and seafood are commercially available, but a very few "biological use by date" or "chemical best before date" indicators for dairy products had been reported to the best of our knowledge indicating research possibilities in this area.

5.3 Ripening indicators

Cheese ripening indicator could be defined as the use of any technique/process/sensor for spotting metabolites (majorly volatiles) or chemical breakdown by-products of glycolysis, proteolysis and lipolysis to quantify the maturity or age of any cheese variety. The earliest attempt in cheese segment included the use of amido black dye for detecting the age of Cheddar and lactose-hydrolyzed cheddar cheese. Dye binding values were correlated with the free amino acid content [84]. Electric nose (or e-nose) had been used for headspace fingerprinting of packaged ripened cheese (Crescenza) volatiles and the data obtained was found to be helpful for its shelf-life measurement [85]. Tavaria, Ferreira, & Malcata, (2004) [86] quantified major ripening descriptors like free fatty acids, acetic, isobutyric and isovaleric acid concentration during 180 days ripening period of Serra da Estrela cheese. These volatile fatty acids furnished information about the optimal consumption time of cheese which could also be successfully used as ripening indicator. Industrially successful models based on infrared reflectance spectra, attributed to the changes in absorbance patterns of alcohol and amide groups have been used to predict the ripening stages and sensory characteristics of Cheddar [87] and Camembert cheese [88] with a minute error of one day.

5.4 Time temperature indicators (TTIs)

The shelf-life of any food commodity as mentioned on the package in terms of "biological use by date" or "chemical best before date" is subject to its temperature exposure history owing to temperature dependence of microbial growth, enzyme activity and chemical reactions. Time temperature indicators (TTIs) convey information about the temperature exposure of the food commodity over a period of time [89]. TTIs mainly finds their applications in temperature sensitive food products that are stored or distributed in chilled conditions like milk, cheese, ice-cream, yoghurt, meat, fish etc. Shellhammer & Singh (1991) [90] used enzyme-based full history TTI (I-POINT®) on cottage cheese to correlate temperature variation with cheese quality parameters and reported that the

TTIs response was significantly affected by pH, titratable acidity and standard plate count of cheese samples. However, attempts of TTI usage in cheese are few and include shelf-life evaluation of Taleggio cheese [91] and Caprino type cheese [92] using TTIs. Potential of diacetylenic monomers as active ingredient in TTIs based on polymerization reaction for monitoring cheese maturity had also been suggested [93]. A study on evolution of proteolytic activity products in Azeitao cheese with fluctuating temperature revealed prominent presence of two free amino acids (valine and leucine) and two biogenic amines (tyramine and putrescine), which may serve as temperature change indicators for the development of microbial TTI for ripened cheese [94].

5.5 Radio frequency identification devices (RFID)

Cheese traceability at batch level is maintained using self-adhesive casein labels, written records, and in advanced cases information is stored in a local database. However, such systems are inefficient considering food safety, counterfeiting risks, voluminous cheese production, warehouse optimization and cost involved in production [95]. So, application of RFID tags at 'farm to fork' levels of cheese industry could provide reliable solutions as it stores more information and assess at longer distances [12]. Regattieri, Gamberi, & Manzini (2007) [96] developed a RFID based traceability systems for hard cheese (Parmigiano Reggiano) which detects the history of the product over entire supply chain. Every minute information starting from feed input, production details to detailed pedigree of a cheese piece is available, thus even facilitating consumers to authorize cheese origin and prevent cheese imitation. The final cost of such RFID tags on customer was calculated to be 0.5%. Similarly, improved traceability of long-ripened cheeses (Bra Tenero, Bra Duro, Raschera and Toma Piemontese) with automatic movement recording during production, handling in ripening room and warehouse, delivery, packing and selling was achieved using tags operating at low (125 kHz), high (13.56 MHz) and ultra-high (865 MHz) frequency [12]. RFID tags with an ability to store data related to 200 variables of cheese production not only improved the quality and yield control of the production plant but also possessed robustness against different temperature, humidity, acid and frictional forces [97]. Papetti et al. (2012) [98] designed a web based "infotracing system" for Italian cheese (*Caciottina massaggiata di Amaseno*) using RFID tags. On linking maturity level of cheeses with quality information (chemical, sensory and spectrophotometric data), RFID system was found to be reliable and compatible with production process. An additional application is "Smart Shelf" which consists of network of RFID antennae for identifying a product's location. It had been successfully validated for tracking expiry dates of processed cheese [99].

5.6 Physical shock indicators

Physical shock indicators are of prime importance for status quo of any fragile product during its rough handling or carriage. Cheeses are often exported across the globe with highest probability of mishandling by personnel during any step of distribution channels or improper selection of transportation channel. Physical shock indicators could be developed using diffusion mechanism, where a fluid leaks and collects irreversibly in another impermeable package, thus indicating the force or pressure to which package content had been exposed. To the best of our knowledge and literature mining no physical shock indicator for cheese and food packaging had been reported. Convex-concave type of metallic structure could also be used to identify the forces to which any cheese packages are exposed over long distances.

6. Development Trends and Research Directions for Smart Packaging of Cheese

6.1 Active packaging systems

Packaging could also be used for facilitating the reduction of cholesterol and lactose in cheeses using cholesterol reductase and lactase enzymes. Cholesterol reductase enzyme converts cholesterol to undigested form (coprosterol), reducing its absorption in intestine. An innovative ethylene-vinyl alcohol copolymer (EVOH) plastic encompassing 30% beta-cyclodextrins reduced the cholesterol concentration by 23% in UHT milk [100]. Such type of active plastic films could be incorporated with β-galactosidase enzyme (lactase) and explored for the development of lactose free whey cheeses due to increased incidences of lactose intolerance across the globe [101].

Citric acid, ferrous salt/ascorbic acid, cellulose triacetate and activated carbon/clays/zeolites are most commonly used off-odor absorbers finding their use in fish, cereals, fruits and poultry products [3]. Off-flavor and odor scavengers prevent cross contamination of pungent odor and aids in improving the overall acceptance of cheeses. However, it is imperative that the constituents scavenged should not be spoilage indicators or essential for flavor development. Some ketones, aldehydes and esters are associated with fruity flavor of cheeses which may be undesirable for some customers [102]. Aldehyde and ester scavengers in cheese packaging can be helpful in improving its sensorial quality. The identified volatile compounds from the headspace of cheese packages revealed the possibilities for development of absorption system and stabilization of sensory qualities of semi-soft ripened cheese [103].

The earliest documented and patented step to achieve the tack ability of a multilayered polyester film over cheese surface was the electrical discharge or flame treatment of the inner surface [104]. Such films were temporarily adherent and easily peel able while opening cheese package. Presently, these anti-stick films can find their vast application for packaging individual slices of processed cheese or Mozzarella cheese spheres thus, reducing sticking losses.

Carbon dioxide and ethanol not only inhibit bacteria, yeasts, molds but also reduces oxidation and could be used individually or in combination for cheese packaging systems to inhibit microbial growth and pack shrinkage [105]. Cheese is most commonly packed with higher CO_2 concentration using MAP technique but CO_2 dissolves in the product leading to package collapse [6]. Package collapse could be overcome by inserting CO_2 emitters in standard MAP cheese trays with perforated false bottom. The controlled release of ethanol in cheese packs could be obtained by encapsulating in a carrier material [65]. Ethicap®, a commercialized ethanol emitter absorbed in silica pads and embedded in sachets made from ethylene vinyl acetate copolymer prevented the growth of molds and yeast, thereby enhancing the shelf-life of soft cheeses [106]. However, objectionable off-flavors involved with higher concentration of CO_2 and ethanol are concerning and supplementary flavor mixtures may be required.

An innovative single use package having the ability to absorb oxygen, carbon dioxide and water vapor, comprising of calcium hydroxide which emits water due to CO_2 absorption, thus activating transition metal (iron oxide) based oxygen scavenger has been developed. Such containers would be suitable for hard cheeses like Taleggio, which emits large amount of CO_2 during ripening and require slight oxygen for maintaining the growth of live cultures [107].

Self-cooling packaging technique is based on an endothermic chemical reaction involving the dissolution of ammonium chloride or ammonium nitrate in water and heat pump technology using water as the heat transmission medium. Such type of packaging systems may remunerate the cold chain conditions, especially where supply channel is inefficient [3]. Initially, thermal sensitive cheese varieties may be shipped using secondary or tertiary thermal management system. Greenbox Thermal Management Systems™ utilizes organic phase change nanomaterial labeled as PureTemp®, to provide specifically designed distribution carriage systems with an ability to maintain temperature precisely for longer durations of supply [108]. It consists of a reusable, recyclable and completely biodegradable boxes in box arrangement with exterior layer of corrugated plastic. Such type of self-cooling containers may be really helpful for exporting cheeses over longer distances without any thermal abuse and quality deterioration.

6.2 Intelligent packaging systems

Emmental and Gouda cheese possess typical and desired regular round holes (eyes) owing to the production of large amount of carbon dioxide during lactate metabolism [109]. Dye based CO_2 indicators based on color intensity that is correlated with amount of CO_2 released could be used to monitor advances in ripening and signpost the accomplishment of optimal ripening. Recently, a novel consumable adhesive CO_2 indicator strip consisting of phenol red dye and tetrabutylammonium hydroxide coated onto silica nanoparticles was developed by Wang, Yusufu, & Mills, (2019) [110]. The color response was dependent on temperature and thickness of polymer barrier films. Such type of indicators could be explored for the development of CO_2 indicator or freshness indicator for modified atmosphere packaged cheese and cheese-based products.

Temperature sensitive networks based on chitosan-poly-(N-isopropylacrylamide) for controlled release were developed by Alvarez-Lorenzo et al. (2005) [111], which can be used in active cheese packaging materials for precise emission of any active component. Films changing their gas permeability in response to degree of temperature and exposure duration may be frequently used during storage and distribution of respiring cheeses like Camembert and Gouda. BreatheWay® membrane technology (Apio Inc., California), based on side chain crystallizable (SCC) polymers provides the solution for gas permeability control according to change in temperature. The change in polymer properties like chain length and side chains can be used for attaining required oxygen and carbon dioxide permeabilities in cheese packages [3].

7. Conclusion

With the focal point being shifted to consumer convenience, quality and safety, active and intelligent packaging tools may help customers with informed choice. As the world is witnessing increased consumption of cheese, these packaging tools have potential market growth. The expansion of smart packaging technologies in cheese industry remains at a nascent stage. Recent research publications on smart packaging of meat, fish, fruits and vegetables suggest innovative ideas which could be conceptualized for cheese in near future. Smart packaging tools need to be of low cost and multiple benefits. The partnership of active and intelligent packaging can be used to complement each other's actions. Existing challenges could be overcome by multidisciplinary approaches for the development of smaller, more powerful and cost-effective smart packaging systems. Biotechnology, nanotechnology, food science, sensor technology and information technology could be combined for overcoming the shortcomings. Biosensor and hybrid devices for cheese packaging remains untouched in terms of its development and commercialization. It could be expected that with the continuous advances in intelligent packaging and growing modified atmosphere packaged dairy products market, the demand for such type of intelligent packaging systems is expected to rise.

References

1. Mihindukulasuriya SDF, Lim LT. Nanotechnology development in food packaging: A review. Trends in Food Science & Technology. 2014;40:149-167
2. Lee SJ, Rahman AM. Intelligent packaging for food products. In: Innovations in Food Packaging. Academic Press. 2014; p. 171-209
3. Robertson GL. Food Packaging: Principles and Practice. 3rd ed. CRC Press, Boca Raton, USA. 2013
4. Mohan CO, Ravishankar CN, Gopal TS. Active packaging of fishery products: a review. Fishery Technology. 2010;47:1-18
5. Mordor Intelligence [Internet]. 2020. Active and intelligent packaging market- Growth, Trends and Forecast (2020-2025). Available from: https://www.mordorintelligence.com/industry-reports/active-and-intelligent-packaging-market-industry. [Accessed: 2020-03-31]
6. O'Callaghan KA, Kerry JP. Consumer attitudes towards the application of smart packaging technologies to cheese products. Food Packaging and Shelf Life. 2016;9: 1-9
7. FAOSTAT-Food and Agricultural Organization of the United Nations Statistics Division. [Internet] (2019). Available from: http://faostat.fao.org/ [Accessed: 2020-05-10]
8. IDF. Bulletin of the International Dairy Federation 501/2019. The World Dairy Situation. 2019
9. Hanusova K, Stastna M, Votavova L, Klaudisova K, Dobias J, Voldrich M, Marek M. Polymer films releasing nisin and/or natamycin from polyvinyldichloride lacquer coating: nisin and natamycin migration, efficiency in cheese packaging. Journal of Food Engineering. 2010;99:491-496
10. Jayadevan GR. A strategic analysis of cheese and cheese products market in India. Indian Journal of Dairy Research. 2013;2:247-250
11. Schneider Y, Kluge C, Weiss U, Rohm H. Packaging Materials and Equipment. In Law BA, Tamime AY, editors. Technology of cheesemaking. 2nd ed. Blackwell Publishing

House, Malden, USA. 2010. p. 413-439

12. Barge P, Gay P, Merlino V, Tortia C. Item-level radio-frequency identification for the traceability of food products: application on a dairy product. Journal of Food Engineering. 2014;125:119-130
13. Sinigaglia M, Bevilacqua A, Corbo MR, Pati S, Del Nobile MA. Use of active compounds for prolonging the shelf life of mozzarella cheese. International Dairy Journal. 2008;18:624-630
14. Pocas MdeF, Pintado M. Packaging and the shelf life of cheese. In: Robertson GL (Ed) Food Packaging and Shelf life. A Practical Guide. CRC Press, Boca Raton, Florida, USA. 2010; p. 103-125
15. Deshwal GK. Development and evaluation of intelligent oxygen sensor applied to modified atmosphere packaged Mozzarella cheese. M.Tech Dissertation, ICAR-National Dairy Research Institute; 2017
16. Rodriguez-Aguilera R, Oliveira JC, Montanez JC, Mahajan PV. Effect of modified atmosphere packaging on quality factors and shelf-life of surface mold ripened cheese: Part I constant temperature. LWT-Food Science and Technology. 2011;44:330-336
17. Kapoor R, Metzger LE. Process cheese: Scientific and technological aspects—A review. Comprehensive Reviews in Food Science and Food Safety. 2008;7:194-214
18. Realini CE, Marcos B. Active and intelligent packaging systems for a modern society. Meat Science. 2014;98:404-419
19. Conceicao Goncalves MPJ, Dos Santos Pires AC, Soares NDFF, Araujo EA. Use of allyl isothiocyanate sachet to preserve cottage cheese. Journal of Foodservice. 2009;20:275-279
20. Unalan IU, Arcan I, Korel F, Yemenicioglu A. Application of active zein-based films with controlled release properties to control Listeria monocytogenes growth and lipid oxidation in fresh Kashar cheese. Innovative Food Science & Emerging Technologies. 2013;20:208-214
21. Faccia M, Gambacorta G, Natrella G, Caponio F. Shelf life extension of Italian mozzarella by use of calcium lactate buffered brine. Food Control. 2019;100:287-291
22. Huvaere K, Skibsted LH Flavonoids protecting food and beverages against light. Journal of the Science of Food and Agriculture. 2015;95:20-35
23. Otero V, Becerril R, Santos JA, Rodriguez-Calleja JM, Nerin C, Garcia-Lopez ML. Evaluation of two antimicrobial packaging films against Escherichia coli O157: H7 strains in vitro and during storage of a Spanish ripened sheep cheese (Zamorano). Food Control. 2014;42:296-302
24. Ramos OL, Pereira JO, Silva SI, Fernandes JC, Franco MI, Lopes-da-Silva JA, Malcata FX. Evaluation of antimicrobial edible coatings from a whey protein isolate base to improve the shelf life of cheese. Journal of Dairy Science. 2012;95:6282-6292
25. Lillevang SK, Mortensen G, Kristensen MN, Pedersen HS, Friis TL, Sorensen NK. Microbial oxygen absorber. U.S. Patent Application No. 14/307,979. U.S. Patent and Trademark Office. Washington, DC. 2014

26. Lillevang S, Mortensen G, Kristensen M, Pedersen H, Friis T, Sorensen N. Microbial oxygen absorber. U.S. Patent Application No. 11/602,328. U.S. Patent and Trademark Office. Washington, DC. 2007
27. Pantaleao I, Pintado MME, Pocas MFF. Evaluation of two packaging systems for regional cheese. Food Chemistry. 2007;102:481-487
28. Desobry S, Hardy J. Camembert cheese water loss through absorbent packaging. Journal of Food Science. 1994;59:986-989
29. Swartz WE, Hartford CG. Method for preventing spoilage, rancidity, and off-color by using a tricalcium phosphate ultraviolet light inhibitor. U.S. Patent No. 5,948,458. Washington, DC: U.S. Patent and Trademark Office. 1999
30. Marbler CA, Parmentier R. Packaging material. U.S. Patent No. 5,958,534. U.S. Patent and Trademark Office. Washington, DC. 1999
31. Biji KB, Ravishankar CN, Mohan CO, Gopal TS. Smart packaging systems for food applications: a review. Journal of Food Science and Technology. 2015;52:6125-6135
32. Brown SR, Forauer EC, D'Amico DJ. Effect of modified atmosphere packaging on the growth of spoilage microorganisms and Listeria monocytogenes on fresh cheese. Journal of Dairy Science. 2018;101:7768-7779
33. Mexis SF, Chouliara E, Kontominas MG. Quality evaluation of grated Graviera cheese stored at 4 and 12 C using active and modified atmosphere packaging. Packaging Technology and Science. 2011;24:15-29
34. Panfil-Kuncewicz H, Lis A, Majewska M. The use of oxygen absorbers for packaging ripened cheese. Polish Journal of Natural Sciences. 2015;30:285-295
35. Sanguinetti AM, Secchi N, Del Caro A, Stara G, Roggio T, Piga A. Effectiveness of active and modified atmosphere packaging on shelf life extension of a cheese tart. International Journal of Food Science & Technology. 2009;44:1192-1198
36. Arena F, Dainelli D, McAlister Jr LB. Oxygen scavenging films. U.S. Patent Application No. 15/574,635. U.S. Patent and Trademark Office. Washington, DC. 2018
37. Veronique COMA. Bioactive packaging technologies for extended shelf life of meat-based products. Meat science. 2008;78:90-103
38. Ozdemir M, Sadikoglu H. A new and emerging technology: Laser-induced surface modification of polymers. Trends in Food Science & Technology. 1998;9:159-167
39. CORDIS-Community Research and Development Information Service. [Internet] (2005) Biopack: Proactive biobased cheese packaging. Available from: https://cordis.europa.eu/project/id/QLK5-CT-2000-00799. [Accessed: 2020-08-15]
40. Hala MF, Ebtisam ED, Sanaa I, Badran MA, Marwa AS, Said ME. Manufacture of low fat UF-soft cheese supplemented with rosemary extract (as natural antioxidant). Journal of American Science. 2010;6:570-579
41. Bonilla J, Sobral PJ. Gelatin-chitosan edible film activated with Boldo extract for improving microbiological and antioxidant stability of sliced Prato cheese. International Journal of Food Science & Technology. 2019;54:1617-1624

42. Soto-Cantu CD, Graciano-Verdugo AZ, Peralta E, Islas-Rubio AR, Gonzalez-Cordova, A, Gonzalez-Leon A, Soto-Valdez H. Release of butylated hydroxytoluene from an active film packaging to asadero cheese and its effect on oxidation and odor stability. Journal of Dairy Science. 2008;91:11-19

43. Mushtaq M, Gani A, Gani A, Punoo HA, Masoodi FA. Use of pomegranate peel extract incorporated zein film with improved properties for prolonged shelf life of fresh Himalayan cheese (Kalari/kradi). Innovative Food Science & Emerging Technologies. 2018;48:25-32

44. Shin YJ, Song HY, Seo YB, Song KB. Preparation of red algae film containing grapefruit seed extract and application for the packaging of cheese and bacon. Food Science and Biotechnology. 2012;21:225-231

45. Huvaere K, Nielsen JH, Bakman M, Hammershoj M, Skibsted LH, Sorensen J, Dalsgaard TK. Antioxidant properties of green tea extract protect reduced fat soft cheese against oxidation induced by light exposure. Journal of Agricultural and Food chemistry. 2011;59:8718-8723

46. Rashidinejad A, Birch EJ, Sun-Waterhouse D, Everett DW. Effects of catechin on the phenolic content and antioxidant properties of low-fat cheese. International Journal of Food Science & Technology. 2013;48:2448-2455

47. Lee DS. Carbon dioxide absorbers for food packaging applications. Trends in Food Science & Technology. 2016;57:146-155

48. Jakobsen M, Risbo J. Carbon dioxide equilibrium between product and gas phase of modified atmosphere packaging systems: Exemplified by semihard cheese. Journal of Food Engineering. 2009;92:285-290

49. Fellows PJ. Food Processing Technology: Principles and Practice. (3rd ed.). Woodhead Publishing Ltd. Cambridge. 2009

50. Crump J. Method of rapid carbon dioxide absorption. U.S. Patent Application No. 13/480,222. U.S. Patent and Trademark Office. Washington, DC. 2012

51. Mortensen G, Bertelsen G, Mortensen BK, Stapelfeldt H. Light-induced changes in packaged cheeses-a review. International Dairy Journal. 2004;14:85-102

52. Yousif E, Haddad R. Photodegradation and photostabilization of polymers, especially polystyrene. SpringerPlus, 2013;2:398

53. Kristoffersen T, Stussi DB, Gould IA. Consumer-Packaged Cheese. I. Flavor Stability1. Journal of Dairy Science. 1964;47:496-501

54. Coughlin G, Schambony S. New UV absorber for PET packaging: better protection with less discoloration. Journal of Plastic Film & Sheeting. 2008;24:227-238

55. Var I, Erginkaya Z, Guven M, Kabak B. Effects of antifungal agent and packaging material on microflora of Kashar cheese during storage period. Food Control. 2006;17:132-136

56. Govaris A, Botsoglou E, Sergelidis D, Chatzopoulou PS. Antibacterial activity of oregano and thyme essential oils against Listeria monocytogenes and Escherichia coli

O157: H7 in feta cheese packaged under modified atmosphere. LWT-Food Science and Technology. 2011;44:1240-1244

57. Balaguer MP, Lopez-Carballo G, Catala R, Gavara R, Hernandez-Munoz P. Antifungal properties of gliadin films incorporating cinnamaldehyde and application in active food packaging of bread and cheese spread foodstuffs. International Journal of Food Microbiology. 2013;166:369-377
58. Han JH. Antimicrobial food packaging. In: Ahvenainen, R. (Ed.), Novel Food Packaging Techniques. CRC Press, Boca Raton, Florida, USA. 2003; p. 50-70
59. Singh A, Khamrai M, Samanta S, Kumari K, Kundu PP. Microbial, Physicochemical, and Sensory Analyses-Based Shelf Life Appraisal of White Fresh Cheese Packaged into PET Waste-Based Active Packaging Film. Journal of Packaging Technology and Research. 2018;1-23
60. Winther M, Nielsen PV. Active packaging of cheese with allyl isothiocyanate, an alternative to modified atmosphere packaging. Journal of Food Protection. 2006;69:2430-2435
61. Moreira MDR, Pereda M, Marcovich NE, Roura SI. Antimicrobial effectiveness of bioactive packaging materials from edible chitosan and casein polymers: assessment on carrot, cheese, and salami. Journal of Food Science. 2011;76:M54-M63
62. Dos Santos Pires AC, de Fatima Ferreira Soares N, de Andrade NJ, da Silva LHM, Camilloto GP, Bernardes PC. Development and evaluation of active packaging for sliced mozzarella preservation. Packaging Technology and Science: An International Journal. 2008;21:375-383
63. Cui H, Wu J, Li C, Lin L. Improving anti-listeria activity of cheese packaging via nanofiber containing nisin-loaded nanoparticles. LWT-Food Science and Technology. 2017;81:233-242
64. Gorrasi G, Bugatti V, Tammaro L, Vertuccio L, Vigliotta G, Vittoria V. Active coating for storage of Mozzarella cheese packaged under thermal abuse. Food Control. 2016;64:10-16
65. DSM [Internet] (2020). Available from: https://www.dsm.com/food-specialties/en_US/products/dairy/package.html [Accessed: 2020-06-18]
66. Lopez-Rubio A, Almenar E, Hernandez-Munoz P, Lagaron JM, Catala R, Gavara R. Overview of active polymer-based packaging technologies for food applications. Food Reviews International. 2004;20:357-387
67. Koontz JL. Special delivery: Controlled release of active ingredients from food and beverage packaging. Italian Packaging Technology Award (IPTA) Paper Competition. Department of Food Science and Technology, Virginia. 2006
68. Oliveira MA, Bastos MS, Magalhaes HC, Garruti DS, Benevides SD, Furtado RF, Egito AS. α, β-citral from Cymbopogon citratus on cellulosic film: Release potential and quality of coalho cheese. LWT-Food Science and Technology. 2017;85:246-251
69. Gerber LC, Koehler FM, Grass RN, Stark WJ. Incorporating microorganisms into polymer layers provides bioinspired functional living materials. Proceedings of the National Academy of Sciences. 2012;109:90-94

70. Emmi [Internet] (2020). Available from: https://www.emmiusa.com/cheese/fondu/. [Accessed: 2020-08-15]
71. 71.Romero V, Borneo R, Passalacqua N, Aguirre A. Biodegradable films obtained from triticale (X Triticosecale Wittmack) flour activated with natamycin for cheese packaging. Food Packaging and Shelf Life. 2016;10:54-59
72. Mills A. Oxygen indicators and intelligent inks for packaging food. Chemical Society Reviews. 2005;34:1003-1011
73. Mills A, Hazafy D. Nanocrystalline SnO2-based, UVB-activated, colorimetric oxygen indicator. Sensors and Actuators B: Chemical. 2009;136:344-349
74. O'Mahony FC, O'Riordan TC, Papkovskaia N, Kerry JP, Papkovsky DB. Non-destructive assessment of oxygen levels in industrial modified atmosphere packaged cheddar cheese. Food Control. 2006;17:286-292
75. Deshwal GK, Panjagari NR, Badola R, Singh AK, Minz PS, Ganguly S, Alam T. Characterization of biopolymer-based uv-activated intelligent oxygen indicator for food-packaging applications. Journal of Packaging Technology and Research. 2018;2:29-43
76. Hempel AW, Gillanders RN, Papkovsky DB, Kerry JP. Detection of cheese packaging containment failures using reversible optical oxygen sensors. International Journal of Dairy Technology. 2012;65:456-460
77. Pirsa S, Karimi Sani I, Pirouzifard MK, Erfani A. Smart film based on chitosan/Melissa officinalis essences/pomegranate peel extract to detect cream cheeses spoilage. Food Additives & Contaminants: Part A. 2020;37:634-648
78. Chen JH, Zall RR. Packaged milk, cream and cottage cheese can be monitored for freshness using polymer indicator labels. Dairy and food sanitation (USA). 1987
79. Sunesen LO, Lund P, Sorensen J, Holmer G. Development of volatile compounds in processed cheese during storage. LWT-Food Science and Technology. 2002;35:128-134
80. Cattaneo TMP, Giardina C, Sinelli N, Riva M, Giangiacomo R. Application of FT-NIR and FT-IR spectroscopy to study the shelf-life of Crescenza cheese. International Dairy Journal. 2005;15:693-700
81. Zaragoza P, Ros-Lis JV, Vivancos JL, Martinez-Manez R. Proof of concept of using chromogenic arrays as a tool to identify blue cheese varieties. Food Chemistry. 2015;172:823-830
82. Bento LM, Pereira MCS, Chaves KDS, Stefani R. Development and evaluation of a smart packaging for the monitoring of ricotta cheese spoilage. MOJ Food Processing & Technology. 2015;1:9-11
83. Stratton JE, Hutkins RW, Taylor SL. Biogenic amines in cheese and other fermented foods: a review. Journal of Food Protection. 1991;54:460-470
84. Kroger M, Weaver JC. Use of protein dye-binding values as indicators of the "chemical age" of conventionally made Cheddar cheese and hydrolyzed-lactose Cheddar cheese. Journal of Food Science. 1979;44:304-305

85. Benedetti S, Sinelli N, Buratti S, Riva M. Shelf life of Crescenza cheese as measured by electronic nose. Journal of Dairy Science. 2005;88:3044-3051

86. Tavaria FK, Ferreira AS, Malcata FX. Volatile free fatty acids as ripening indicators for Serra da Estrela cheese. Journal of Dairy Science. 2004;87:4064-4072

87. Downey G, Sheehan E, Delahunty C, O'Callaghan D, Guinee T, Howard V. Prediction of maturity and sensory attributes of Cheddar cheese using near-infrared spectroscopy. International Dairy Journal. 2005;15:701-709

88. Martin-del-Campo ST, Picque D, Cosio-Ramirez R, Corrieu G. Middle infrared spectroscopy characterization of ripening stages of Camembert-type cheeses. International Dairy Journal. 2007;17:835-845

89. Tsironi T, Stamatiou A, Giannoglou, M, Velliou E, Taoukis PS. Predictive modelling and selection of time temperature integrators for monitoring the shelf life of modified atmosphere packed gilthead seabream fillets. LWT-Food Science and Technology. 2011;44:1156-1163

90. Shellhammer TH, Singh RP. Monitoring chemical and microbial changes of cottage cheese using a full-history time-temperature indicator. Journal of Food Science, 1991;56:402-405

91. Toppino PM, Cigoli A, Civardi G, Riva M. Shelf life of packed Taleggio cheese. 1: Conventional parameters evaluation and use of TTI devices [Time Temperature Indicators-Lombardy]. Scienza e Tecnica Lattiero Casearia (Italy). 2002

92. Riva M, Toppino PM, Cattaneo TMP. Use of TTI devices to monitor and evaluate the shelf-life of classic and light caprino-type cheese. Scienza e Tecnica Lattiero Casearia (Italy). 1999

93. Baughman RH, Hall LJ, Kozlov M, Smith DE, Prusik T. Time-temperature indicators comprising crystallized diacetylenic indicator compounds. U.S. Patent No. 8,813,675. U.S. Patent and Trademark Office. Washington, DC. 2014

94. Pinho O, Ferreira IM, Mendes E, Oliveira BM, Ferreira M. Effect of temperature on evolution of free amino acid and biogenic amine contents during storage of Azeitao cheese. Food Chemistry. 2001;75:287-291

95. Abdelnour A, Buchin F, Kaddour D, Tedjini S. Improved Traceability Solution Based on UHF RFID for Cheese Production Sector. IEEE Journal of Radio Frequency Identification. 2018;2:68-72

96. Regattieri A, Gamberi M, Manzini R. Traceability of food products: General framework and experimental evidence. Journal of Food Engineering. 2007;81:347-356

97. Perez-Aloe R, Valverde JM, Lara A, Castano F, Carrillo JM, Gonzalez J, Roa I. Use of RFID tags for data storage on quality control in cheese industries. In: Data Storage. IntechOpen. 2010. p. 213-226

98. Papetti P, Costa C, Antonucci F, Figorilli S, Solaini S, Menesatti P. A RFID web-based infotracing system for the artisanal Italian cheese quality traceability. Food Control. 2012;27:234-241

99. Bornhovd C, Lin T, Haller S, Schaper J. Integrating automatic data acquisition with business processes experiences with SAP's auto-ID infrastructure. In: Proceedings of the Thirtieth international conference on Very large data bases. VLDB Endowment. 2004; p. 1182-1188

100. Lopez-de-Dicastillo C, Catala R, Gavara R, Hernandez-Munoz P. Food applications of active packaging EVOH films containing cyclodextrins for the preferential scavenging of undesirable compounds. Journal of Food Engineering. 2011;104:380-386

101. Pulinas L, Spanu C, Idda I, Ibba I, Nieddu G, Virdis S, Scarano C, Piras F, Spano N, Sanna G, De Santis EPL. Production of farmstead lactose-free Pecorino di Osilo and ricotta cheeses from sheep's milk. Italian Journal of Food Safety. 2017;6:33-39

102. Hong Q, Liu XM, Hang F, Zhao JX, Zhang H, Chen W. Screening of adjunct cultures and their application in ester formation in Camembert-type cheese. Food Microbiology. 2018;70:33-41

103. Lecanu L, Ducruet V, Jouquand C, Gratadoux JJ, Feigenbaum A. Optimization of headspace solid-phase microextraction (SPME) for the odor analysis of surface-ripened cheese. Journal of Agricultural and Food Chemistry. 2002;50:3810-3817

104. Kane W. Surface treated cheese package and method. U.S. Patent Application No. 3,784,711. U.S. Patent and Trademark Office. Washington, DC. 1974

105. Bilska A. Packaging systems for animal origin food. Scientific Journal of Logistics. 2011;7:35-44

106. Singh P, Abas Wani A, Saengerlaub S. Active packaging of food products: recent trends. Nutrition & Food Science. 2011;41:249-260

107. Crump JW, Chau CC, McKedy GE, Pyne DS, Powers TH, Solovyov SE, Hurley TJ. Oxygen, water vapor and carbon dioxide absorption in a single use container. U.S. Patent Application No. 14/3200,192. U.S. Patent and Trademark Office. Washington, DC. 2015

108. Coldchain Technology Services. [Internet] (2020) Available from: https://coldchain-tech.com/consulting/thermal-systems/greenbox.html [Accessed: 2020-06-26]

109. Acerbi F, Guillard V, Aliani M, Guillaume C, Gontard N. Novel methodology for the in situ assessment of CO2 production rate and its application to anaerobic ripened cheese. Food Research International. 2015;78:295-301

110. Wang C, Yusufu D, Mills A. A smart adhesive' consume within'(CW) indicator for food packaging. Food Packaging and Shelf Life. 2019;22:100395

111. Alvarez-Lorenzo C, Concheiro A, Dubovik AS, Grinberg NV, Burova TV, Grinberg VY. Temperature-sensitive chitosan-poly (N-isopropylacrylamide) interpenetrated networks with enhanced loading capacity and controlled release properties. Journal of Controlled Release. 2005;102:629-641

Chapter-9

Demystifying Emulsifiers as Additives Through Molecular Gastronomy: A Contribution to Rise Consumer's Sovereignty

1. Introduction

Despite all the literature available concerning the food additives, the availability of easy reading yet scientifically sounding information about the nature, chemical composition, role, and safety of the additives is scarce. It appears that, in a world where fake information is quickly spread in social media and food additives start to be perceived as harmful, the development of tools that permit the demystification of their use and therefore their acceptance, is essential as they, at least, contribute to the expansion of food's the shelf life and fight waste.

Following the work that our group has undertaken, to raise the consumer's knowledge about additives, we desire to present and explain in a simple manner some physical/chemical properties of emulsifiers, EU codes, in the 400–499 range, namely fatty acids esters and sucrose esters of fatty acids. How those additives are used in food preparation in order to achieve and maintain certain desirable characteristics, how they contribute to obtain a better result in final preparation and how they can be used in modern cuisine will be illustrated and reviewed recurring to recipes of molecular gastronomy. Preparations coming from the discipline of molecular gastronomy have been chosen since they are based on laboratory related procedures and they use only a few ingredients, including an additive.

This review chapter is a means to attain a final objective which is to develop tools for the demystification of food additives targeted to wider target people audience than researchers as, for example, undergraduate students of chemical, biological and nutrition sciences, students of gastronomy, chefs and high school teachers, in the hope of encouraging them to spread genuine scientific information based, as far as possible, on the current state-of-art. This will contribute to the raising of the level of knowledge of those additives thus allowing the consumer to make informed decisions.

Emulsifying agents are discussed first, the chapter begins with a simple review of emulsification and interfacial tension. Then the E471 and E473 class of additives will be presented (chemical constitution and relevant physical properties). For each class the most common uses in food processing will be discussed as well as their role in a particular recipe. Finally, a brief summary concerning their safety will be made.

2. Interfacial Tension and Emulsions

A mixture is formed when it is thermodynamically more stable than the states given by its separate components. That occurs when the interaction energies between molecules of the mixture are higher than the sum of the interaction energies of the primary constituents [1]. This condition is usually satisfied when the molecular constituents that are to be mixed have cohesive forces of similar magnitude. Otherwise, they do not tend to mix, even when shaken and whipped, and phase separation occurs spontaneously [2]. Anyhow, components must contact in the interfaces and interfaces are particular: surface phenomena such as surface tension (or interfacial tension) applies. The term interfacial tension usually relates to the liquid/liquid and liquid/solid phase boundaries; the term surface tension applies to the liquid/gaseous interface and the term surface free energy applies for the solid/gaseous. The surface free energy affects wettability [3].

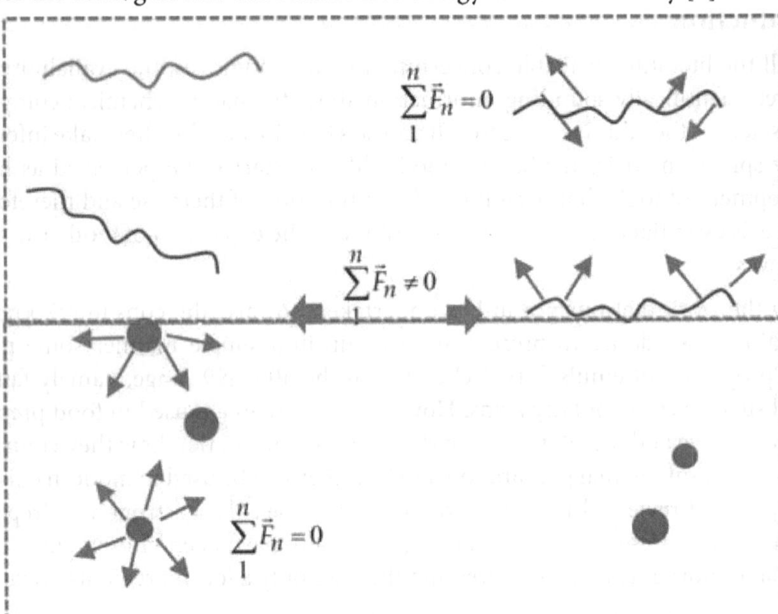

Figure 1. Schematic representation of the differences on the net forces in two liquid that are not mixable, e.g., polar water molecules (in red circles) and non-polar oil molecules (blue curved lines): in the bulk of each liquid the net forces between molecules are balanced while on the interface the net forces acting on the molecules are not, as consequence of the differences on the mutual interaction energies. Such molecules tend to undergo a displacement to the respective bulk solution and new molecules are displaced to the inner solution to replace them. This dynamic is kept in order to maintain the interface and the energy required to do it, per surface unit, is called the interfacial tension.

Interfacial tension and surface tension are defined as the work that must be done to maintain the unit of interface or free surface, respectively [1]. By way of an example, consider the interface of two immiscible liquids (oil and water), see Figure 1. The molecules placed in the boundary surface are interacting with molecules of both liquids, as such, the sum of interactive forces that act in the boundary molecule is unbalanced. That means that there is a net force that tends to push those molecules into their respective bulk. Thus, in order to keep the surface or the interface, a network of molecules is formed against the pull, as so, work must be done in order to maintain or increase the size of the interface. As consequence, when a water solution is vigorously stirred with an oil solution, the droplets formed after stirring tend to aggregate and phase separation occur. This happens because the system will minimise its energy when it reaches the smaller surface of contact [3].

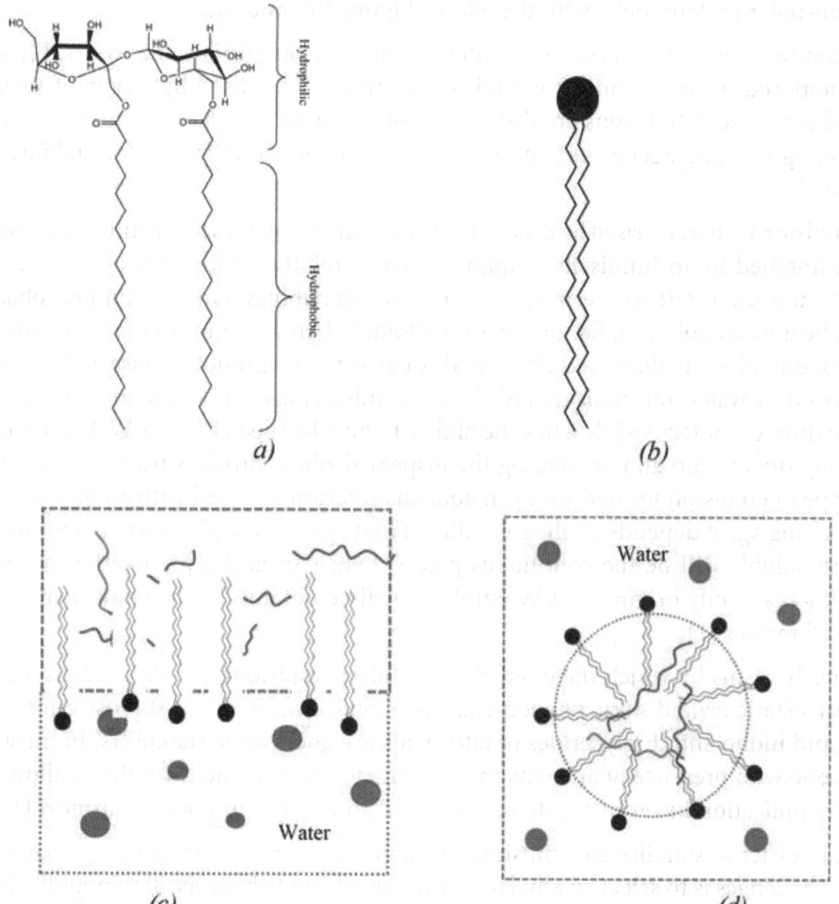

Figure 2. (a) Structure of a sucrose di-ester (derived from stearic acid). The polar part of the molecule is identified as hydrophilic and the non-polar as hydrophobic (or lipophilic), (b) usual scheme representation of the molecule, (c) the surface-active agent has the ability to stabilise the interface: The polar part of the molecule interacts preferentially with the polar liquid (water) and the non-polar preferentially with the oil; (d) when the mixture is shaken, droplets of oil disperse into the water phase (continuous phase).

The additives that will be here presented can act as surfactants that is, they have the capability of modify the surface tension/interfacial tension/surface free energy of the heterogeneous systems. As so, they can act as emulsifiers and/or humectants. This ability comes from their chemical constitution: they are amphipathic molecules, *e.g.*, they are unbalanced in their charge distribution, which means that one can recognise a part of the molecule that is polar (hydrophilic) and another that is non-polar (hydrophobic), see Figure 2(a) for an example (a sucrose ester). Those kinds of molecules are named as surfactants agents due to their ability to modify the interface/surface tension value. If this value is lowered by action of the surfactant agent, then they stabilise the interface. The mechanism by which this is possible is related to the amphiphilic nature of the molecule: the polar part of the molecule will interact preferentially with the polar liquid (water) and the non-polar preferentially with the oil, see Figure 2(c) and (d).

Two common two-phase systems in food products are emulsions and foams. Emulsions are composed of two immiscible liquids that may be stabilised by a type of surfactant named emulsifier [3]. Foams are dispersions of gas bubbles in a liquid continuous phase containing foaming agents, i.e., surface-active substances that ensure the stability of the system [2].

An emulsion is characterised by the existence of small droplets that constitute the dispersed phase imbibed in an immiscible liquid, which forms the continuous phase, see Figure 2(d). Foams are a state where air (gas) is confined as bubbles into a continuous phase that could be a water solution, fat, or a hydrocolloid. When the emulsion is called oil–water (O/W), the oil is the dispersed phase and water is the continuous phase and the reverse happens for a water–oil emulsion (W/O). An emulsifier agent is a tension-active molecule (or mixture of molecules) that has the ability to be adsorbed at the O/W/W/O interface, lowering the tension and preventing the dispersed-phase droplets from aggregating [4]. The type of emulsion formed when an aqueous solution is mixed with an oil and with an emulsifying agent depends on the solubility of that agent: "The phase where the emulsifier is most soluble will be the continuous phase", hence, even if a preparation has 60% of constituents of oily origin an O/W emulsion will be obtained if the emulsifier chosen is soluble in water [1].

The mechanisms by which the emulsifiers stabilise emulsions and dispersions are based in electrostatic or/and steric phenomena. Non-ionic emulsifiers, as sucrose esters of fatty acids and mono and di-glycerides of fatty acids are good steric stabilisers. In most cases, they cope with presence of any ions in solution and do not react with them, allowing for a wide application in various systems, especially in water of unknown hardness [1, 5].

Thus, in order to stabilise an emulsion, the appropriate emulsifier must be selected. One of the challenges is to select an emulsifier with an adequate hydrophilic/lipophilic balance (HLB) suitable for the system. The HLB is a measure of the hydrophilicity of an emulsifier: higher HLB values indicate a higher hydrophilicity. Emulsifiers with HLB values of 3.5–6 tend to be the most suitable for W/O emulsions, and those with HLB values from 8–18 are best suited for O/W emulsions. Surfactants with values ranging from 7–9 are good wetting agents [1, 5].

Most commercial emulsifiers, especially glycerides of fatty acids and sucrose esters, are mixtures of mono-, di- and tri-esters. Emulsifier HLB values are largely dependent on the monoester content; a greater monoester content leads to a higher HLB value [5]. The HLB value also depends on the chain length(s) of the fatty acids attached to the emulsifier; short fatty acid chains lead to higher HLB values [6].

Another factor to take into account is the molecular weight of the surfactant: emulsifiers with low molecular weight are usually more rapidly adsorbed onto the phase interface and prevent droplet coalescence during homogenisation, they are a preferable choice for making micro-emulsions, as those used in food processing [7, 8].

2.1 The critical micelle concentration

A micelle is considered a supramolecular aggregate shape of molecules with spherical that is formed by surfactant molecules. Micelles in water are aggregates with the hydrophilic part of the molecule contacting with surrounding solvent and the hydrophobic tail regions in the micelle centre. The micelle content can be air (see Figure 3) or a fat. In colloidal and surface chemistry, the critical micelle concentration (CMC) is defined as "the concentration of surfactants above which micelles form and all additional surfactants added to the system will form micelles".

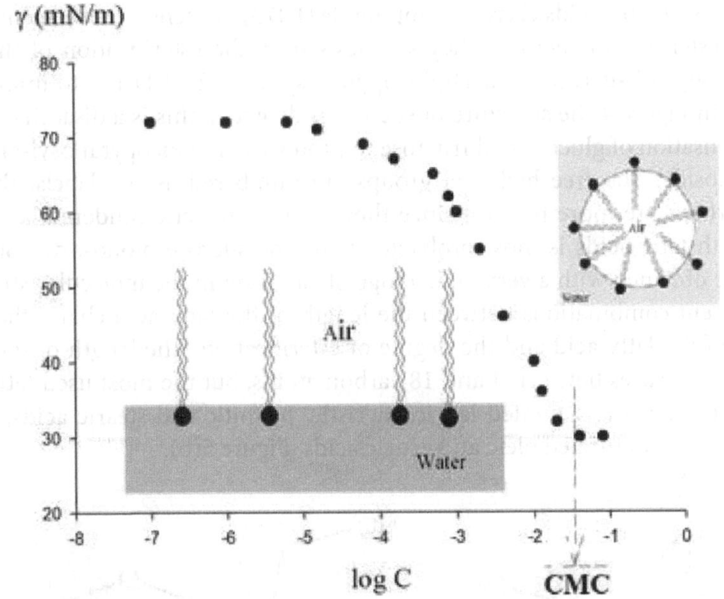

Figure 3. A graphical representation of the surface tension (γ) of water solutions of a fatty acid ester of a glucopyranoside at various concentrations. The arrow indicates the value of the critical micelle concentration (CMC): before the CMC there is a strong change of the surface tension; after the CMC that value is minimum and stays constant and the air micelle in water is stabilised. Adapted from [9, 10].

Figure 3 depicts a graphical representation of the surface tension (γ) of water solutions of a fatty acid di-ester of a glucopyranoside at various concentrations. When the concentration of the surfactant is low, the whipping of the solution will not give "stable bubbles", but this will occur after the surfactant concentration reaches the CMC (value indicated by the arrow). As seen, before the CMC there is a strong change of the surface tension and after the CMC that value reaches a minimum and stays independent of the concentration, giving stability to the air droplets. For the mono and di-glycerides of fatty acids and as well for sucrose esters the stability of the micelles is given by the strong hydrogen bonds that the sugar moiety can make with water [9, 10].

The CMC depends on the temperature, the pressure and it also changes with the composition of the water solution, namely with the presence of other surface-active compounds.

3. E473 and E471 Classes of Additives

This revision will concern about two classes of food additives, the sucrose esters of fatty acids (E473) and the mono- and diglycerides of fatty acids (E471). Both are widely used nowadays in food industry and thus present in the preparations available in supermarket shelves. The sucrose esters of fatty acids are used to obtain mostly O/W emulsion (but not only) and mono- and diglycerides of fatty acids are used to obtain W/O emulsions. Both are good agents for form stable foams.

3.1 Sucrose Esters of fatty acids (sucrose esters), E473

Sucrose esters of fatty acids class of compounds (E473) are tensioactive agents that have fatty acids esterified in sucrose. They synthesised by the esterification of the available hydroxyl groups of sucrose (O-α-D-glucopyranosyl-(1 → 2)-β-D-fructofuranoside) and fatty acids. In Figure 4 the structure of sucrose is depicted: this is a disaccharide formed from condensation of glucose and fructose to produce α-D-glucopyranosyl-(1 → 2)-β-D-fructofuranoside. The free hydroxyl groups are numbered as red labels; the primary hydroxyl groups are more reactive since they are not so steric hindered, as such, their reaction with fatty acids is most probable giving the sucrose mono-, di-, or tri-esters. They can be obtained with a very wide range of variation in the molecular structure due to the different combinations between the length of the fatty acid chain, the degree of saturation of the fatty acid and the degree of esterification. The length of the fatty acid chain generally varies between 4 and 18 carbon atoms, but the most used fatty acids for esterification are i) the saturated lauric, myristic, palmitic and stearic acids, see Figure 5(a), and ii) the unsaturated oleic and erucic acids, Figure 5(b).

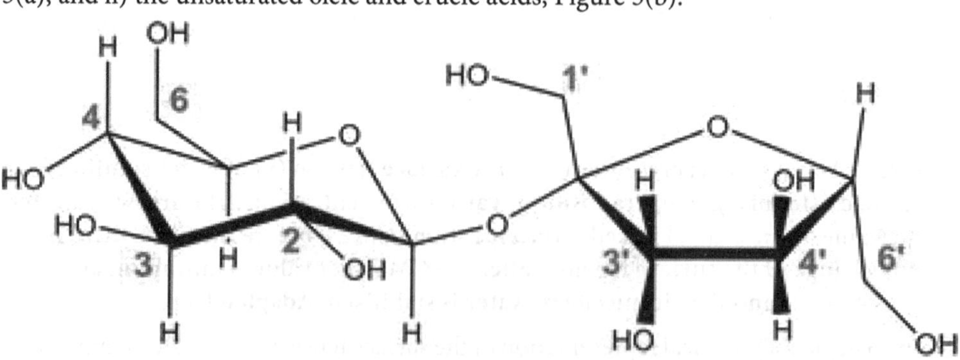

Figure 4. The structure of produce α-D-glucopyranosyl-(1 → 2)-β-D-fructofuranoside (sucrose). The labelled atoms are carbons atoms containing hydroxyl groups available for sterifications. The primary C1', C6' and C6 are the most reactive.

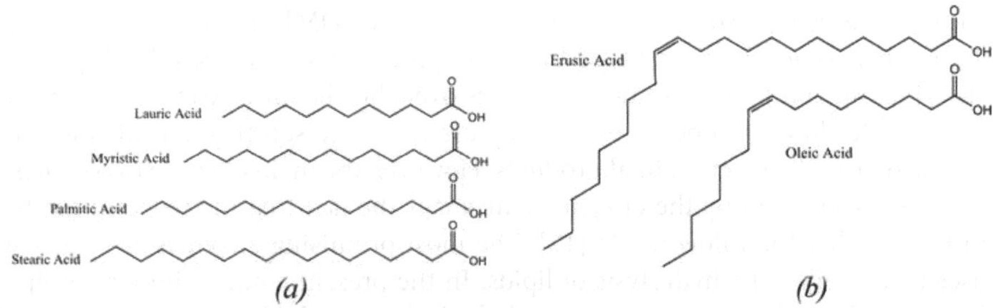

Figure 5. The most used fatty acids for esterification into sucrose (a) saturated fatty acids (b) unsaturated fatty acids.

Sucrose esters can be synthesised either chemically or enzymatically. In industrial scale they are obtained by trans-esterification (see Figure 6): the reactants are sucrose, a triglyceride (TG), potassium carbonate as catalyst in *N,N*-dimethyformamide (DMF, **CAS:** 68–12-2) or dimethylsulfoxide (DMSO, **CAS:** 67–68-5) as solvents. This reaction gives a product that is a mixture of the sucrose monoester (> 50%), sucrose di-esters (~10%) higher esters, sucrose and TAG [11]. This processing method has issues connected with the removal of the solvents that have high boiling points [426.2 ± 0.7 K (~153°C) for DMF and 463 ± 1 K (~189°C) for DMSO] and the separation of the monoester from the other by-products. Sucrose monoesters may be also produced in large scale by the reversible trans-esterification process, see Figure 6, using sucrose and a fatty acid methyl ester as reagents and sucrose esters and methanol will be obtained as products. This this a reversible process, so, the methanol should be removed during the reaction. Under the optimal conditions this protocol can provide material containing upwards of 70% of the monoester [11].

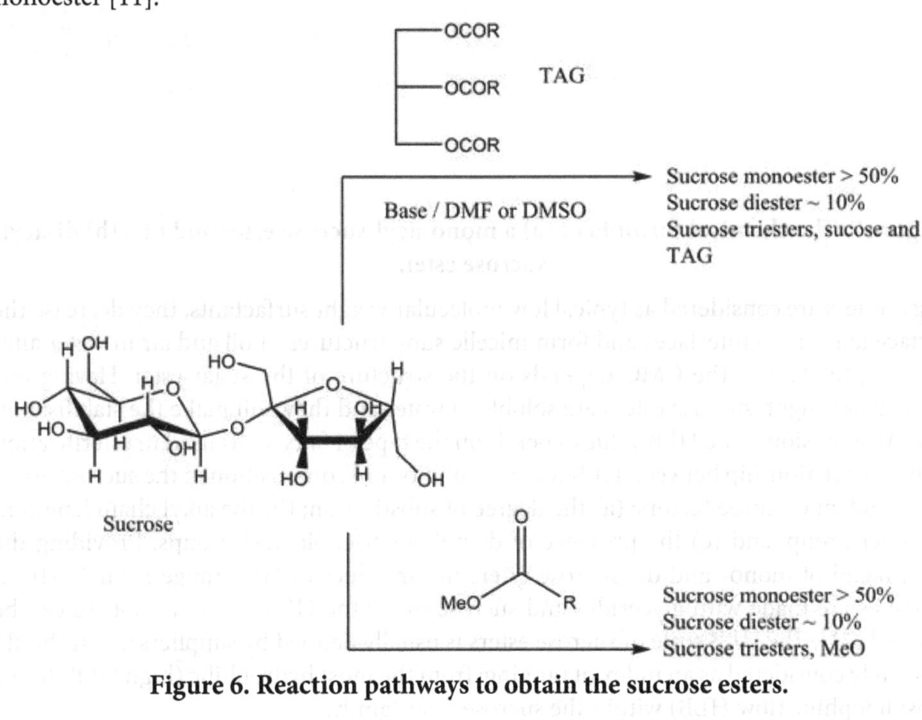

Figure 6. Reaction pathways to obtain the sucrose esters.

Chemical synthesis uses solvents such as DMSO and/or DMF that display toxicity [12, 13, 14]. In alternative, enzymatic based reaction procedures have been developed more recently [15, 16], and present some advantages over the chemical reactions: i) they can take place under lower temperatures ii) they use less toxic solvents and iii) they allow better control over the reaction final products. However, the main obstacle to commercial sucrose ester production by the enzymatic means is the fact that the reaction has to be run in batch rather than flow mode [17]. The most promising enzymes are a group of esterases that catalyse the hydrolysis of lipids. In the present context, important lipases are the triacylglycerol lipases that selectively hydrolyse insoluble triacylglycerol's at the substrate-water interface [18, 19].

Most of the physical–chemical characteristics of a determined sucrose ester are related to the kind of fatty acids used for esterification. In Figure 7 the chemical formula of a di-ester of sucrose is show, where the esterification occurred in the 1 and 6' position and the fatty acid used was the stearic acid.

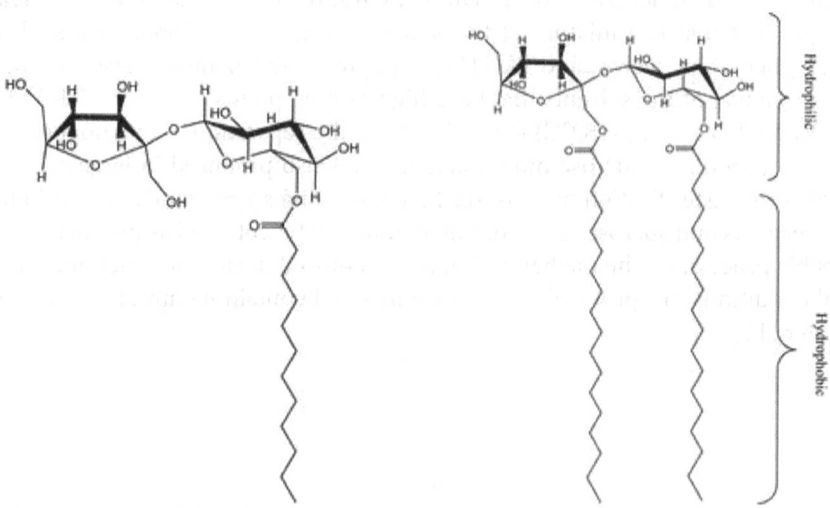

(a) (b)

Figure 7. The chemical formula of (a) a mono-acyl-sucrose ester and of a (b) di-acyl sucrose ester.

Sugar esters are considered as typical low molecular weight surfactants, they decrease the surface tension on interfaces and form micelle sub-structures of oil and air in water after reaching the CMC. The CMC depends on the structure of the sugar ester. Having low molecular weight, sucrose esters are soluble in water and they will make the stabilisation of O/W emulsions. The HLB value depends on the type of fatty acids used in esterification [20]. The relationship between HLB water-solubility and composition in the sucrose esters is dependent on three factors: (a) the degree of substitution; (b) the alkyl chain length in the ester group; and (c) the presence of dienoic or trienoic acyl groups. Providing the right blend of mono- and di- sucrose esters linear values of HLB range from 10–16. If mixtures are made with glycerides and sucrose esters the HLB range from 4–16 can be covered [21]. The HLB scale of sucrose esters is usually defined by suppliers and it should be merely considered as an index of ranking: from the most hydrophilic (high HLB) to the most lipophilic (low HLB) within the sucrose ester family.

Some authors [22] worked on the clarification of the real HLB for the sucrose esters. They concluded that the experimental HLB of sucrose monoesters, would be around 11–12 for short fatty chains (6 to 12 carbons) and around 10–11 for long fatty chains (14 to 18 carbons). Those values are slightly different than the values presented by the producers that range from 2–16. Sucrose esters powders may come with residual amounts of fatty acids and those, by their turn, may appear in the neutral form or in the protonated form. Fatty acids in neutral state are also able to act as surfactants and they are good to stabilise the O/W emulsions.

The easiest parameter to evaluate the thermal stability of sucrose esters is the melting point, it can vary within the 40–60°C range depending on the type of fatty acids that have been esterified. Nevertheless, the emulsions can be heated to about 180°C (they decompose above 220°C) without compromise their emulsification properties of the surfactants (however a colour change may be observed in high temperatures if sugars are present due to caramelisation). The thermal properties of sucrose esters depend on the various HLB values: high or medium HLB values tend to be vitrified by melting. Since sucrose esters are usually a mixture, apart from crystallisation, they also present amorphous structures which slowly crystallise in time [23].

Sucrose esters are stable in the pH range of 4 to 8, so they can be used as an additive in most foods. At pH higher than 8, saponification (hydrolysis of the ester bond to release the original sucrose and the salt of fatty acids) might occur. Hydrolysis could also occur at pH lower than 4.

3.1.1 Common Uses for Sucrose esters

Monoesters of sucrose esters are most commonly studied, although applications of di-esters or higher esters have also a wide range of applications. The sugar esters can be used in white sauces, dairy products that are alternative to milk, aerated dairy desserts, like ice cream, mousses, bakery, confectionery, preparation of cereal based products [18].

Thus, no wonder that the use of sucrose esters as surfactants in the food industry has been increasing exponentially. Apart from the emulsification power that will be discussed next, they also present advantages concerning about their biodegradability: after ingestion, the sucrose esters are hydrolysed by digestive enzymes into sucrose and their component fatty acids that are further metabolised in the usual way [5].

In a general approach the use of sucrose esters as additive in food preparations will allow the formation of O/W emulsions. The presence of the sucrose ester will replace the oil (fat) continuous phase by smaller globules/oil droplets imbibed into a water phase. The encapsulation of the oil will protect it against oxidation, which occurs in the presence of oxygen and light. This will improve the stability of the prepared food and will increase the shelf life. They can also act as humectants as they can make the stabilisation of the water/ air interface, as so, they are popular in preparations of foams. In addition of acting as emulsifiers and texturizers, they present several other advantages such as the protection of proteins from flocculation, the prevention of sugar crystallisation and they form stable structures resulting from interaction with starch.

3.1.1.1 Preparations based in oil/water emulsions

Typical O/W emulsions are mayonnaise and dressings. Sucrose esters are used in their preparation as an alternative to egg yolk. Sucrose esters with a high HLB value are soluble in cold water so the preparation can be made at room temperature. They also do permit the formation of stable emulsions with very small droplets; this is only possible due to their ability of significantly lowering the tension in the oil–water interface. Another important feature is that emulsions can be made with homogeneous droplet radius; this will avoid coalescence of the small droplets into to bigger, ones once the former have higher Laplace pressure.

Visually the emulsions made with sucrose esters appear like a white sauce and they impart a creamy sensation. A large number of small oil droplets give the impression of a higher fat content, thus, the creamy mouth feel and flavour perception of the preparations and will mimic the sensation of taste provided by fats. In addition, the small droplets will scatter the light into various directions increasing the brightness of the preparation and giving a whiter appearance. In the preparation of mayonnaise, the emulsion must be made before adding the acidic components in order to prevent the hydrolysis of the ester that can occur at values below pH 4.

Sucrose esters can replace milk proteins in the preparation of white sauces and improve the stability of the emulsion. There are sucrose esters (or mixture of sucrose esters) in the market that have higher O/W emulsification power than proteins. The small size of the oil droplets obtained and their homogeneity, provide high stability against flocculation and the emulsions may not require a high viscous continuous phase. The viscosity of the continuous phase is, generally, obtained by adding starch (in this case it will work as a stabiliser). The advantage in gastronomic terms is to have sauces that are less sticky, more satiny and with good power to release flavours. Furthermore, the preparation does not decompose by Maillard reaction[1] during pasteurisation or sterilisation. Drinks based on soy, almond, rice and coconut are popular alternative products to milk. The addition of sucrose esters in their processing allows the stabilisation of the oil emulsion in a continuous hydrophilic phase that usually displays low viscosity. Again, "the milk alternative product" will appear whiter. Usually the mono-esterified esters are chosen since they interact better with proteins that are pH sensitive due to their isoelectric point. The tension-active agent will interact with these proteins, stabilising preparations against flocculation [28].

Pre-preparations of fat-soluble substances for further water dilution: fat soluble substances such as omega-3 and some natural dyes may be previously mixed with sucrose esters and kept as a concentrated emulsion for further incorporation into water-dispersed form in beverages, dairy products, and confectionery.

3.1.1.2 Preparations based in oil/water emulsions with significant sugar content

In preparations with sugar like chewy soft candy and ices, sucrose esters are also widely used nowadays. The reason for that is related to the capacity of sugar esters to control sugar crystallisation. Sucrose esters promote the formation of secondary grains during crystal growth process and prevent the agglomeration by making a homogenisation of the crystal size. The delay on sucrose crystallisation may be due to the decrease of the molecular mobility [29] or to the disrupting the crystal lattice due to molecular interactions between sucrose and the additive [30]. The polar group of the sucrose ester interferes with crystallisation after cooling precluding the crystal growth: during cooling the non-polar fatty acid of the sucrose ester will move from the polar sugar solution to the less polar sugar crystal. As a result, the sugar crystals will be surrounded by a layer of fatty acids that will prevent further crystal growth see Figure 8. The re-solubilisation and re-crystallisation of the sugar will be inhibited. The sugar ester layer around the micro-crystalline nuclei stabilises the outer shell of water and prevents water evaporation from the sugar mixture (in fondants). This will result in a gritty mouth feel.

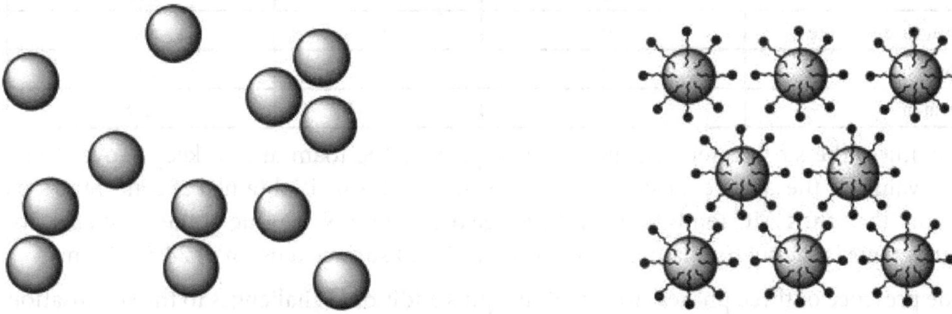

Figure 8. Effect of sucrose ester into the crystal growth of sugar (a) without the surfactant; (b) with surfactant: the crystals are surrounded by the fatty acid chain that protect the micro-crystals from further growth.

Bakery products with sucrose esters as additives are very usual. The esters form a flexible network with gluten making the preparations more resistant to mechanical stress during processing. In addition, they contribute to maximise gas retention due to their properties of stabilisation of the air/solid interface (see 3.2.1.3). Sucrose esters are also able to interact with starch: they have the ability to complex with the linear and helical amylose and the branched amylopectin. In consequence the temperature and enthalpy of gelatinization will change [31]. The capacity of interaction with starch varies with the length of the monoester and the degree of saturation. The sucrose fatty acids interact preferentially with amylose and form a helical complex during gelatinization. This reaction prevents retrogradation of the starch and increases the long-life term [32].

3.1.1.3 Aerated oil/water emulsions preparations

Sucrose esters are good humectant agents since they significantly decrease the surface tension value of water solutions or hydrocolloids. They help into aeration and promote foam stabilisation.

Foams correspond to the macroscopic manifestation of an air/liquid or air/solid interface, air appearing as bubbles. The dispersive phase can be a water solution, fat, protein or hydrocolloids. Water has a high surface tension thus, without a surfactant, it is difficult to retain air inside. The addition of a sugar ester to the preparation reduces the surface tension in the interphase and makes the foam stable. Table 1 gives the values for surface tension of several sucrose surfactants [33, 34].

Table 1. Values for hydrophilic/lipophilic balance (HLB) and surface tension (Du Noüy method) for several sucrose esters surfactants and for water at T = 20°C [33]. Concentration above the critical micelle concentration.

	% of monoester	HLB value	Surface tension 10^{-3} (N/m)
Sucrose laurate	70	15	28.5
Sucrose palmitate	75	16	34.0
Sucrose stearate	70	15	34.5
Sucrose stearate	50	11	36.7
Sucrose stearate	30	6	46.8
water			72.8

The lower the surface tension the easier to prepare the foam and to keep it over time. The value for the surface tension of air/water interfaces that is obtained at concentrations higher than the CMC tends to be in the range of 24–44 mN/m. One of the most effective surfactants is the sucrose laurate with HLB of 15 and surface tension of 28.5 mN/m.

The presence of three phases (water/oil/air) puts additional challenges to the stabilisation of the preparation. First, the presence of fatty materials destabilises the aqueous foams because the water film around the air bubble will be squeezed by the presence of the fat: this will decrease its thickness leading to the collapse of the air bubble. If an emulsion is made previously to aeration the surfactant must be used to stabilise the fat droplets first in order to preclude the interference of the fat with the water layer around the bubble. For this purpose, the use of a surfactant with medium HLB value (6–11) is the better choice since they can make stable oil-in-water emulsions. However, to ensure a longer shelf life without syneresis and retraction, preparations must contain stabilising and gelling agents.

The second challenge deals with the presence of proteins. Proteins are surface active agents and compete with the sucrose esters in the surface of air/water and in the interface of oil/water droplets and, as seen, generally the formation of the oil/water emulsion precedes the aeration process. In preparations of creams those proteins are egg and milk proteins. It seems that egg proteins are more effectively absorbed in the oil/water interface than milk proteins [35], but low weight surfactants are able to displace the proteins from the fat layer drops, lowering the steric stabilisation made by the macromolecules, especially if they are daily proteins [36]. This will increase the probability of the coalescence of the emulsion during whipping. Those effects have been studied at least one decade ago and it seems that the content of sucrose ester is a key to obtain a good stability for the foam: in concentrations lower than 0.1% the amount of sucrose ester that is absorbed into the surface oil films is insignificant while in concentrations higher than 0.1125% most of the proteins are displaced by the surfactant in the oil/water interface [37], that is, little protein is left around the oil droplets of the emulsion due to the preferential adsorption of sucrose ester over milk proteins (Figure 9).

Additionally, the shear stress given by whipping promotes a higher desorption of milk proteins on the oil/water interface, that is, the egg proteins are better for preparation of stable foams, but in order to get a stable foam, partially coalescence of fat droplets should occur [38, 39].

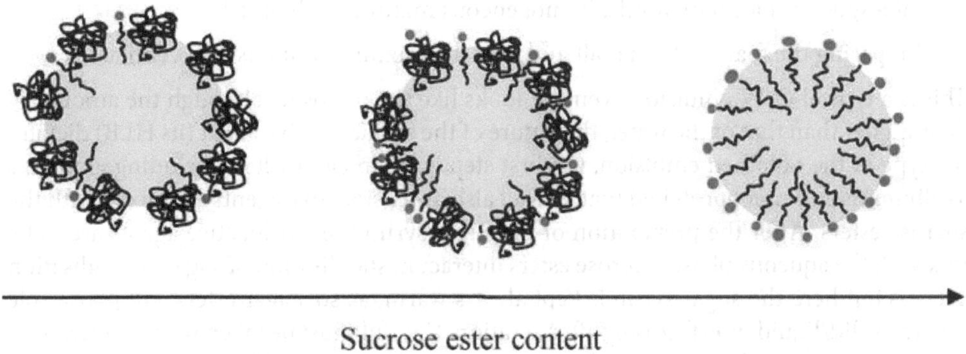

Figure 9. There is a competitive absorption in the oil–water interface between milk proteins: in consequence a mixed surfactant protein film is formed around oil droplets. This last film is weaker than that obtained when only protein is present and will contribute to the destabilisation of the emulsion. The higher the content of sucrose ester in the preparation, the higher the degree of substitution.

3.1.2 Examples for the use of sucrose esters in molecular gastronomy

3.1.2.1 Preparation of a butter cream

The ingredients for the preparation of a butter cream are: egg whites (4 units); sugar (267 g) and water (60 g); unsalted butter (454 g); salt (1.5 g) and sucrose esters (5.4 g) (Figure 10).

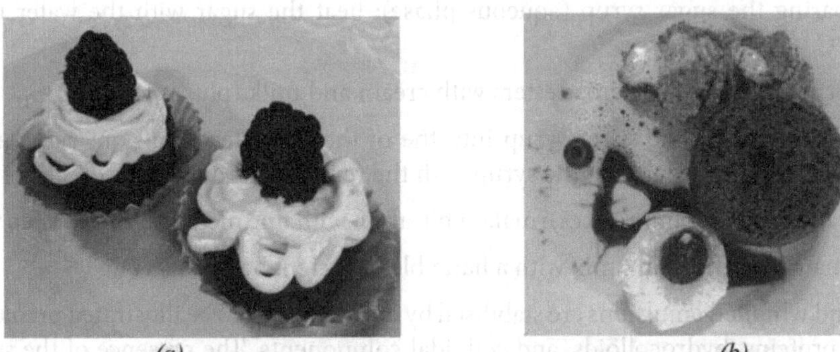

(a) *(b)*

Figure 10. Butter cream (a) and cream bubbles (b) preparations using sucrose esters, suggestion of presentation.

Preparation consists in:

1. Mixing the surface-active agents: whisk the whites and sucrose esters and beat in an electric mixer initially at low speed (1) and then at medium speed (3).
2. Preparing the sugar syrup (aqueous phase): heat the sugar with the water to T = 115°C.
3. Incorporating of surfactants in the aqueous phase: slowly mix the glucose syrup into the egg whites until incorporated, increase the speed to medium-high and beat for 10 minutes. Allow to cool.

4. Preparing the O/W emulsion: add the butter slowly at room temperature (2 tablespoons at a time), until a homogeneous mixture is obtained.
5. Preparing the foam: add the salt and beat the mixture until it is well ventilated.

This is a typical O/W emulsion even if it looks like the opposite; although the amount of fat is higher than that of the water, the nature of the surface-active agent (its HLB) dictates the type of the obtained emulsion. On first step, protein egg whites (including globulins as albumins and mucoproteins) that can act also as tensioactive agents are mixed with the sucrose esters. After the preparation of the sugar syrup the tensioactive agents are to be mix with the aqueous phase. Sucrose esters interact in stabilisation of sugar crystallisation process but here the sugar syrup is kept always warm, as so, sugar esters will play a role into the solid/liquid interface only after aeration. Careful must be taken with temperature: not too low to avoid sugar crystallisation; not too high to avoid proteins denaturation (sugar esters protect protein against from flocculation). The oil phase is added slowly at room temperature in order to form the emulsion. In step 5 the whipping allows the air to be retain in the pre-obtained emulsion, that will be stabilised by the sucrose esters in the air/(O/W) emulsion.

3.1.2.2 Cream bubbles

Ingredients: Cream (250 g); milk (125 g); water (87.5 g); water for sugar syrup (37.5 g); sugar (37.5 g); sucrose ester (5 g) and spices (seeds of 1 piece cardamom, seeds of 1 piece of vanilla, ginger flower).

Preparation consists in:

1. Preparing the sugar syrup (aqueous phase): heat the sugar with the water to T = 115°C.
2. Incorporation of the sucrose esters with cream and milk (plus spices).
3. Incorporation of the sugar syrup into the of the oil/water emulsion: using a hand blender make a mixture of the syrup with the remaining boiled ingredients.
4. Filtration of the solids (spices): make a filtration with fine mesh sieves and keep warm.
5. Formation of the foam: mix with a hand blender right before serving.

Foams and whipped emulsions are stabilised by dairy proteins. The illustrated preparation has fats, proteins, hydrocolloids, and colloidal components. The presence of the sucrose ester contributes, together with the proteins, to stabilising the interacting components of the preparation as an emulsion when they are first boiled and mix (homogenisation). Then, at last minute, the aeration process to make the foam should be implemented (aeration).

Proteins (namely dairy proteins) and the sucrose ester present in the bulk preparation compete at interfaces air/water and oil/water. At high concentrations, the water-soluble sucrose ester is more effectively adsorbed at those interfaces and displace the proteins. This mechanism occurs depending on the ratio protein/surfactant. As so, proteins they are to be apportioned between the bulk phase (water) and interfaces (oil/water).

The mix of the sucrose ester with proteins in the interfacial layers makes the surface film to be weaker when compared only with proteins, contributing to the destabilisation of the emulsion. This is important to the aeration process: when a cream is whipped, shear forces are acting into an aqueous dispersive system with a high interface area that encloses, oil and gaseous phases brought together. The destabilisation of the emulsion will make the incorporating of air easier during the foam manufacture and at same time the aqueous phase behaves like a reservoir of active material surface, say, protein and surfactant. This is beneficial for building a network of partially coalesced fat globules around the air bubbles and for stabilising new formed air-water interfaces.

3.1.3 Safety and maximum daily intake

Sucrose esters were approved and registered by European Food Safety Authority or EFSA under the E number of 473. The authorization for the use of sucrose esters of fatty acids (E 473) was made in 1995 in various preparations but not for colour stabilisation and sweetening preparations but, having a wide range of applications, sucrose esters are used in many industrial preparations and the European Food Safety Authority (EFSA) requested to the food industries extra data about the use of the sucrose esters in baked goods and flavoured drinks (chocolate milk, cocoa, eggnog, drinking yogurt and whey based drinks) in 2004 [40].

The and the panel responsible for the evaluation concluded that the daily the exposure to the additive exceeded the acceptable daily intake of 40 mg/kg/bw/day.

Meanwhile health concerns on high consumption of the sugar esters of fatty acids raised since it is used in several categories of food of preparations where data was not provided from producers. As so, the current exposes could be higher than that advised by the panel.

In 2010, the panel recommend the collection of reported used form industry for fine bakery and flavoured drinks and in the light of new data, ANS Panel indeed concluded that the use of sucrose esters of fatty acids (E 473) may lead to an exposure higher than the 40 mg/kg/bw/day [41].

The uncertainty regarding the amount of E473 that could be ingested by the population, due to the wide use of these additives, led the commission to launch of a public call for data aiming at collecting reported use levels from industry or analytical data and a motorization of the presence of the surfactant on the label of foods in 2014. A refined analysis was made based on the preference of the population for consuming certain food categories and brands (flavoured drinks, fine bakery, only dairy-based and almond drinks). The panel concluded that the refined exposure in the brand-loyal scenario could be estimated as 54 mg/kg/bw/per day for babies (12–35 months) and as 124 mg/kg/bw/per day for children (3–9 years), but as not all categories were included this value may be overestimated [42].

The panel in 2018 recommended a higher detail in data collection in order to provide the scientific experts with data for a more realistic estimation. Thus, the actual studies point that there is an over cumulative intake of this additive by the European population due to quantity of preparations available in marked that use sucrose esters of fatty acids [42].

3.2 Mono- and diglycerides of fatty acids (E471)

Monoglycerides and diglycerides are lipid molecules composed of a single fatty acid or two fatty acids, respectively, esterified with hydroxyl groups of the glycerol [43, 44]. The food additive itself, consists of a mixture of glycerol mono-, di- and tri-esters of fatty acids derived from edible oils and fats [45, 46]. The amount of mono- and di-esters should be at least 70% according to the European Regulation (2012), and may also contain a small quantity of free glycerol (maximum 7%) and fatty acids [45]. Therefore, it is a mixture of substances which may contain different mono- and diglycerides of fatty acids, depending on the fatty acid sources. The fatty acids in mono- and diglycerides may be saturated or unsaturated, being the lauric, linoleic, myristic, oleic, palmitic, and stearic fatty acids the most prevalent. [46].

A revision concerning the synthesis of monoglycerides was published some years ago [46]. Their synthesis can be made according to schematic reactions presented in Figure 11. As seen, mono- and diglycerides of fatty acids may be produced by chemical methods, such as inter-esterification (glycerolysis) of fatty acid esters with glycerol and direct esterification of glycerol with fatty acids [47, 48, 49]. Some methyl esters of the fatty acids used are depicted in Figure 12. Both reactions occur under alkaline conditions at a high temperature (200–250°C), and yield a mixture of mono-, di- and triglycerides in addition to a reduced fraction of free glycerol [46, 50]. Alternatively, mono- and diglycerides of fatty acids may be produced by enzymatic hydrolysis of triglycerides [51, 52]. The advantages of enzymatic synthesis are higher yields and mild reaction conditions, resulting in products of higher quality and lower energy consumption [53].

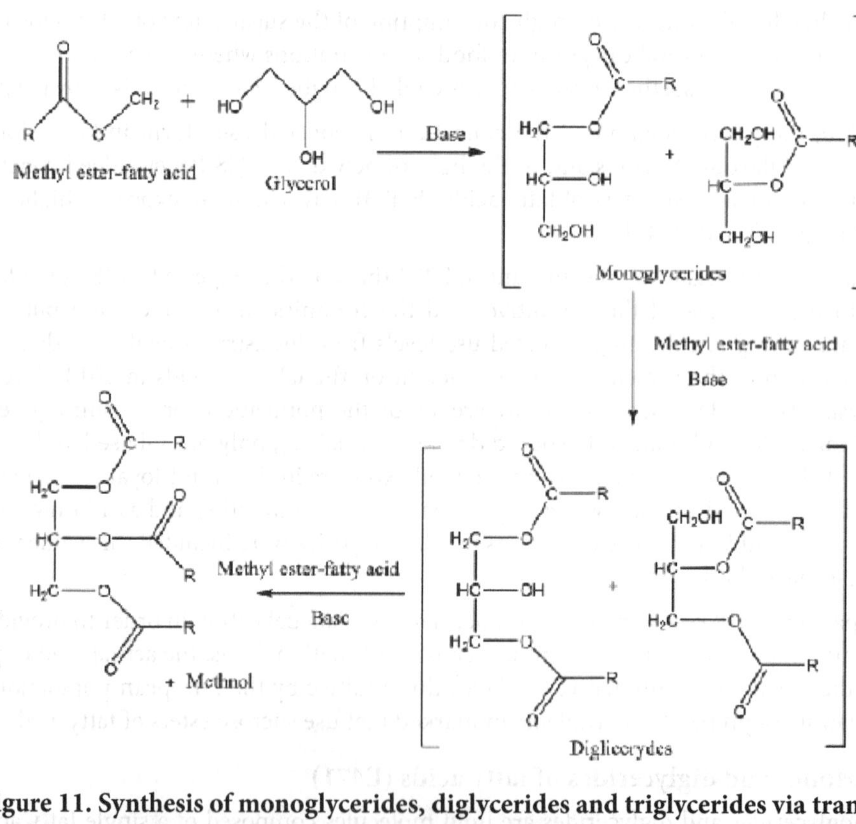

Figure 11. Synthesis of monoglycerides, diglycerides and triglycerides via trans-esterification (glycerolysis) using methyl-fatty acids and glycerol in a basic reaction medium (adapted from [46]).

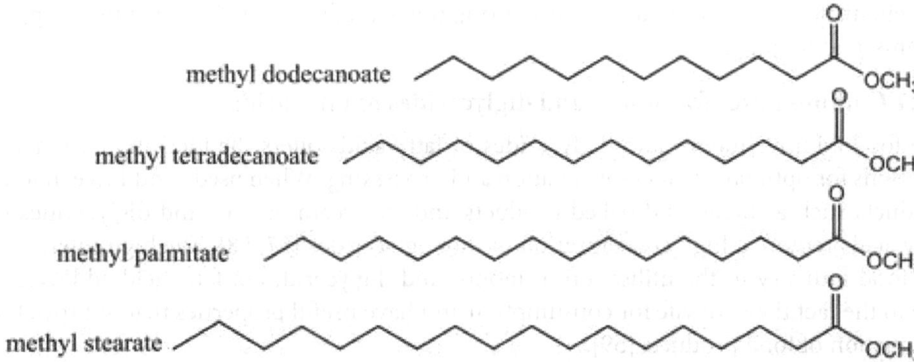

Figure 12. Examples of methyl esters fatty acids used in the esterification of glycerol.

Regardless of the chosen method, it is usually necessary to separate and to concentrate these glycerides [48]. Monoglycerides may be separated from the diglycerides, triglycerides, and glycerol by molecular distillation. This a method to separate and purify thermally unstable compounds and substances with low vapour pressure and higher molecular weight, without significant thermal decomposition [48]. Molecular distilled monoglycerides contain an equilibrium of 1-monoglycerides and 2-monoglycerides dependent on the temperature used during the process. The monoester content is usually 90–95% in commercially distilled monoglycerides [50]. Distilled monoglycerides have a better dispersibility in water than the mono- and diglycerides of fatty acids because they present high purity and well-defined molecular structure, forming a liquid crystalline mesomorphic phase with ordered bilayers of the fatty acid chains, separated by water layers associated with the polar groups [50]. This feature is useful in food products where interactions with water-soluble ingredients (e.g., starch components) or the aerating properties of fat-free products are of importance [54]. When distilled monoglycerides are heated to their melting point with water, a gel is formed in which the water layers alternate with lipid bilayers [55]. The melting point depends on the chain length of the fatty acid and on the purity of the monoglyceride [50].

Mono- and diglycerides of fatty acids are amphipathic molecules, which can be represented as depicted in synthetic pathways of Figure 11. They can be absorbed in the water/oil surfaces once the hydrophilic head has a high affinity for water, and a lipophilic tail to the oil [49]. The structure of mono- and di-glycerides of fatty acids classifies them as surfactants, i.e., surface-active agents or molecules that migrate to the interface between two phases (solid, liquid, or gas).

The functional properties of mono- and diglycerides of fatty acids are determined by the HLB of the mixture that will depend on i) the type of fat used as the base ingredient, i.e., if the original fat is saturated or unsaturated, ii) the length of the hydrocarbon chain fatty acids and iii) the percentage of monoglycerides. The mono- and di-glycerides of fatty acids are non-ionic oil soluble surfactants, predominantly lipophilic and only sparingly soluble in water, having a low HLB. Therefore, they stabilise W/O emulsions such as margarines and spreads, by forming reversed micelles in oil [50, 56]. They are also used to inhibit fat crystallisation in some O/W emulsions because this improves the stability of the food product to refrigeration conditions, for example, in dressings. Moreover, oil-soluble emulsifiers can be used in combination with water-soluble emulsifiers to facilitate protein displacement from fat droplets during the manufacturing of ice creams, whipped creams, and toppings.

3.2.1 Common uses for mono- and diglycerides of fatty acids

The food additive mono- and diglycerides of fatty acids offers the food industry many solutions for optimal product formulation and processing. When used as additive in food products such as bread and baked products and ice cream, mono- and diglycerides of fatty acids require a high concentration of monoglycerides [57, 58]. The high interest of the food industry in the utilisation of mono- and diglycerides of fatty acid additives is due to the fact they are safe for consumption and have useful properties that improve the production of food products [59].

Food-grade mono- and diglycerides of fatty acids have been studied as surface-active substances or foaming agents. Diglycerol monomyristate influences the foamability and stability of olive oil foams, depending on the surfactant concentration, and the size of its solid particles [60]. Moreover, the addition of a mixture of distilled mono- and diglycerides of edible saturated fatty acids from vegetable origin with a total content in monoglycerides larger than 90% result in high stable foams of rapeseed oil, in which a dense layer of surfactant crystals at the oil-air interface protects the air bubbles, thus preventing the dissolution and coalescence of the system [61]. In addition, there is reported evidence of Pickering stabilisation of gas bubbles by monoglyceride crystals at the air–oil interface. Gunes *et al.* observed that the air bubbles produced in sunflower oil with a monoglyceride (98% saturated) were covered by a layer of monoglyceride crystals preventing rapid dissolution and coalescence of the oil foam [62]. In addition, whipping is an important mechanical process for the effective coverage of the bubble surface with the stabilising layer of fat crystals as compared to foams obtained by depressurization [62].

3.2.2 Examples for the use of mono- and diglycerides of fatty acids in molecular gastronomy

3.2.2.1 Mono- and diglycerides of fatty acids to prepare "olive oil foam"

To prepare "olive oil foam", two ingredients are necessary: extra virgin olive oil (200 g) and mono- and diglycerides flakes (200 g) Figure 13.

Figure 13. A olive oil foam prepared with diglycerides flakes.

Preparation consists in:

1. Heat the olive oil with the mono- and diglycerides of fatty acids flakes until the mixture reaches the temperature of 60°C. Stir until the flakes dissolve.
2. Pour the mixture into a heat resistant whipping siphon and charge as instructed by the manufacturer.
3. Serve warm. To use it later, keep the whipping siphon at room temperature.

3.2.3 Safety and maximum daily intake

According to the European Regulation the presentation of the food additive E-471 varies from a pale yellow to pale brown oily liquid to a white or slightly off-white hard waxy solid. The solids may be in the form of flakes, powders, or small beads [63].

Mono- and diglycerides of fatty acids are the most widely used food emulsifiers, accounting for around 70% of the world production of food emulsifiers [3, 56]. Commercial mono- and diglycerides of fatty acids usually contain 45–55% monoglycerides, 38–45% diglycerides, 8–12% triglyceride and 1–7% free glycerol [50]. Mono- and diglycerides of fatty acids are also produced in the gastrointestinal tract during the hydrolysis of the food derived triglycerides by gastric and pancreatic lipases [64]. Mono- and diglycerides of fatty acids have been evaluated by the Joint FAO/WHO Expert Committee on Food Additives in 1974. The commission recommended a specific risk assessment for its use in food products for infants under the age of 12 weeks [65]. In 2017, the safety of mono- and diglycerides of fatty acids when used as a food additive was re-evaluated by the EFSA Panel on Food Additives and Nutrient Sources added to Food for ages above 12 weeks of age [66]. Since the hydrolysis of mono- and diglycerides of fatty acids is likely to occur in the gastrointestinal tract with production of free glycerol and free fatty acids. For this reason, the Panel also evaluated the safety of the food additives glycerol (E 422) and fatty acids (E 570). No evidence for adverse effects was reported in toxicity studies that assessed these additives. Neither genotoxic nor carcinogenic effects have been observed [67]. On the other hand, exposure to mono- and diglycerides of fatty acids may be compared with the fat consumed, due to the similarity in their metabolic breakdown. US dietary guidelines recommend an overall fat consumption of 30% of total energy intake and saturated fat consumption of 10% of total energy intake [68]. High saturated fat intake has been associated with higher risk of coronary heart disease and mortality [69]. Although not all saturated fats have the same biological activity, the saturated fatty acids palmitic acid and myristic acid, often used in E-471 additives, may have the most negative impact on serum lipidic levels [70]. However, EFSA Panel reports a small contribution of mono- and diglycerides of fatty acids to the daily fat intake, representing around 0.8–3.5% of the recommendation [67]. EFSA Panel concluded that there was no safety concern regarding the use of these food additives for the general population and that there was no need for a numerical acceptable daily intake [66]. In addition, since the dietary exposure to emulsifiers has not increased over the 10-year period there is no reason to suspect that the dietary exposure may cause a safety concern [47].

As mentioned before, the additives here discussed act as surfactant food additives. Independently of the assessment and evaluation, it is worthy to be mention that there are studies concerning their possible effects on intestinal barriers and microbiota that probably will keep the experts watchful. Recent studies suggest that high doses of emulsifier consumption may contribute to the rising incidence of several diseases involving the gastro-intestinal tract [71, 72, 73]. The rise of scientific reports and population studies will help risk assessors make adequate re-evaluations concerning the authorised food emulsifiers.

4. Conclusions

The utilisation of sucrose esters of fatty acids and mono- and diacylglycerides of fatty acids as food additives has been increasing exponentially. Those molecules act as surfactants and, depending of their mixtures and nature of the esterified fatty acids, they can present a wide range of HLB values. As so, they are useful for promoting the stabilisation of oil–water; water–oil and water/oil-gas emulsions. They are a common ingredient in industrial bakery, white sauces, alternative dairy to milk, aerated dairy desserts, ice creams, mousses, confectionery and preparation of cereals based products. They are obtained by chemical or enzymatic synthesises from natural reagents as fatty acids, sucrose and glycerol by esterification present advantages concerning about their biodegradability: after ingestion, they are hydrolysed by digestive enzymes into sucrose/glycerol and their component fatty acids that are further metabolised in the usual way. Despite this, concerns about their safety are rising since the daily intake doses may become higher than the maximum safe daily intake doses. Evaluation panels made studies about refined exposure in the brand-loyal scenario concluding that the daily intake doses could be overestimated specially in babies and children.

References

1. McClements DJ. Food Emulsions: Principles, Practices, and Techniques. Third. London: CRC Press; 2015
2. Green AJ, Littlejohn KA, Hooley P, Cox PW. Formation and stability of food foams and aerated emulsions: Hydrophobins as novel functional ingredients. Current Opinion in Colloid and Interface Science. 2013;18(4):292-301. DOI:10.1016/j.cocis.2013.04.008
3. Miller R. Emulsifiers: Types and Uses. In: Encyclopedia of Food and Health. 1st ed. Elsevier Ltd.; 2016. p. 498-502. DOI:10.1016/B978-0-12-384947-2.00249-X
4. Jafari SM, He Y, Bhandari B. Optimization of nano-emulsions production by microfluidization. European Food Research and Technology. 2007;225(5):733-741. DOI:10.1007/s00217-006-0476-9
5. Whitehurst RJ, editor. Emulsifiers in Food Technology. Wiley-Blackwell; 2008
6. Tan CP, Nakajima M. Effect of polyglycerol esters of fatty acids on physicochemical properties and stability of β-carotene nanodispersions prepared by emulsification/ evaporation method. Journal of the Science of Food and Agriculture. 2005;85(1):121-126. DOI:https://doi.org/10.1002/jsfa.1947
7. Chambi H, Grosso C. Effect of surfactants on the functional properties of gelatin–polysaccharide-based films. European Food Research and Technology. 2011;232(1):63-69. DOI:10.1007/s00217-010-1361-0

8. Zhang H, Feng F, Li J, Zhan X, Wei H, Li H, et al. Formulation of food-grade microemulsions with glycerol monolaurate: effects of short-chain alcohols, polyols, salts and nonionic surfactants. European Food Research and Technology. 2008;226(3):613-619. DOI:10.1007/s00217-007-0606-z

9. Krawczyk J. Temperature impact on the water-air interfacial activity of n-octyl and n-dodecyl-β-d-glucopyranosides. Colloids and Surfaces A: Physicochemical and Engineering Aspects. 2017;533:61-67. DOI:https://doi.org/10.1016/j.colsurfa.2017.08.015

10. Krawczyk J. Aggregation properties of sucrose fatty acid esters and some other sugar-based surfactants at different temperatures. Journal of Molecular Liquids. 2018;271:610-620. DOI:https://doi.org/10.1016/j.molliq.2018.09.013

11. Plat T, Linhardt RJ. Syntheses and applications of sucrose-based esters. Journal of Surfactants and Detergents. 2001;4(4):415-421. DOI:10.1007/s11743-001-0196-y

12. Galvao J, Davis B, Tilley M, Normando E, Duchen MR, Cordeiro MF. Unexpected low-dose toxicity of the universal solvent DMSO. The FASEB Journal. 2014;28(3):1317-1330. DOI:https://doi.org/10.1096/fj.13-235440

13. World Health Organization. N,N-Dimethylformamide. Geneva; 2001

14. Lynch D. NTP technical report on the toxicity studies of N,N-Dimethylformamide (CAS No. 68-12-2) Administered by Inhalation to F344/N Rats and B6C3F1 Mice. 1992

15. Plou FJ, Cruces MA, Ferrer M, Fuentes G, Pastor E, Bernabé M, et al. Enzymatic acylation of di- and trisaccharides with fatty acids: choosing the appropriate enzyme, support and solvent. Journal of Biotechnology. 2002;96(1):55-66. DOI:https://doi.org/10.1016/S0168-1656(02)00037-8

16. Tarahomjoo S, Alemzadeh I. Surfactant production by an enzymatic method. Enzyme and Microbial Technology. 2003;33(1):33-37. DOI:https://doi.org/10.1016/S0141-0229(03)00085-1

17. Franssen MCR, Steunenberg P, Scott EL, Zuilhof H, Sanders JPM. Immobilised enzymes in biorenewables production. Chemical Society Reviews. 2013;42(15):6491-6533. DOI:10.1039/C3CS00004D

18. Teng Y, Stewart SG, Hai Y-W, Li X, Banwell MG, Lan P. Sucrose fatty acid esters: synthesis, emulsifying capacities, biological activities and structure-property profiles. Critical Reviews in Food Science and Nutrition. 2020;1-21. DOI:10.1080/10408398.2020.1798346

19. Neta N do AS, Santos JCS dos, Sancho S de O, Rodrigues S, Gonçalves LRB, Rodrigues LR, et al. Enzymatic synthesis of sugar esters and their potential as surface-active stabilizers of coconut milk emulsions. Food Hydrocolloids. 2012;27(2):324-331. DOI:https://doi.org/10.1016/j.foodhyd.2011.10.009

20. Fang S, Zuo X-B, Xu H-N, Meng Y-C, Liu Y-M. Effect of sucrose fatty acid esters with different hydrophilic-lipophilic balance values on pasting and rheological properties of waxy rice flour. Food Science and Biotechnology. 2016;25(3):721-727. DOI:10.1007/s10068-016-0125-4

21. Gupta RK, James K, Smith FJ. Sucrose esters and sucrose ester/glyceride blends as emulsifiers. Journal of the American Oil Chemists' Society. 1983;60(4):862-869. DOI:10.1007/BF02787451

22. Ontiveros JF, Pierlot C, Catté M, Molinier V, Salager J-L, Aubry J-M. A simple method to assess the hydrophilic lipophilic balance of food and cosmetic surfactants using the phase inversion temperature of C10E4/n-octane/water emulsions. Colloids and Surfaces A: Physicochemical and Engineering Aspects. 2014;458:32-39. DOI:https://doi.org/10.1016/j.colsurfa.2014.02.058

23. Szűts A, Pallagi E, Regdon G, Aigner Z, Szabó-Révész P. Study of thermal behaviour of sugar esters. International Journal of Pharmaceutics. 2007;336(2):199-207. DOI:https://doi.org/10.1016/j.ijpharm.2006.11.053

24. Finot P-A. Historical Perspective of the Maillard Reaction in Food Science. Annals of the New York Academy of Sciences. 2005;1043(1):1-8. DOI:https://doi.org/10.1196/annals.1333.001

25. Morales FJ, van Boekel MAJS. A Study on Advanced Maillard Reaction in Heated Casein/Sugar Solutions: Colour Formation. International Dairy Journal. 1998;8(10):907-915. DOI:https://doi.org/10.1016/S0958-6946(99)00014-X

26. Kwak E-J, Lim S-I. The effect of sugar, amino acid, metal ion, and NaCl on model Maillard reaction under pH control. Amino Acids. 2004;27(1):85-90. DOI:10.1007/s00726-004-0067-7

27. Friedman M. Food Browning and Its Prevention: An Overview. Journal of Agricultural and Food Chemistry. 1996;44(3):631-653. DOI:10.1021/jf950394r

28. Makino S, Ogimoto S, Koga S. Sucrose Monoesters of Fatty Acids: Their Properties and Interaction with Proteins. Agricultural and Biological Chemistry. 1983;47(2):319-326. DOI:10.1080/00021369.1983.10865614

29. Saleki-Gerhardt A, Zografi G. Non-Isothermal and Isothermal Crystallization of Sucrose from the Amorphous State. Pharmaceutical Research. 1994;11(8):1166-1173. DOI:10.1023/A:1018945117471

30. Gabarra P, Hartel RW. Corn Syrup Solids and Their Saccharide Fractions Affect Crystallization of Amorphous Sucrose. Journal of Food Science. 1998;63(3):523-528. DOI:https://doi.org/10.1111/j.1365-2621.1998.tb15778.x

31. Yamashita S, Matsumiya K, Kogo Y, Takamatsu K, Matsumura Y. Emulsifiers efficiently prevent hardening of pancakes under refrigerated conditions via inclusion complexes with starch molecules. Food Hydrocolloids. 2020;100:105432. DOI:https://doi.org/10.1016/j.foodhyd.2019.105432

32. Breyer LM, Walker CE. Comparative Effects of Various Sucrose-Fatty Acid Esters upon Bread and Cookies. Journal of Food Science. 1983;48(3):955-958. DOI:https://doi.org/10.1111/j.1365-2621.1983.tb14939.x

33. Zhang X, Song F, Taxipalati M, Wei W, Feng F. Comparative Study of Surface-Active Properties and Antimicrobial Activities of Disaccharide Monoesters. PLOS ONE. 2014;9(12):e114845

34. Yanke L, Shufen Z, Qinghui W, Jinzong Y. Study on Surface Activity and Critical Aggregation Concentration of Sucrose Esters Containing Different Isomers of Mono-, Di- and Polyesters. Tenside Surfactants Detergents. 41(1):26-30. DOI:https://doi.org/10.3139/113.100205

35. van Kempen SEHJ, Schols HA, van der Linden E, Sagis LMC. Effect of Variations in the Fatty Acid Chain of Oligofructose Fatty Acid Esters on Their Foaming Functionality. Food Biophysics. 2014;9(2):114-124. DOI:10.1007/s11483-013-9324-1

36. Garofalakis G, Murray BS. Dilatational rheology and foaming properties of sucrose monoesters in the presence of β-lactoglobulin. Colloids and Surfaces B: Biointerfaces. 2001;21(1):3-17. DOI:https://doi.org/10.1016/S0927-7765(01)00179-5

37. Tual A, Bourles E, Barey P, Houdoux A, Desprairies M, Courthaudon J-L. Effect of surfactant sucrose ester on physical properties of dairy whipped emulsions in relation to those of O/W interfacial layers. Journal of Colloid and Interface Science. 2006;295(2):495-503. DOI:https://doi.org/10.1016/j.jcis.2005.09.010

38. Goff HD, Jordan WK. Action of Emulsifers in Promoting Fat Destabilization During the Manufacture of Ice Cream. Journal of Dairy Science. 1989;72(1):18-29. DOI:https://doi.org/10.3168/jds.S0022-0302(89)79075-5

39. Melsen J, Walstra P. Stability of recombined milk fat globules. Netherlands Milk and Dairy Journal. 1989;43:63-78

40. European Food Safety Authority. Opinion of the Scientific Panel on food additives, flavourings, processing aids and materials in contact with food (AFC) on Sucrose esters of fatty acids E 473 and sucroglycerides E 474. EFSA Journal. 2004;2(11):106. DOI:https://doi.org/10.2903/j.efsa.2004.106

41. EFSA Panel on Food Additives and Nutrient Sources added to Food (ANS). Scientific Opinion on the safety of sucrose esters of fatty acids prepared from vinyl esters of fatty acids and on the extension of use of sucrose esters of fatty acids in flavourings. EFSA Journal. 2010;8(3):1512. DOI:https://doi.org/10.2903/j.efsa.2010.1512

42. EFSA Panel on Food Additives and Nutrient Sources added to Food, Younes M, Aggett P, Aguilar F, Crebelli R, Dusemund B, et al. Refined exposure assessment of sucrose esters of fatty acids (E 473) from its use as a food additive. EFSA Journal. 2018;16(1):e05087. DOI:https://doi.org/10.2903/j.efsa.2018.5087

43. Nicholson RA, Marangoni AG. Diglycerides. In: Melton L, Shahidi F, Varelis PBT-E of FC, editors. Oxford: Academic Press; 2019. p. 70-73. DOI:https://doi.org/10.1016/B978-0-08-100596-5.22343-3

44. Chen CH, Terentjev EM. Monoglycerides in Oils [Internet]. Second Edi. Edible Oleogels: Structure and Health Implications. AOCS Press; 2011. 173-201 p. DOI:10.1016/B978-0-9830791-1-8.50011-5

45. European Commission. Commission Regulation (EU) No 231/2012 of 9 March 2012 laying down specifications for food additives listed in Annexes II and III to Regulation (EC) No 1333/2008 of the European Parliament and of the Council. Official Journal of the European Union. 2012;55

46. Food and Drug Administration. Title 21: Food and Drugs. In: Code of Federal Regulations. 2021

47. Shah R, Kolanos R, DiNovi MJ, Mattia A, Kaneko KJ. Dietary exposures for the safety assessment of seven emulsifiers commonly added to foods in the United States and implications for safety. Food Additives and Contaminants - Part A Chemistry, Analysis, Control, Exposure and Risk Assessment. 2017;34(6):905-917. DOI:10.1080/19440049.2017.1311420

48. Fregolente PBL, Pinto GMF, Wolf-Maciel MR, Filho RMI. Monoglyceride and diglyceride production through lipase-catalyzed glycerolysis and molecular distillation. Applied Biochemistry and Biotechnology. 2010;160(7):1879-1887

49. Nitbani FO, Tjitda PJP, Nurohmah BA, Wogo HE. Preparation of fatty acid and monoglyceride from vegetable oil. Journal of Oleo Science. 2020;69(4):277-295

50. Norn V. Emulsifiers in Food Technology. Vol. 9780470670, Emulsifiers in Food Technology: Second Edition. Wiley-Blackwell; 2015. 1-342 p

51. Fregolente PBL, Fregolente LV, Pinto GMF, Batistella BC, Wolf-Maciel MR, Filho RM. Monoglycerides and Diglycerides Synthesis in a Solvent-Free System by Lipase-Catalyzed Glycerolysis. Applied Biochemistry and Biotechnology. 2008;146(1):165-172. DOI:10.1007/s12010-008-8133-3

52. Garba U, Singanusong R, Jiamyangyuen S, Thongsook T. Preparing spray-dried cholesterol free salad dressing emulsified with enzymatically synthesized mixed mono- and diglycerides from rice bran oil and glycerol. Journal of Food Science and Technology. 2020; DOI:10.1007/s13197-020-04611-2

53. Yang T, Rebsdorf M, Engelrud U, Xu X. Monoacylglycerol synthesis via enzymatic glycerolysis using a simple and efficient reaction system. Journal of Food Lipids. 2005;12(4):299-312. DOI:https://doi.org/10.1111/j.1745-4522.2005.00025.x

54. Goff HD. Controlling ice-cream structure by examining fat: protein interactions. Australian Journal of Dairy Technology. 2000;55(2):78-81

55. Ojijo NKO, Neeman I, Eger S, Shimoni E. Effects of monoglyceride content, cooling rate and shear on the rheological properties of olive oil/monoglyceride gel networks. Journal of the Science of Food and Agriculture. 2004;84(12):1585-1593

56. Chen L. Emulsifiers as food texture modifiers [Internet]. Vol. 1, Modifying Food Texture: Novel Ingredients and Processing Techniques. Elsevier Ltd.; 2015. 27-49 p. DOI:10.1016/B978-1-78242-333-1.00002-4

57. Cauvain SP, Young LS. The stability and shelf life of bread and other bakery products [Internet]. Food and Beverage Stability and Shelf Life. Woodhead Publishing Limited; 2011. 657-682 p. DOI:10.1533/9780857092540.3.657

58. Cropper SL, Kocaoglu-Vurma NA, Tharp BW, Harper WJ. Effects of Locust Bean Gum and Mono- and Diglyceride Concentrations on Particle Size and Melting Rates of Ice Cream. Journal of Food Science. 2013;78(6):811-816

59. Suman M, Silva G, Catellani D, Bersellini U, Caffarra V, Careri M. Determination of food emulsifiers in commercial additives and food products by liquid chromatography/atmospheric-pressure chemical ionisation mass spectrometry. Journal of Chromatography A. 2009;1216(18):3758-3766

60. Shrestha RG, Shrestha LK, Solans C, Gonzalez C, Aramaki K. Nonaqueous foam with outstanding stability in diglycerol monomyristate/olive oil system. Colloids and Surfaces A: Physicochemical and Engineering Aspects. 2010;353(2-3):157-165

61. Brun M, Delample M, Harte E, Lecomte S, Leal-Calderon F. Stabilization of air bubbles in oil by surfactant crystals: A route to produce air-in-oil foams and air-in-oil-in-water emulsions. Food Research International. 2015;67:366-375. DOI:10.1016/j.foodres.2014.11.044

62. Gunes DZ, Murith M, Godefroid J, Pelloux C, Deyber H, Schafer O, et al. Oleofoams: Properties of Crystal-Coated Bubbles from Whipped Oleogels-Evidence for Pickering Stabilization. Langmuir. 2017;33(6):1563-1575

63. European Commission. Commission Regulation (EU) N° 231/2012 of 9 March 2012 laying down specifications for food additives listed in Annexes II and III to Regulation (EC) No 1333/2008 of the European Parliament and of the Council. Official Journal of the European Union. 2012;L83:1-295

64. Bauer E, Jakob S, Mosenthin R. Principles of Physiology of Lipid Digestion. Asian-Australas J Anim Sci. 2005;18(2):282-295. DOI:10.5713/ajas.2005.282

65. Joint FAO/WHO Expert Committee on Food Additives. Toxicological evaluation of some food additives including anticaking agents, antimicrobials, antioxidants, emulsifiers and thickening agents. FAO Nutrition Meetings Report Series. 1974;53A

66. EFSA Panel on Food Additives and Nutrient Sources added to Food, Younes M, Aggett P, Aguilar F, Crebelli R, Dusemund B, et al. Re-evaluation of mono- and di-glycerides of fatty acids (E 471) as food additives. EFSA Journal. 2017;15(11)

67. EFSA Panel on Food Additives and Nutrient Sources added to Food, Younes M, Aggett P, Aguilar F, Crebelli R, Dusemund B, et al. Re-evaluation of mono- and di-glycerides of fatty acids (E 471) as food additives. EFSA Journal. 2017;15(11):e05045. DOI:https://doi.org/10.2903/j.efsa.2017.5045

68. U.S. Department of Health and Human Services, U.S. Department of Agriculture. 2015-2020 Dietary Guidelines for Americans. 2015

69. Clifton PM, Keogh JB. A systematic review of the effect of dietary saturated and polyunsaturated fat on heart disease. Nutrition, Metabolism and Cardiovascular Diseases. 2017;27(12):1060-1080. DOI:10.1016/j.numecd.2017.10.010

70. Micha R, Mozaffarian D. Saturated Fat and Cardiometabolic Risk Factors, Coronary Heart Disease, Stroke, and Diabetes: a Fresh Look at the Evidence. Lipids. 2010;45(10):893-905. DOI:10.1007/s11745-010-3393-4

71. Mine Y, Zhang JW. Surfactants enhance the tight-junction permeability of food allergens in human intestinal epithelial Caco-2 cells. Int. Arch. Allergy Immunol. 2003; 130:135-142. DOI:10.1159/000069009

72. Glynn A, Igra AM, Sand S, Ilbäck NG, Hellenäs KE, Rosén J, Aspenström-Fagerlund B. Are additive effects of dietary surfactants on intestinal tight junction integrity an overlooked human health risk? – a mixture study on Caco-2 monolayers. Food Chem. Toxicol. 2017; 106 (Pt A):314-323. DOI:10.1016/j.fct.2017.05.068

73. Csáki KF, Sebestyén E. Who will carry out the tests that would be necessary for proper safety evaluation of food emulsifiers? Food Science and Human Wellness. 2019: 8(2): 126-135. https://doi.org/10.1016/j.fshw.2019.04.001

Chapter-10

Molecular Breeding of Sweetpotato Carotenoids

1. Introduction

Sweetpotato [*Ipomoea batatas* (L.) Lam.], also described as "sweet potato," belongs to the family Convolvulaceae and occupies the seventh position among the food crops of the world after wheat, rice, maize, potato, barley, and cassava [1, 2]. The largest genus in the family Convolvulaceae is *Ipomoea,* consisting of 600–700 species, among which, only *I. batatas* is cultivated widely as a food crop around the world [3, 4]. In comparison with other tuber crops, sweetpotato comprises higher contents of carbohydrates, many minerals, and more protein estimates than other vegetables [5, 6]. It also contains much higher levels of provitamin A, vitamin C, and minerals than those of rice or wheat [7]. Hundred grams of raw sweetpotato contain 1.57 g of protein, 20.12 g of carbohydrates, 3 g of total dietary fiber, 41.8 g of total sugars, 30 mg of calcium, 0.61 mg of iron, 25 mg of magnesium, 47 mg of phosphorous, 337 mg of potassium, 55 mg of sodium, 0.3 mg of zinc, 2.4 mg of vitamin C, 0.5 mg of niacin, 0.2 mg of vitamin B6, 14,187 IU of vitamin A (VA), 0.2 mg of vitamin E, 11 μg of vitamin B-9, and 8509 μg of β-carotene (β,β-carotene) [8]. The starch in sweetpotato is easy to digest. Therefore, it is a valuable constituent in the preparation of excellent weaning meals [9]. It is a source of food supply to combat malnutrition in the developing nations, since the tuberous roots (tubers) are enriched with starch and dietary fiber, along with carotenoids, anthocyanin, ascorbic acid, potassium, calcium, iron, and other bioactive ingredients [10, 11, 12, 13]. For people of South East Asia and Africa, this crop is the main source of β-carotene [10, 14]. The tubers of the Japanese cultivars are diverse concerning carotenoids accumulation [15]. Sweetpotato may exert diverse health positive effects, since it contains high amounts of numerous phytochemicals in roots or leaves [6, 16]. *I. batatas* cultivars with color-fleshed tubers have been reported for their excellent bioactivities, such as antimutagenic [17], free radical scavenging [18], hepatoprotactive, reduction of liver injury [19, 20], anticancer [21, 22, 23], antioxidative activities [23, 24, 25], antimicrobial activity, antihypertension, anti-inflammatory, antidiabetic, anticaries effect, ultraviolet protection [23], and chemopreventive activities [26]. Previous reports also suggest that its tubers may be useful for treating peptic ulcers [27]. The genome of *I. batatas* is structurally complex and has a size of 4.8–5.3 pg/2C nucleus [28]. Due to the existence of polyploidy, sweetpotato is a hexaploid species (2n = 6x = 90) that has a basic chromosome number of 15, [3] with a huge genome size of 2200–3000 Mbp [29].

The genetic studies on this species are exhausting, since it is difficult to generate seeds and to evaluate the effects of polyploidy on the genome [30, 31]. Complex structure of its genome also manifests self and cross-incompatibility, causing barrier for genetic studies on important agronomical characters [3, 32]. Its tubers exhibit various colors, such as white, yellow, orange and purple orange, and yellow and orange-fleshed lines, were shown to contain β-carotene as the predominant carotenoid [11, 12, 13, 33, 34]. Annual yield of sweetpotato is currently exceeding the value of 105 million metric tons, 95% of which is shared by the developing countries. China is the world's leader among all in sweetpotato consumption that counts about 66% of the total global consumption. China is followed by Nigeria and Tanzania, though, each of these last two countries shares only 4% of the total global consumption [35]. By applying the conventional breeding, biofortification of sweetpotato involved the selection of orange-fleshed varieties, to combat vitamin A deficiency among the developing nations [36]. In Japan, high β-carotene-accumulating varieties, such as "Benihayato," "J-Red," and "Sunny-Red," were initially developed by Japanese breeders at the Kyushu-Okinawa Agricultural Research Center (formerly the Kyushu National Agricultural Experiment Station), Miyakonojo, Miyazaki, Japan [37, 38, 39]. The enhancement of sweetpotato with provitamin A carotenoids (PVACs) has also been the area of research focus for the HarvestPlus (a company headquartered in Washington, DC, USA, involved in the development of nutritious food crops through biofortification and promotion of such crops) since the launch of its projects on biofortification [40]. Genetic modification of sweetpotato by using the transgenic tools and in order to improve the nutritional quality offers huge scope, and numerous research reports have already been published on genetic modification of sweetpotato using molecular gene engineering technologies [41]; however, it is the immense need to overcome hidden hunger, specially the one related with the insufficiency of provitamin A carotenoids among the poorly fed but rapidly growing populations in the developing countries by molecular breeding of sweetpotato varieties on sustainable basis.

2. Carotenoids and their Distribution

Carotenoids, the visible colors of life, are the 40-carbon isoprenoids synthesized naturally by fungi, bacteria, algae, and cyanobacteria [42, 43, 44] and conspicuously by green plants including bryophytes [45] and higher plants [44, 46, 47]. Being intracellular, carotenoids are commonly located in the membranes of chloroplasts, mitochondria, or endoplasmic reticulum [48]. Approximately 750 carotenoids have been reported so far [49, 50].

Maoka et al. (unpublished) analyzed carotenoids that were extracted from the orange tubers of the cultivar W71. It was consequently found that there were β-carotene-5,8,5`,8`-diepoxide (13.8% of the total carotenoids), β-carotene-5,6,5`,8`-diepoxide (9.2%), β-carotene-5,8-epoxide (4.6%), β-cryptoxanthin (3.2%), β-cryptoxanthin-5`,6`-epoxide (2.2%), lutein (2%), and zeaxanthin (trace amounts), in addition to β-carotene (59.3%). The biosynthetic pathway of these carotenoids is proposed in Figure 1. The carotenoids with 5,6-epoxy-β-ring or with 5,8-epoxy-β-ring are unique to sweetpotato. The tubers of the Japanese cultivar "Benimasari" were also found to accumulate not only the unique carotenoids, such as β-carotene-5,8,5`,8`-diepoxide (40.5% of the total carotenoids), β-carotene-5,8-epoxide (6.5%), β-cryptoxanthin-5`,8`-epoxide (10.5%), and β-carotene (10.5%), but also typical carotenoids that included 5,6-dihydroxy-β-ring (named ipomoeaxanthins) [15].

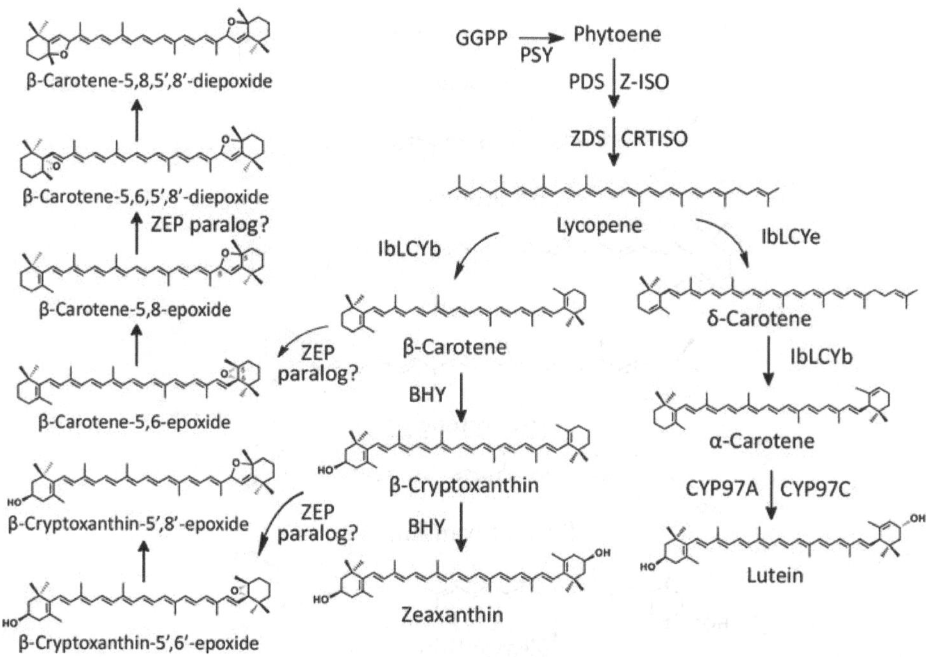

Figure 1. Proposed carotenoid biosynthetic pathway in sweetpotato orange tubers [10].

2.1 Role of carotenoids in animals

Animals, with very few exceptions [51, 52] are unable to synthesize them [53, 54]; however, carotenoids are accumulated by crustaceans, crabs, fish, crayfish, prawns, mammals, and in insects such as butterflies. The animal and human diet must include carotenoids as essential nutrients [44].

In marine animals, astaxanthin (3,3`-dihydroxy-β,β-carotene-4,4`-dione) has been reported as the most commonly stored carotenoid pigment [55]. It is responsible for the red/pink coloration of crustaceans [56, 57] and the flesh of salmonoids [58]. Astaxanthin has received much attraction for its likely role in preventing cardiovascular diseases and aging caused by UV light in human body [59]. Chemical structures of some major dietary carotenoids are shown in Figure 2.

Both α-carotene and β-carotene have provitamin A activity and are converted to retinol in the human body [60, 61, 62, 63, 64]. Carotenes such as the lycopene and β-carotene play a potential role in human nutrition and act as protectants against diseases, such as lycopene protects against cardiovascular [65], aging-related diseases, macular degradation of eye [66, 67], and certain types of cancers including gastrointestinal, cervix, breast, and prostate cancer [47, 61, 68, 69, 70, 71, 72, 73].

Beneficial effects of dietary carotenes, α-carotene, and β-carotene on human health related to enhancement of immune system and minimizing the risk of cancer are due to their antioxidant potential [69, 74]. β-Carotene, α-carotene, and β-cryptoxanthin are provitamin A carotenoids (PVACs) and hence they are the main precursors of vitamin A (VA) in the human body [75].

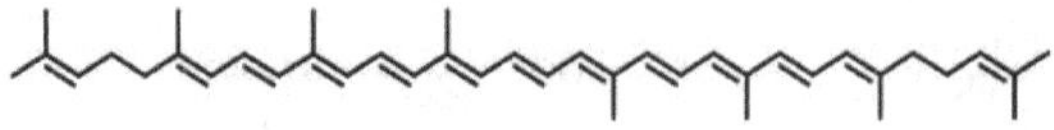

Lycopene

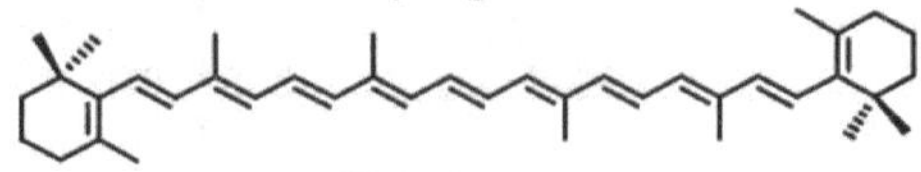

β-Carotene

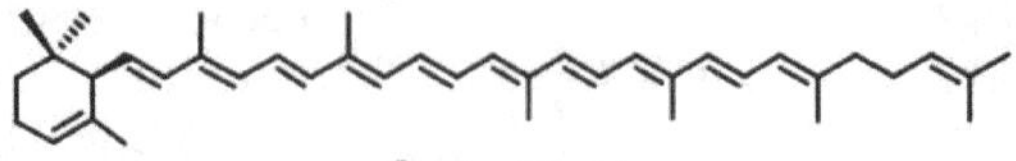

δ-Carotene

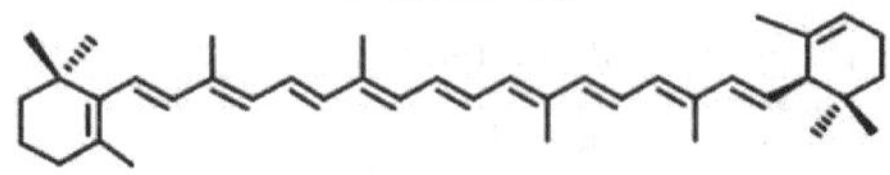

α-Carotene

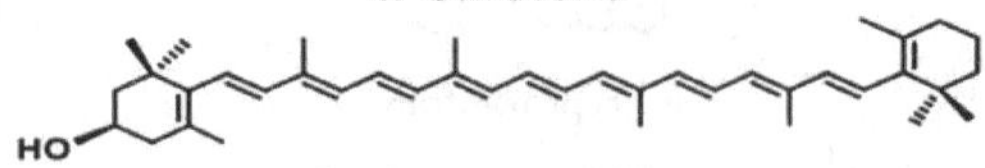

β-Cryptoxanthin

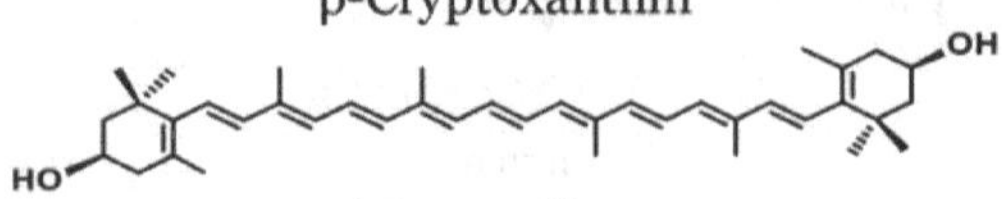

Zeaxanthin

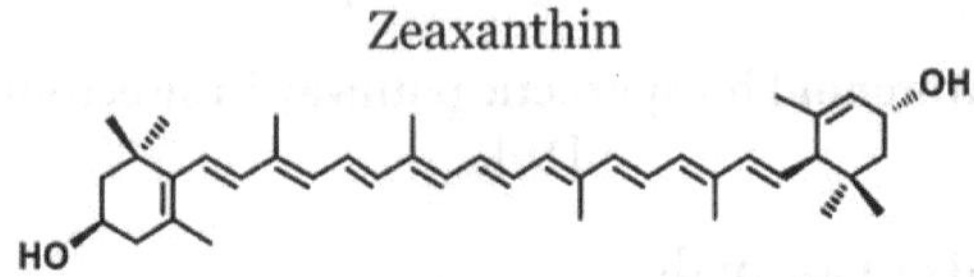

Lutein

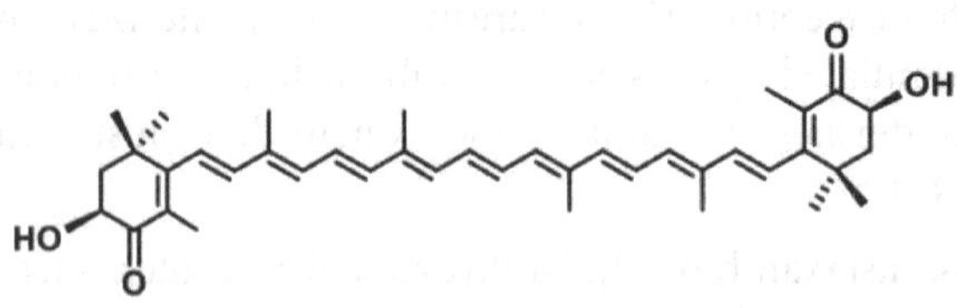

Astaxanthin

Figure 2. Structures of some major dietary carotenoids.

2.2 Carotenoids use in disease prevention

Carotenoids, especially astaxanthin, have been reported to enhance both the nonspecific and specific immune system and protect cell membranes and cellular DNA from mutation [44, 76]. Intake of fruits and vegetables rich in carotenoids mainly lycopene, α-carotene, β-carotene, β-cryptoxanthin, zeaxanthin, and lutein lowers the risk of morbidity and mortality by cardiovascular diseases and atherosclerosis [69]. Epidemiological studies have reported that lycopene can lower the risk of prostate cancer [77] and in its ability to quench singlet oxygen; it is 2- to 10-fold stronger than β-carotene and α-tocopherol, respectively [78]. Clinical studies have reported that the lycopene-enriched foods are protective against oxidative DNA damage in leukocytes in vitro [79] and prostate tissue in vivo [80]. β-Carotene is useful in reducing the risk of ischemic heart disease and myocardial infarction [81]. In the macular region of human eye including eye lens, two xanthophylls, lutein and zeaxanthin, exist in high concentrations and are regarded very important carotenoids for eye health. Reports suggest that these two carotenoids protect eye from high energy UV light and are excellent reactive oxygen species scavengers [82]. The role of lutein and zeaxanthin as macular pigments and their function in eye health has been reported in previous studies [83]. It has been anticipated that phytoene and phytofluene which are colorless precursors of other carotenoids possess light absorption in UV-A and UV-B range and protect skin by their photo-protective characteristics [84, 85]. Astaxanthin is also known as the super antioxidant. Since, it contains particular molecular configuration, making it extremely powerful antioxidant consequently, protecting cells against oxidation by quenching singlet oxygen and dissipating the energy as heat. It has the strong potential for scavenging free radicals and effectively breaks peroxide chain reactions [86, 87]. Studies have showed that the low-density lipoprotein (LDL) high cholesterol levels in mice decreased when supplemented with astaxanthin. Neither β-carotene nor canthaxanthin produced the same effect. Astaxanthin or other carotenoids can decrease the oxidation of the lipid carriers and thereby reduce the risk of atherosclerosis [88]. It also has positive effects in case of antitumor activity [89].

2.3 Industrial uses of carotenoids

All carotenoids show antioxidants activities appearing in a variety of colors in red, yellow, and orange; therefore, carotenoids are used as natural pigments in food, food supplements, nutraceuticals, pharmaceuticals, and cosmetic industry and various biotechnological purposes [90, 91]. Global carotenoids market touched $1.5 billion ($1500 million) in 2017 with a projection of $2.0 billion by 2022 [92]. In a previous report, the global market for carotenoids was $766 million in 2007. The expected projection for the year 2015 was $919 million with a compound annual growth rate (CAGR) of 2.3%. In 2007, β-carotene alone shared the market value at $247 million; this segment was predicted to be worth $285 million by 2015 with CAGR of 1.8% [91, 93]. In horticultural crops, they appear as a trait of attractiveness, adding value to the marketing potential of fruits and vegetables [94, 95]. Green algae Haematococcus pluvialis, which is the natural source of astaxanthin, has been reported for huge amounts ranging from 10,000 to 40,000 ppm (mg/kg) of astaxanthin in addition to other important carotenoids such as β -carotene, lutein, and canthaxanthin [58, 96]. Industrially, astaxanthin has been utilized as a feed supplement for cultured fish and shellfish [97, 98]. Other diverse biological functions of astaxanthin include an involvement in cancer prevention [99], enhancer of immune responses [100], and a free radical quencher [58, 101]. It is evident, therefore, that astaxanthin is a biomolecule with huge biofunction potential both to the pharmaceutical and food industries [58].

3. Sweetpotato Carotenoids

The carotenoids present in the sweetpotato leaves can scavenge free radical agents as singlet-oxygen quenchers [102, 103, 104, 105]. In a recent analytical report [105], the total phenol, carotenoid, anthocyanin, and flavonoids contents of the sweetpotato leaves ranged from 2.0 to 22.5 (g/100 g DW), 0.9 to 23.4 (β-carotene equivalents/100 g; BET/100 g), 2.2 to 24.5 (color value/g DW), and 62.8–272.2 (catechin equivalents; μg/g), respectively [105]. Consumption of sweetpotato in Asia ranges from its use as additional food of minute status to a very vital supplementary food to rice and/or other root and tuber crops [106]. It is cooked or used to make cakes, chapatis, mandazia, bread, buns, and cookies [107]. In the United States and some other developed countries, sweetpotato is strictly used as a luxury food. In Japan, it is used in novel plant products and/or nutraceuticals [108]. By using absorption spectroscopy, Ishiguro et al. [109] analyzed carotenoids from eight cultivars of yellow-fleshed sweetpotato and compared them in terms of their carotenoids. By HPLC analyses, they revealed some 17 different carotenoids from yellow- and orange-fleshed sweetpotato. In yellow-fleshed sweetpotato, the major carotenoids included β-carotene-5,8, 5′,8′-diepoxide (32–51%) and β-cryptoxanthin 5,8-epxide (11–30%), whereas β-carotene with amounts ranging from 80 to 92% were dominant in the orange-fleshed cultivars. For other orange cultivars, e.g. W71 and "Benimasari," carotenoid composition in the tubers has already been described along with a comprehensive metabolic pathway [10, 15]. Kammona et al. [110] analyzed and compared the carotenoid composition in some Malaysian orange, yellow, purple, and white sweetpotato tubers. They reported the highest total carotenoid contents from orange sweetpotato followed by yellow, purple, and white sweetpotato. Among the individual carotenoids analyzed, β-carotene existed in all types ranging from 91.95 ± 2.05 μg/g DW in white sweetpotato to 376.03 ± 11.05 μg/g DW in orange sweetpotato tubers. Traces of zeaxanthin were reported with values 5.44 ± 3.23 μg/g DW and 20.47 ± 2.03 μg/g DW in yellow and white sweetpotato, respectively. Lutein was available only in orange sweetpotato at trace amount of 0.91 ± 1.03 μg/g DW. Purple sweetpotato contained only β-carotene (113.86 ± 14.17 μg/g DW) with absence of other carotenoids [110].

Islam et al. [111] performed HPLC analyzes of *trans-* and *cis*-β-carotene from raw and boiled sweetpotato which included three orange-fleshed, three yellowish-cream-fleshed, and one white-fleshed varieties of sweetpotato. The deep-orange-fleshed variety Kamalasundari (BARI SP-2) showed the highest amounts of β-carotene among all the varieties followed by yellow varieties. On the other hand, from one of the two white-fleshed varieties, only trace amounts of β-carotene were obtained with no amounts at all from the other one. Their results proposed that the orange-fleshed varieties of sweetpotato contain the highest amounts of β-carotene in raw as compared to those which were boiled.

Despite huge economic value, sweetpotato has not received due importance as compared with common staple crops such as wheat, maize, and rice. World increasing hidden hunger, especially in developing countries, needs new foods and nutrition sources on sustainable bases. In this regard, sweetpotato not only offers immense nutritional, medicinal, industrial, and potential benefits but is also a new horizon in modern industrial biotechnological uses for biofunction development through the latest molecular tools and technologies of molecular plant breeding.

3.1 Isolation and functional identification of carotenoids biosynthesis genes

The heterologous complementation expression system in *Escherichia coli* offers unique tool for functional analysis of isolated new carotenoids biosynthesis genes from different organisms [112]. Carotenoid biosynthetic pathway in microorganisms, such as *Erwinia uredovora* and *Erwinia herbicola* (reclassified as *Pantoea ananatis* and *Pantoea agglomerans*, respectively), is specified by a gene cluster, encoding biosynthetic enzymes that function in a pathway starting with the synthesis of geranylgeranyl pyrophosphate (*GGPP*) and ending in the synthesis of zeaxanthin glucosides [113, 114]. Complete carotenoid gene clusters or part of it from *E. uredovora* or *E. herbicola* have been introduced into *E. coli*, which is otherwise a nonpigmented bacterium, and such transformed *E. coli* engineered in a way that they accumulate a range of colorful carotenoids [114, 115]. Since carotenoids are derived from isoprenoid precursors, *E. coli* can accumulate carotenoids by coupling an endogenous isoprenoid biosynthetic pathway with enzymes encoded by transformed genes of carotenogenic organisms such as *E. uredovora*. Hence, the biosynthetic pathway can be reconstructed in vivo even if the enzymes are of such diverse origin as those encoded by bacteria and plants [116, 117, 118]. The expression of carotenoid genes in *E. coli* has been useful for identifying function of gene products [118, 119, 120], the manipulation of the pathway [121, 122], investigating transcriptional regulators of carotenoids biosynthesis genes [123], and the isolation of new genes encoding enzymes of the carotenoid biosynthetic pathway [124] or enzymes catalyzing the synthesis of carotenoid precursors [125].

Misawa et al. [55] isolated and functionally identified the carotenoids biosynthesis genes cluster that included *crtB* (phytoene synthase), *crtI* (phytoene desaturase), *crtW* (β-carotene ketolase), and *crtZ* (β-carotene hydroxylase) from *Agrobacterium aurantiacum* (reclassified as *Paracoccus* sp. strain N81106). The functional identification of the isolated gene cluster led them to propose astaxanthin biosynthetic pathway for the first time.

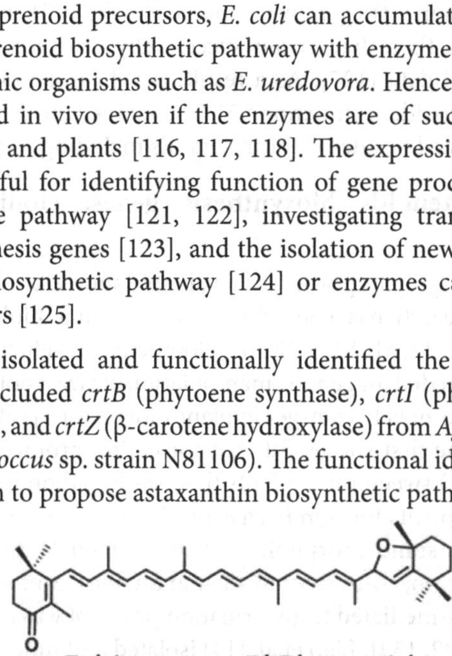

Echinenone-5′,8′-epoxide

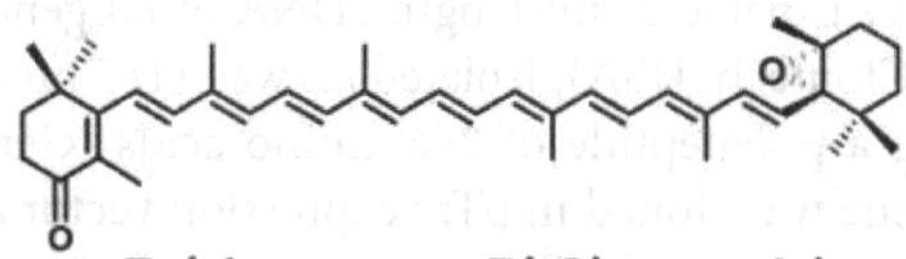

Echinenone-5′,6′-epoxide

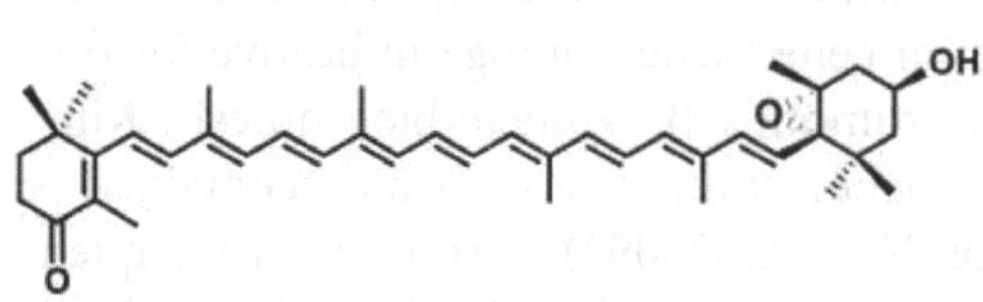

3′-Hydroxyechinenone-5′,6′-epoxide

Figure 3. Novel carotenoids produced in the tuber of the transgenic sweetpotato engineered with the crtW gene of bacterial origin.

Misawa et al. [114] isolated and functionally identified the carotenoid biosynthesis genes, such as *crtE* (GGPP synthase), *crtX* (Zeaxanthin glucosyltransferase), *crtY* (lycopene β-cyclase), *crtI* (phytoene desaturase), *crtB* (phytoene synthase), and *crtZ* (β-carotene hydroxylase), form *E. uredovora* by analyzing carotenoids accumulated in *E. coli* transformants in which these genes were expressed. By analysis of accumulated carotenoids in the transformed *E. coli* by these individual genes, they found that carotenoids in this pathway appeared to be close to those in higher plants rather than to those in bacteria. Although HPLC is a routine analytical tool to analyze various metabolic products from plants, highly developed and comprehensive metabolome analytical techniques with respect to particular tissues now offer precise analytical approaches such as nuclear magnetic resonance (NMR; COSY and NOESY) and accurate mass spectrometry (MS) techniques [47, 126]. A foreign *crtW* gene was expressed in the W71 cultivar of sweetpotato, and carotenoids generated there have been successfully analyzed by UV-vis, ESI-MS, ^{1}H-NMR, and CD spectral data [127]. As a result, novel carotenoids, shown in Figure 3, i.e. echinenone 5`,8`-epoxide, echinenone 5`,6`-epoxide, and 3`-hydroxyechinenone 5`,6`-epoxide, were identified besides ketocarotenoids including astaxanthin.

3.2 Sweetpotato carotenoids biosynthesis genes, cloning, and genetic engineering

Although, sweetpotato is highly important as a valuable source of carotenoids especially β-carotene, very little research has been done on molecular biological aspects of its carotenoid biosynthesis [10, 14, 31, 128]. The development of an efficient and reproducible transformation system is needed for genetic manipulation of sweet potato to either improve the crop or establish it as a novel "transgenic plant bioreactor" [129]. Otani et al. [130] developed and reported the first successful transformation protocol for the production of transformed (transgenic) sweetpotato plants that was based on the formation of hairy roots using leaf disks as explants for *Agrobacterium rhizogenes*. However, the regenerated transgenic plants showed some morphological abnormalities such as short storage root and internodes. Later on, to overcome such anomalies, a modified and successful *Agrobacterium tumefaciens*-mediated transformation protocol was developed via somatic cell embryogenesis [131, 132, 133]. Liao et al. [14] isolated and functionally characterized an isopentenyl diphosphate isomerase (*idi*) gene from sweetpotato cultivar YUSU 303 from Southeast China. They isolated a full-length cDNA of *idi* gene by SMART™ RACE cDNA Amplification Kit (Clontech, USA). Isolated *idi* was 1155 bp with an open reading frame of 892 bp encoding a polypeptide of 296 amino acids (GenBank accession No. DQ150100). Isolated *idi* gene was cloned in pTrc expression vector and was fed to *E. coli* which contained pAC-BETA plasmid for β-carotene accumulation. *E. coli* were cultured and carotenoids were analyzed by color complementation. Cultures of *E. coli* which were transformed with *idi* gene turned orange indicative for β-carotene and suggested its potential activity in promoting β-carotene biosynthesis. Kim et al. [134] isolated a partial sequence of phytoene synthase (*PSY*) which contained 354 bp from a cultivar Shinhwangmi (accession No. HQ828092). It showed 94% sequence identity with a *PSY* isolated from Ipomoea species Kenyan (GenBank accession No. AB499050.1). However, no gene function of isolated *PSY* from sweetpotato could be reported. Ling et al. [135] isolated a lycopene ε-cyclase (*LCYe*) gene from sweetpotato cultivar Nongdafu 14 from China. However, they did not functionally characterize it. They isolated a full-length cDNA of *idi* gene by GeneRacer TM Kit (Invitrogen Carlsbad, CA, USA). Isolated *LCYe* was 1805 bp with an open reading frame of 1236 bp encoding a polypeptide of 411 amino acids. Quantitative real-time PCR analysis showed that *IbLCYe* expression levels were desirably higher in roots as compared to those in leaves. Isolated *LCYe* gene was expressed in tobacco cultivar Winconsin 38. Carotenoids from transgenic tobacco plants were extracted and analyzed by HPLC which revealed transgenes accumulating more β-carotene as compared to control plants. Kim et al. [128] isolated a partial lycopene β-cyclase (*IbLCYb*) from a cultivar Yulmi of sweetpotato. They synthesized primers by using a partial sequence of *IbLCYb* from database with accession number JX393306 and amplified a partial cDNA of *IbLCYb* by RT-PCR. By using isolated *IbLCYb*, an IbLCYb-RNAi vector was constructed and then used to transform white-fleshed sweetpotato. Transformed sweetpotatoes were cultured and analyzed for accumulated carotenoids. Their results showed a total increase in the carotenoids contents along with increase in resistance against salt stress in transgenic sweetpotato as compared to the control. Significant levels of carotenoids genes expression were observed in all plant parts with highest expression in leaves to lowest in the fibrous roots. But in case of transgenic calli, expressions of *IbLCYb* were dramatically reduced and found high in non-transgenic calli. Lycopene was not produced both by transgenic and non-transgenic sweetpotato. In an another experiment, Kim et al. [134] cloned a partial cDNA encoding β-carotene hydroxylase (*BHY*) from storage roots of sweetpotato cultivar Shinhwangmi and constructed an RNA-i-IbCHY-β vector for transformation of white-fleshed cultivar Yulmi and evaluation of inhibition effects of β-carotene hydroxylase (*BHY*) in transgenic lines. Downregulation of *IbBHY* gene expression altered the content and degree of carotenoids between transgenic and non-transgenic cells with an increase in the β-carotene and total carotenoids in transgenic sweetpotato cells along with an increase in their antioxidation potential.

3.3 Metabolic engineering of the carotenoid biosynthetic pathway to enhance carotenoid contents in higher plants

The pathway engineering approach using a variety of carotenoid biosynthesis genes is becoming a potential approach as one of the most effective methods to generate large quantities of structurally diverse carotenoids [59, 136, 137]. Astaxanthin (3,30-dihydroxy-4,40-diketo-β-carotene) is a high-value ketocarotenoid that is biosynthesized only by a few organisms typically at low levels. This red pigment (produced through chemical synthesis) has been used in large amounts in aquaculture. Currently, natural astaxanthin is employed as a health boosting food and is investigated for the treatment of a number of human diseases including cancers [138]. The limited renewable sources and growing demand for natural astaxanthin have attracted tremendous interest in its engineering into heterologous hosts, especially plants with the ability of sequestering 10- to 50-fold higher carotenoids than microorganisms, to produce the high-value pigment, during the past decade [139, 140]. The most promising approach reaching high astaxanthin yields was by chloroplast transformation using a bacterial ketolase gene [141]. Plastid genome transformation of lettuce (*Lactuca sativa*) has similarly been site-specifically modified with the addition of three transgenes, which encoded β,β-carotenoid 3,3`-hydroxylase (*crtZ*) and β,β-carotenoid 4,4Misawa et al. [114] isolated and functionally identified the carotenoid biosynthesis genes, such as *crtE* (GGPP synthase), *crtX* (Zeaxanthin glucosyltransferase), *crtY* (lycopene β-cyclase), *crtI* (phytoene desaturase), *crtB* (phytoene synthase), and *crtZ* (β-carotene hydroxylase), form *E. uredovora* by analyzing carotenoids accumulated in *E. coli* transformants in which these genes were expressed. By analysis of accumulated carotenoids in the transformed *E. coli* by these individual genes, they found that carotenoids in this pathway appeared to be close to those in higher plants rather than to those in bacteria. Although HPLC is a routine analytical tool to analyze various metabolic products from plants, highly developed and comprehensive metabolome analytical techniques with respect to particular tissues now offer precise analytical approaches such as nuclear magnetic resonance (NMR; COSY and NOESY) and accurate mass spectrometry (MS) techniques [47, 126]. A foreign *crtW* gene was expressed in the W71 cultivar of sweetpotato, and carotenoids generated there have been successfully analyzed by UV-vis, ESI-MS, ^{1}H-NMR, and CD spectral data [127]. As a result, novel carotenoids, shown in Figure 3, i.e. echinenone 5`,8`-epoxide, echinenone 5`,6`-epoxide, and 3`-hydroxyechinenone 5`,6`-epoxide, were identified besides ketocarotenoids including astaxanthin.

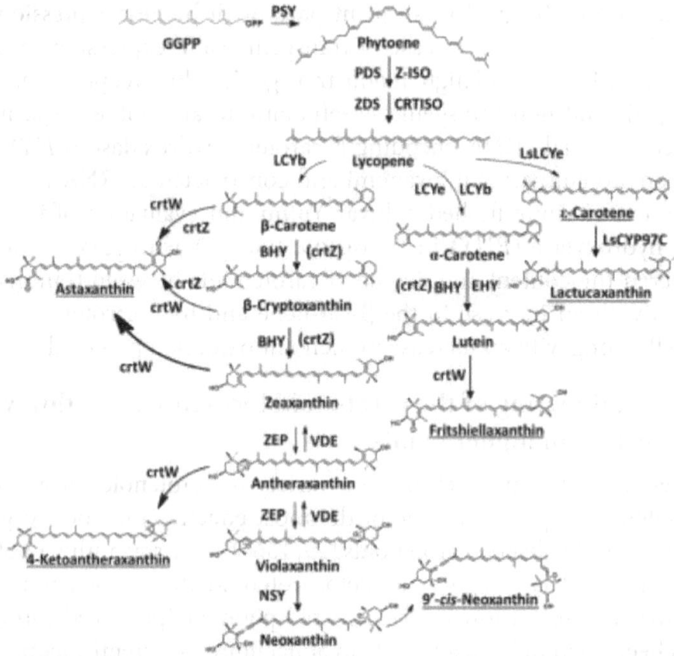

Figure 4. Carotenoid biosynthetic pathway in higher plants. A summarized illustration for the introduction and function of heterologous crtW and crtZ genes expressed in tobacco [141] and lettuce [142] leaves. The carotenoids shown in black represent native carotenoids accumulated by both tobacco and lettuce, and black and underlined are those reported from both of the transgenic tobacco and lettuce. Carotenoids underlined red are reported from lettuce only, where GGPS, is geranylgeranyl pyrophosphate synthase, PSY is phytoene synthase, PDS is phytoene-desaturase, ZDS is ζ-carotene desaturase, CRTISO is carotenoid isomerase, LCYb is lycopene β-cyclase, LCYe is lycopene ε-cyclase, LsLCYe is lettuce LCYe, LsCYP97C is lettuce heme-containing cytochrome P450-type carotene ε-ring hydroxylase, BHY is non-heme di-iron-type carotene β-ring hydroxylase, EHY is carotene ε-ring hydroxylase, ZEP is zeaxanthin-epoxidase, VDE is violaxanthin de-epoxidase, and NSY is neoxanthin synthase.

Through pathway engineering that utilizes the marine bacterial carotenoid 4,4`-ketolase (4,4`-oxygenase) gene named *crtW,* unique keto-carotenoids such as astaxanthindiglucoside, 2,2`-dihydroxyastaxanthin, and 2,2`-dihydroxycanthaxanthin have been produced in *Escherichia coli* [146, 147] and 4-ketoantheraxanthin in tobacco (*Nicotiana tabacum*) plants [148]. Breitenbach et al. [149] also synthesized α-echinenone (4-keto-α-carotene) in rice callus using *crtW*. Recently, 4-keto-zeinoxanthin was produced in *E. coli* cells by introducing the bacterial *crtW* gene and carotenogenic genes from liverwort [150].

3.4 Metabolic engineering of the carotenoids biosynthetic pathway in sweetpotato

Metabolic engineering of carotenoid biosynthetic pathway using a combinatorial approach has led to the efficient production of interesting carotenoids of high commercial value and pharmaceutical potential [44, 59, 151].

Starting with transgenic approach, prerequisite is to have a sound knowledge on the metabolic pathways regulating the carotenoid biosynthesis and their accumulation. Due to efforts of many scientists, the carotenoid metabolic pathway and the function of the biosynthetic enzymes involved in carotenoids biosynthesis have been elaborated well [152]. It was reported that sweetpotato contained not only β-carotene but also several epoxy carotenoids unique to the sweetpotato tubers, e.g. β-carotene-5, 8-epoxside and β-carotene-5, 8, 5`8`-diepoxside [15]. Therefore, it was assumed that the new structural carotenoids with epoxy and keto groups can be produced by expressing the ketolase *crtW* gene in sweetpotato tubers. Recently, marine bacterial genes that include the *crtW* gene encoding carotenoid 4,4`-ketolase [148] was introduced into sweetpotato cultivar W71 under the control of the CaMV promoter. Consequently, novel carotenoids with epoxy and keto groups 1, 2, and 3 were obtained along with a series of ketocarotenoids. The structural elucidation of these novel epoxy-keto carotenoids along with biosynthetic pathway in sweetpotato was also proposed [127]. A tabulated summary of recent developments in molecular breeding of sweet potato by genetic, metabolic, and pathway engineering approaches is presented in Table 1.

Table 1. Role of metabolic engineering in carotenoids enhancement in sweetpotato*.

Gene engineered	Promoter used	Carotenoids enhancement	Ref.
ZDS	Cauliflower mosaic virus (CaMV) 35S	3.96–2.37 increase in β-carotene and lutein and, 2.23-fold increase in total carotenoids accompanied with enhanced salt tolerance	[153]
PSY	Tuber-specific primer	6.3-fold increase in carotenoid, 19-fold increase in β-carotene	[154]
LCYb	Cauliflower mosaic virus (CaMV) 35S	1.4–1.8 times higher in β-carotene, increased tolerance to drought stress	[128]
LCYe		5.44-fold to 6.59-fold Increase in β-carotene and 1.77–2.75 times increase in total carotenoids	[155]
BHY	Cauliflower mosaic virus (CaMV) 35S	Twofold increase in total carotenoids and 16-fold increase in β-carotene	[152]
CrtW	Cauliflower mosaic virus (CaMV) 35S	Novel carotenoids with epoxy and keto groups were produced including a series of ketocarotenoids	[127]
CrtO	Cauliflower mosaic virus (CaMV) 35S	10–12% increase in total carotenoid	[156]
BHY	Silencing of BHY	117 μg/g (dry weight) increase in total carotenoids and 34.43 lg/g (dryweight) in β- carotene	[134]
IbOr-R96H	cauliflower mosaic virus (CaMV) 35S promoter	19.6- and 186.2-fold higher total carotenoid and β-carotene contents, respectively	[157]

"White Star" (WS) and W71, which produce white- and orange-fleshed tubers, respectively, are important sweetpotato cultivars, since they are amenable to *Agrobacterium*-mediated transformation [10, 127, 158, 159]. Chemical analysis of the carotenoids and isolation and functional characterization of the carotenoids biosynthesis genes of these two cultivars was reported in more details by Khan et al. [10] for the first time. One of the initial works that led to the sweetpotato genetic improvement for enhancing provitamin A amounts was done by Kim et al. [160] which involved isolation and functional analysis of the orange (*Or*) gene, from orange-fleshed sweetpotato controlling the carotenoids accumulation in the transgenic calli of sweetpotato. White-fleshed sweetpotatoes were transformed by the orange (*Or*) gene, which resulted in the 10-fold increased accumulation of β-carotene and total carotenoids. Later, identical results were presented by Park et al. [161] who observed that the overexpression of *IbOr* gene boosted the carotenoid composition in purple-flesh sweetpotato cultivar. In higher plants, the biosynthesis of carotenoids from lycopene involves the enzymatic activity of lycopene ∈-cyclase (*LCYe*) gene, via β-branch-specific biosynthetic pathway yielding β-carotene. By downregulating the expression of *IbLCYe* through RNA interference (RNAi) technology, higher amounts of β-carotene were recorded [162]. To increase the β-carotene contents in sweetpotato, researchers have also made use of the molecular markers along with the evaluation and screening of available germplasm.

This combined approach is thought very useful in selecting the desirable parents for breeding new sweetpotato varieties with the higher levels of β-carotenes [163]. To analyze the gene diversity and evolutionary relationships among various cultivars of sweetpotato, Hwang et al. [164] have applied the use of Simple Sequence Repeats (SSRs). Their results showed that polycross-derived cultivars have higher levels of genetic diversity suggesting the application of polycross breeding that overcomes the challenges of cross-incompatibility. For breeding high β-carotene sweetpotato varieties, Quantitative Trait Loci (QTLs) for β-carotene content in a cross sweetpotato were reported for the first time by Cervantes-Flores et al. [165] which led to the understanding of the inheritance pattern and is considered the foundation of the development of marker-assisted breeding techniques for breeding high β-carotene (provitamin A) accumulating sweetpotato cultivars.

Orange-fleshed sweetpotato, which is a genetically modified crop, is now well accepted by consumers [166, 167] and has appeared as a sound supply of provitamin A. To achieve the daily provitamin A needs, mere 125 g of fresh orange-fleshed sweetpotato roots from most varieties are enough [35].

4. Conclusion

The rapidly increasing world population demands sustainable supply of ample quantities of quality food, especially under the changing climatic conditions. Food insecurity accompanied with already existing malnutrition among the developing countries is a grand challenge of the day. Foods rich in phytonutrients not only contribute toward enhancing the health but also reduce the risk of many diseases including early aging. In this regard, genetic improvement of the major staple crops such as sweetpotato needs additional strategies of the molecular plant breeding to overcome the genetic complexity. The application of metabolic engineering supplemented with the omics and the recently developed gene editing tools and technologies are the potential strategies to be adopted which promise scope for improving the quantity of phytonutrients, especially the carotenoids in sweet potato. It will contribute to prevent the malnutrition and the diseases linked with foods deficit in quality nutrients. There is a dire need to apply multiple gene engineering approaches for multi-phytonutrients improvement to meet the need.

5. Recent Development and Future Scope

Recent developments in carotenoids gene manipulations have helped to make insight that engineering sweetpotato with IbOr gene manipulations would be a potential strategy to improve the total carotenoids and specially the β-carotene through enhancing sink strength in storage roots of sweet potato. Moreover, site-directed mutagenesis supplanted with the genome editing tools such as CRISPR-Cas9 such as CRISPR-Cas9 and its different modifications will further lead to a fruitful biofortification of sweet potato for nutritional enhancement through carotenoids improvement.

References

1. Ahn YO, Kim SH, Kim CY, Lee JS, Kwak SS, Lee HS. Exogenous sucrose utilization and starch biosynthesis among sweetpotato cultivars. Carbohydrate Research. 2010;345(1):55-60
2. Rodriguez-Bonilla L, Cuevas HE, Montero-Rojas M, Bird-Pico F, Luciano-Rosario D, Siritunga D. Assessment of genetic diversity of sweet potato in Puerto Rico. PLoS One. 2014;9(12):e116184
3. Hirakawa H, Okada Y, Tabuchi H, Shirasawa K, Watanabe A, Tsuruoka H, et al. Survey of genome sequences in a wild sweet potato, Ipomoea trifida (HBK) G. Don. DNA Research. 2015;22(2):171-179
4. Austin DF, Huáman Z. A synopsis of Ipomoea (Convolvulaceae) in the Americas. Taxon. Feb 1996;45(1):3-38
5. Shih PH, Yeh CT, Yen GC. Anthocyanins induce the activation of phase II enzymes through the antioxidant response element pathway against oxidative stress-induced apoptosis. Journal of Agricultural and Food Chemistry. 2007;55(23):9427-9435
6. Ji H, Zhang H, Li H, Li Y. Analysis on the nutrition composition and antioxidant activity of different types of sweet potato cultivars. Food and Nutrition Sciences. 2015;6(01):161
7. Wang H, Cao G, Prior RL. Oxygen radical absorbing capacity of anthocyanins. Journal of Agricultural and Food Chemistry. 1997;45(2):304-309
8. USDA. (U.S. Department of Agriculture), National Nutrient Database for Standard Reference, Release 28. Basic Report: Raw, 11507, Sweetpotato, Unprepared. USA: Online, 2016. Available from: https://ndb.nal.usda.gov/ndb/foods/show/ [Accessed: December 16, 2016]
9. Antonio GC, Takeiti CY, de Oliveira RA, Park KJ. Sweet potato: Production, morphological and physicochemical characteristics, and technological process. Embrapa Agroindústria de Alimentos-Artigo em periódico indexado (ALICE). 2011;5(2):1-18

10. Khan MZ, Takemura M, Maoka T, Otani M, Misawa N. Carotenoid analysis of sweetpotato Ipomoea batatas and functional identification of its lycopene β-and ε-cyclase genes. Zeitschrift für Naturforschung C. 2016;71(9-10):313-322

11. Purcell AE, Walter WM. Carotenoids of centennial variety sweet potato, Ipomea batatas. Journal of Agricultural and Food Chemistry. 1968;16(5):769-770

12. Teow CC, Truong VD, McFeeters RF, Thompson RL, Pecota KV, Yencho GC. Antioxidant activities, phenolic and β-carotene contents of sweet potato genotypes with varying flesh colours. Food Chemistry. 2007;103(3):829-838

13. Grace MH, Yousef GG, Gustafson SJ, Truong VD, Yencho GC, Lila MA. Phytochemical changes in phenolics, anthocyanins, ascorbic acid, and carotenoids associated with sweetpotato storage and impacts on bioactive properties. Food Chemistry. 2014;145:717-724

14. Liao Z, Chen M, Yang Y, Yang C, Fu Y, Zhang Q, et al. A new isopentenyl diphosphate isomerase gene from sweet potato: Cloning, characterization and color complementation. Biologia. 2008;63(2):221-226

15. Maoka T, Akimoto N, Ishiguro K, Yoshinaga M, Yoshimoto M. Carotenoids with a 5, 6-dihydro-5, 6-dihydroxy-β-end group, from yellow sweet potato "Benimasari", Ipomoea batatas Lam. Phytochemistry. 2007;68(13):1740-1745

16. Tsuda T, Horio F, Osawa T. Dietary cyanidin 3-O-β-d-glucoside increases ex vivo oxidation resistance of serum in rats. Lipids. 1998;33(6):583

17. Mazza G, Kay CD, Cottrell T, Holub BJ. Absorption of anthocyanins from blueberries and serum antioxidant status in human subjects. Journal of Agricultural and Food Chemistry. 2002;50:7731-7737

18. Tsoyi K, Park HB, Kim YM, Chung JI, Shin SC, Lee WS, et al. Anthocyanins from black soybean seed coats inhibit UVB-induced inflammatory cylooxygenase-2 gene expression and PGE2 production through regulation of the nuclear factor-κB and phosphatidylinositol 3-kinase/Akt pathway. Journal of Agricultural and Food Chemistry. 2008;56(19):8969-8974

19. Wang J, Mazza G. Inhibitory effects of anthocyanins and other phenolic compounds on nitric oxide production in LPS/IFN-γ-activated RAW 264.7 macrophages. Journal of Agricultural and Food Chemistry. 2002;50(4):850-857

20. Pisha E, Pezzuto JM. Fruits and vegetables containing compounds that demonstrate pharmacological activity in humans. Economic and Medicinal Plant Research. 1994;6:189-233

21. Hagiwara A, Yoshino H, Ichihara T, Kawabe M, Tamano S, Aoki H, et al. Prevention by natural food anthocyanins, purple sweet potato color and red cabbage color, of 2-amino-1-methyl-6-phenylimidazo [4, 5-B] pyridine (phip)-associated colorectal carcinogenesis in rats. The Journal of Toxicological Sciences. 2002;27(1):57-68

22. Kurata R, Adachi M, Yamakawa O, Yoshimoto M. Growth suppression of human cancer cells by polyphenolics from sweetpotato (Ipomoea batatas L.) leaves. Journal of Agricultural and Food Chemistry. 2007;55(1):185-190

23. Islam S. Sweetpotato (Ipomoea batatas L.) leaf: Its potential effect on human health and nutrition. Journal of Food Science. 2006;71(2):R13-R21

24. Kano M, Takayanagi T, Harada K, Makino K, Ishikawa F. Antioxidative activity of anthocyanins from purple sweet potato, Ipomoera batatas cultivar Ayamurasaki. Bioscience, Biotechnology, and Biochemistry. 2005;69(5):979-988

25. Cho J, Kang JS, Long PH, Jing J, Back Y, Chung KS. Antioxidant and memory enhancing effects of purple sweet potato anthocyanin and cordyceps mushroom extract. Archives of Pharmacal Research. 2003;26(10):821-825

26. Kamei H, Kojima T, Hasegawa M, Koide T, Umeda T, Yukawa T, et al. Suppression of tumor cell growth by anthocyanins in vitro. Cancer Investigation. 1995;13(6):590-594

27. Panda V, Sonkamble M. Anti-ulcer activity of Ipomoea batatas tubers (sweet potato). Functional Foods in Health and Disease. 2012;2(3):48-61

28. Ozias-Akins P, Jarret RL. Nuclear DNA content and ploidy levels in the genus Ipomoea. Journal of the American Society for Horticultural Science. 1994;119(1):110-115

29. Yan L, Gu YH, Tao X, Lai XJ, Zhang YZ, Tan XM, et al. Scanning of transposable elements and analyzing expression of transposase genes of sweet potato [Ipomoea batatas]. PLoS One. 2014;9(3):e90895

30. Roullier C, Duputié A, Wennekes P, Benoit L, Fernández Bringas VM, Rossel G, et al. Disentangling the origins of cultivated sweet potato (Ipomoea batatas (L.) Lam.). PLoS One. 2013;8(5):e62707

31. Arizio CM, Tártara SC, Manifesto MM. Carotenoids gene markers for sweetpotato (Ipomoea batatas l. lam): Applications in genetic mapping, diversity evaluation and cross-species transference. Molecular Genetics and Genomics. 2014;289(2):237-251

32. Martin FW. Incompatibility in the sweet potato. A review. Economic Botany. 1965;19(4):406-415

33. Holden JM, Eldrige AL, Beecher GR, Buzzard MI, Bhagwat S, Davis CS, et al. Carotenoid content of U. S. Foods: An update of the database. Journal of Food Composition and Analysis. 1999;12:169-196

34. Burns J, Fraser PD, Bramley PM. Identification and quantification of carotenoids, tocopherols and chlorophylls in commonly consumed fruits and vegetables. Phytochemistry. 2003;62(6):939-947

35. CIP, International Potato Center. 2019. Available from: https://cipotato.org/

36. Institute of Food Technologists. Chapter 4: Nutritionally Improved Sweetpotato. Comprehensive Reviews in Food Science and Food Safety. 2008;7:81-91.(1t)

37. Kukimura H, Yoshida T, Komaki K. New sweet potato cultivars, Benihayato and Satsumahikari, making a new turn for processing. JARQ. 1988;22(1):7-13

38. Yamakawa O, Yoshinaga M, Kumagai T, Hidaka M, Komaki K, Kukimura H, et al. "J-red": A new sweetpotato cultivar. Bulletin of the Kyushu Agricultural Experiment Station (Japan). 1997;33:49-72

39. Yamakawa O, Kumagai T, Yoshinaga M, Ishiguro K, Hidaka M, Komaki K, et al. "Sunny-red": A new sweetpotato (Ipomoea batatas) cultivar for powder. Bulletin of the Kyushu Agricultural Experiment Station (Japan). 1999;35:19-40

40. Tanumihardjo SA, Ball AM, Kaliwile C, Pixley KV. The research and implementation continuum of biofortified sweet potato and maize in Africa. Annals of the New York Academy of Sciences. 2017;1390(1):88-103

41. Liu Q. Improvement for agronomically important traits by gene engineering in sweetpotato. Breeding Science. 2017;67:15-26. DOI: 10.1270/jsbbs.16126

42. Walter MH, Strack D. Carotenoids and their cleavage products: Biosynthesis and functions. Natural Product Reports. 2011;28(4):663-692

43. Okada K, Kasahara H, Yamaguchi S, Kawaide H, Kamiya Y, Nojiri H, et al. Genetic evidence for the role of isopentenyl diphosphate isomerases in the mevalonate pathway and plant development in Arabidopsis. Plant and Cell Physiology. 2008;49(4):604-616

44. Misawa N. Carotenoids. Comprehensive Natural Products. 2010;II:733-753. DOI: 10.1016/b978-008045382-8.00009-5

45. Takemura M, Maoka T, Misawa N. Biosynthetic routes of hydroxylated carotenoids (xanthophylls) in Marchantia polymorpha, and production of novel and rare xanthophylls through pathway engineering in Escherichia coli. Planta. 2015;241(3):699-710

46. Sugawara T, Yamashita K, Asai A, Nagao A, Shiraishi T, Imai I, et al. Esterification of xanthophylls by human intestinal Caco-2 cells. Archives of Biochemistry and Biophysics. 2009;483(2):205-212

47. Fraser PD, Bramley PM. Metabolic profiling and quantification of carotenoids and related isoprenoids in crop plants. In: Plant Metabolomics. Berlin, Heidelberg: Springer; 2006. pp. 229-242

48. Margalith PZ. Production of ketocarotenoids by microalgae. Applied Microbiology and Biotechnology. 1999;51(4):431-438

49. Farré G, Maiam Rivera S, Alves R, Vilaprinyo E, Sorribas A, Canela R, et al. Targeted transcriptomic and metabolic profiling reveals temporal bottlenecks in the maize carotenoid pathway that may be addressed by multigene engineering. The Plant Journal. 2013;75(3):441-455

50. Britton G, Liaaen-Jenson S, Pfander H. (Eds), Compiled by Mercadante AZ, and Egeland ES, Birkhauser Verlag, Basle, Switzerland. Free Radical Research. 2004;38(8):885. DOI: 10.1080/10715760410001727849

51. Moran NA, Jarvik T. Lateral transfer of genes from fungi underlies carotenoid production in aphids. Science. 2010;328(5978):624-627

52. Tonhosolo R, Fabio LD, de Rosso VV, Gazarini ML, Matsumura MY, Peres VJ, et al. Carotenoid biosynthesis in intraerythrocytic stages of Plasmodium falciparum. Journal of Biological Chemistry. 2009;284(15):9974-9985

53. Ruiz-Sola MA, Rodríguez-Concepción M. Carotenoid biosynthesis in Arabidopsis: A colorful pathway. Arabidopsis Book. 2012;10:e0158

54. Cazzonelli CI. Carotenoids in nature: Insights from plants and beyond. Functional Plant Biology. 2011;38(11):833-847

55. Misawa N, Satomi Y, Kondo K, Yokoyama A, Kajiwara S, Saito T, et al. Structure and functional analysis of a marine bacterial carotenoid biosynthesis gene cluster and astaxanthin biosynthetic pathway proposed at the gene level. Journal of Bacteriology. 1995;177(22):6575-6584

56. Matsuno T. In: Krinsky NI, Mathews-Roth MM, Taylor F, editors. "Carotenoids:" In Chemistry and Biology. New York: Plenum Publishing Corp; 1989. pp. 59-74

57. Renstrøm B, Borch G, Skulberg OM, Liaaen-Jensen S. Optical purity of (3S,3'S)-astaxanthin from Haematococcus pluvialis. Phytochemistry. 1981;20(11):2561-2564. DOI: 10.1016/0031-9422(81)83094-4

58. Fraser PD, Miura Y, Misawa N. In vitro characterization of astaxanthin biosynthetic enzymes. Journal of Biological Chemistry. 1997;272(10):6128-6135

59. Misawa N. Pathway engineering for functional isoprenoids. Current Opinion in Biotechnology. 2011;22(5):627-633

60. Zeb A, Mehmood S. Carotenoids contents from various sources and their potential health applications. Pakistan Journal of Nutrition. 2004;3(3):199-204

61. Jaswir I. Carotenoids: Sources, medicinal properties and their application in food and nutraceutical industry. Journal of Medicinal Plants Research. 2011;5:33. DOI: 10.5897/jmprx11.011

62. Thane C, Reddy S. Processing of fruit and vegetables: Effect on carotenoids. Nutrition & Food Science. 1997;97(2):58-65. ISSN: 0034-6659

63. Park SY, Nomura AM, Murphy SP, Wilkens LR, Henderson BE, Kolonel LN. Carotenoid intake and colorectal cancer risk: The multiethnic cohort study. Journal of Epidemiology. 2009;19(2):63-71. DOI: 10.2188/jea.je20080078

64. Carrillo-Lopez A, Yahia EM, Ramirez-Padilla GK. Bioconversion of carotenoids in five fruits and vegetables to vitamin A measured by retinol accumulation in rat livers. American Journal of Agricultural and Biological Sciences. 2010;5(2):215-221

65. Böhm V. Lycopene and heart health. Molecular Nutrition & Food Research. 2012;56(2):296-303

66. Gupta SK, Trivedi D, Srivastava S, Joshi S, Halder N, Verma SD. Lycopene attenuates oxidative stress induced experimental cataract development: An in vitro and in vivo study. Nutrition. 2003;19(9):794-799

67. Chichili GR, Nohr D, Frank J, Flaccus A, Fraser PD, Enfissi EM, et al. Protective effects of tomato extract with elevated β-carotene levels on oxidative stress in ARPE-19 cells. British Journal of Nutrition. 2006;96(4):643-649

68. Peto R, Doll R, Buckley JD, Sporn MB. Can dietary beta-carotene materially reduce human cancer rates? Nature. 1981;290(5803):201-208

69. Fiedor J, Burda K. Potential role of carotenoids as antioxidants in human health and disease. Nutrients. 2014;6(2):466-488

70. Stahl W, Heinrich U, Jungmann H, von Laar J, Schietzel M, Sies H, et al. Increased dermal carotenoid levels assessed by noninvasive reflection spectrophotometry correlate with serum levels in women ingesting Betatene. The Journal of Nutrition. 1998;128(5):903-907

71. Woodall AA, Lee SW, Weesie RJ, Jackson MJ, Britton G. Oxidation of carotenoids by free radicals: Relationship between structure and reactivity. Biochimica et Biophysica Acta (BBA)-General Subjects. 1997;1336(1):33-42

72. Erdman JW Jr, Ford NA, Lindshield BL. Are the health attributes of lycopene related to its antioxidant function? Archives of Biochemistry and Biophysics. 2009;483(2):229-235

73. Giovannucci E. Obesity, gender, and colon cancer. Gut. 2002;51(2):147-147

74. Das A, Yoon SH, Lee SH, Kim JY, Oh DK, Kim SW. An update on microbial carotenoid production: Application of recent metabolic engineering tools. Applied Microbiology and Biotechnology. 2007;77(3):505-512

75. Arscott SA, Tanumihardjo SA. Carrots of many colors provide basic nutrition and bioavailable phytochemicals acting as a functional food. Comprehensive Reviews in Food Science and Food Safety. 2010;9(2):223-239

76. Bendich A. Carotenoids and the immune system. In: Carotenoids. Boston, MA: Springer; 1989. pp. 323-335

77. Nakazawa Y, Sashima T, Hosokawa M, Miyashita K. Comparative evaluation of growth inhibitory effect of stereoisomers of fucoxanthin in human cancer cell lines. Journal of Functional Foods. 2009;1(1):88-97

78. Rao LG, Guns E, Rao AV. Lycopene: Its role in human health and disease. Agro Food. 2003;7:25-30

79. Pool-Zobel BL, Bub A, Müller H, Wollowski I, Rechkemmer G. Consumption of vegetables reduces genetic damage in humans: First results of a human intervention trial with carotenoid-rich foods. Carcinogenesis. 1997;18(9):1847-1850

80. Bowen P, Chen L, Stacewicz-Sapuntzakis M, Duncan C, Sharifi R, Ghosh L, et al. Tomato sauce supplementation and prostate cancer: Lycopene accumulation and modulation of biomarkers of carcinogenesis. Experimental Biology and Medicine. 2002;227(10):886-893

81. Hennekens CH, Buring JE, Manson JE, Stampfer M, Rosner B, Cook NR, et al. Lack of effect of long-term supplementation with beta carotene on the incidence of malignant neoplasms and cardiovascular disease. New England Journal of Medicine. 1996;334(18):1145-1149

82. Edge R, McGarvey DJ, Truscott TG. The carotenoids as anti-oxidants—A review. Journal of Photochemistry and Photobiology B: Biology. 1997;41(3):189-200

83. Abdel-Aal ES, Akhtar H, Zaheer K, Ali R. Dietary sources of lutein and zeaxanthin carotenoids and their role in eye health. Nutrients. 2013;5(4):1169-1185

84. Stahl W, Sies H. β-Carotene and other carotenoids in protection from sunlight. The American Journal of Clinical Nutrition. 2012;96(5):1179S-1184S

85. Engelmann NJ, Clinton SK, Erdman JW Jr. Nutritional aspects of phytoene and phytofluene, carotenoid precursors to lycopene. Advances in Nutrition. 2011;2(1):51-61

86. Kurashige M, Okimasu E, Inoue M, Utsumi K. Inhibition of oxidative injury of biological membranes by astaxanthin. Physiological Chemistry and Physics and Medical NMR. 1990;22(1):27-38

87. Jørgensen K, Skibsted LH. Carotenoid scavenging of radicals. Zeitschrift für Lebensmittel-Untersuchung und Forschung. 1993;196(5):423-429

88. Murillo E. Efecto hipercolesterolémico de la cantaxantina y la astaxantina en ratas. Archivos Latinoamericanos de Nutrición. 1992;42(4):409-413

89. Jyonouchi H, Sun S, Iijima K, Gross MD. Antitumor activity of astaxanthin and its mode of action. Nutrition and Cancer. 2000;36(1):59-65

90. Martín JF, Gudiña E, Barredo JL. Conversion of β-carotene into astaxanthin: Two separate enzymes or a bifunctional hydroxylase-ketolase protein? Microbial Cell Factories. 2008;7(1):1-10

91. Kushwaha K, Saini A, Saraswat P, Agarwal MK, Saxena J. Colorful world of microbes: Carotenoids and their applications. Advanced Biology. 2014;2014:837891

92. The Global Market for Carotenoids. FOD025F. 2018

93. FOD025C. BBC Research. Available from: http://www.bccresearch.com/report/FOD025C.html

94. Azadi P, Otang NV, Chin DP, Nakamura I, Fujisawa M, Harada H, et al. Metabolic engineering of Lilium× formolongi using multiple genes of the carotenoid biosynthesis pathway. Plant Biotechnology Reports. 2010;4(4):269-280

95. Clotault J, Peltier D, Berruyer R, Thomas M, Briard M, Geoffriau E. Expression of carotenoid biosynthesis genes during carrot root development. Journal of Experimental Botany. 2008;59(13):3563-3573

96. Turujman SA, Wamer WG, Wei RR, Albert RH. Rapid liquid chromatographic method to distinguish wild salmon from aquacultured salmon fed synthetic astaxanthin. Journal of AOAC International. 1997;80(3):622-632

97. Fujita T, Satake M, Watanabe T, Kitajima C, Miki W, Yamaguchi K, et al. Pigmentation of cultured red sea bream with astaxanthin diester purified from krill oil. Nippon Suisan Gakkaishi. 1983;49(12):1855-1861

98. Matsuno T. Xanthophylls as precursors of retinoids. Pure and Applied Chemistry. 1991;63(1):81-88

99. Tanaka T, Morishita Y, Suzui M, Kojima T, Okumura A, Mori H. Chemoprevention of mouse urinary bladder carcinogenesis by the naturally occurring carotenoid astaxanthin. Carcinogenesis. 1994;15(1):15-19

100. Jyonouchi H, Zhang L, Tomita Y. Studies of immunomodulating actions of carotenoids. II. Astaxanthin enhances in vitro antibody production to T-dependent antigens without facilitating polyclonal B-cell activation. Nutrition and Cancer. 1993;19:269-280

101. Miki W, Otaki N, Yokoyama A, Izumida H, Shimidzu N. Okadaxanthin, a novel C 50-carotenoid from a bacterium, Pseudomonas sp. KK10206C associated with marine sponge, Halichondria okadai. Experientia. 1994;50(7):684-686

102. Foote CS. Photosensitized oxidation and singlet oxygen: Consequences in biological systems. In: Pryor WA, editor. Free Radicals in Biology Vol II. Vol. 35. New York: Academic Press; 1976. pp. 3-22

103. Hue SM, Boyce AN, Somasundram C. Comparative study on the antioxidant activity of leaf extract and carotenoids extract from Ipomoea batatas var. Oren (sweetpotato) leaves. International Journal of Nutrition and Food Engineering. 2011;5(10):604-607

104. Hue SM, Boyce AN, Somasundram C. Antioxidant activity, phenolic and flavonoid contents in the leaves of different varieties of sweet potato ('Ipomoea batatas'). Australian Journal of Crop Science. 2012;6(3):375-380

105. Islam S. Some bioactive constituents, antioxidant, and antimutagenic activities in the leaves of Ipomoea batatas Lam. Genotypes. Journal of Food Science and Technology. 2016;4(3):70-80

106. Sosinski B, He L, Cervantes-Flores J, Pokrzywa RM, Bruckner A, Yencho GC. Sweetpotato genomics at North Carolina State University. In: International Conference on Sweetpotato. Food and Health for the Future. Vol. 583. Lima, Peru: Acta Horticulture; 2001. pp. 51-60. DOI: 10.17660/ActaHortic.2002.583.4

107. Aguoru CU. Varietal characterisation and taxonomic evaluation of sweet potato (Ipomoea batatas) using macro-and micromorphological evidence. Open Access Library Journal. 2015;2(08):1

108. Wanda CW. Genetic improvement for meeting human nutrition needs. In: Quebedeaux B, Bliss F, editors. Proceedings of the First International Symposium on Horticulture and Human Nutrition, Contributor of Fruits and Vegetables. London: Prentice Hall; 1987. pp. 191-199

109. Ishiguro K, Yoshinaga M, Kai Y, Maoka T, Yoshimoto M. Composition, content and antioxidative activity of the carotenoids in yellow-fleshed sweetpotato (Ipomoea batatas L.). Breeding Science. 2010;60(4):324-329

110. Kammona S, Othman RA, Jaswir IR, Jamal P. Characterisation of carotenoid content in diverse local sweet potato (Ipomoea batatas) flesh tubers. International Journal of Pharmacy and Pharmaceutical Sciences. 2015;7(2):347-351

111. Islam SN, Nusrat T, Begum P, Ahsan M. Carotenoids and β-carotene in orange fleshed sweet potato: A possible solution to vitamin A deficiency. Food Chemistry. 2016;199:628-631

112. Wurtzel ET, Valdez G, Matthews PD, Wurtzel E. Variation in expression of carotenoid genes in transformed E. coli strains. Bioresearch Journal. 1997;1:1-1

113. Sandmann G, Woods WS, Tuveson RW. Identification of carotenoids in Erwinia herbicola and in a transformed Escherichia coli strain. FEMS Microbiology Letters. 1990;71(1-2):77-82

114. Misawa N, Nakagawa M, Kobayashi K, Yamano S, Izawa Y, Nakamura K, et al. Elucidation of the Erwinia uredovora carotenoid biosynthetic pathway by functional

analysis of gene products expressed in Escherichia coli. Journal of Bacteriology. 1990;172(12):6704-6712

115. Misawa N, Yamano S, Linden H, de Felipe MR, Lucas M, Ikenaga H, et al. Functional expression of the Erwinia uredovora carotenoid biosynthesis gene crtl in transgenic plants showing an increase of β-carotene biosynthesis activity and resistance to the bleaching herbicide norflurazon. The Plant Journal. 1993;4(5):833-840

116. Misawa N, Truesdale MR, Sandmann G, Fraser PD, Bird C, Schuch W, et al. Expression of a tomato cDNA coding for phytoene synthase in Escherichia coli, phytoene formation in vivo and in vitro, and functional analysis of the various truncated gene products. The Journal of Biochemistry. 1994;116(5):980-985

117. Li ZH, Matthews PD, Burr B, Wurtzel ET. Cloning and characterization of a maize cDNA encoding phytoene desaturase, an enzyme of the carotenoid biosynthetic pathway. Plant Molecular Biology. 1996;30(2):269-279

118. Chamovitz D, Misawa N, Sandmann G, Hirschberg J. Molecular cloning and expression in Escherichia coli of a cyanobacterial gene coding for phytoene synthase, a carotenoid biosynthesis enzyme. FEBS Letters. 1992;296(3):305-310

119. Cunningham FX, Chamovitz D, Misawa N, Gantt E, Hirschberg J. Cloning and functional expression in Escherichia coli of a cyanobacterial gene for lycopene cyclase, the enzyme that catalyzes the biosynthesis of β-carotene. FEBS Letters. 1993;328(1-2):130-138

120. Sandmann G. [30] carotenoid analysis in mutants from Escherichia coli transformed with carotenogenic gene cluster and Scenedesmus obliquus mutant C-6D. Methods in Enzymology. 1993;214:341-347

121. Hundle B, Alberti M, Nievelstein V, Beyer P, Kleinig H, Armstrong GA, et al. Functional assignment of Erwinia herbicola Eho10 carotenoid genes expressed in Escherichia coli. Molecular and General Genetics MGG. 1994;245(4):406-416

122. Sandmann G, Misawa N. New functional assignment of the carotenogenic genes crtB and crtE with constructs of these genes from Erwinia species. FEMS Microbiology Letters. 1992;90(3):253-257

123. Penfold RJ, Pemberton JM. Sequencing, chromosomal inactivation, and functional expression in Escherichia coli of ppsR, a gene which represses carotenoid and bacteriochlorophyll synthesis in Rhodobacter sphaeroides. Journal of Bacteriology. 1994;176(10):2869-2876

124. Kajiwara S, Kakizono T, Saito T, Kondo K, Ohtani T, Nishio N, et al. Isolation and functional identification of a novel cDNA for astaxanthin biosynthesis from Haematococcus pluvialis, and astaxanthin synthesis in Escherichia coli. Plant Molecular Biology. 1995;29(2):343-352

125. Ohnuma SI, Suzuki M, Nishino T. Archaebacterial ether-linked lipid biosynthetic gene. Expression cloning, sequencing, and characterization of geranylgeranyl-diphosphate synthase. Journal of Biological Chemistry. 1994;269(20):14792-14797

126. Keeler J. Understanding NMR Spectroscopy, 2nd ed. West Sussex, United Kingdom: Wiley; 2010. pp. 184-187. ISBN: 978-0-470-746080

127. Maoka T, Otani M, Khan MZ, Takemura M, Hattan JI, Misawa N. Novel carotenoids produced on the interaction of the foreign carotenoid ketolase CrtW and endogenous epoxycarotenoids unique to sweetpotato tubers. Tetrahedron Letters. 2016;57(42):4746-4748

128. Kim SH, Jeong JC, Park S, Bae JY, Ahn MJ, Lee HS, et al. Down-regulation of sweetpotato lycopene β-cyclase gene enhances tolerance to abiotic stress in transgenic calli. Molecular Biology Reports. 2014;41(12):8137-8148

129. Song GQ, Honda H, Yamaguchi KI. Efficient Agrobacterium tumefaciens-mediated transformation of sweet potato (Ipomoea batatas (L.) Lam.) from stem explants using a two-step kanamycin-hygromycin selection method. In Vitro Cellular & Developmental Biology-Plant. 2004;40(4):359-365

130. Otani M, Mii M, Handa T, Kamada H, Shimada T. Transformation of sweet potato (Ipomoea batatas (L.) Lam.) plants by Agrobacterium rhizogenes. Plant Science. 1993;94(1-2):151-159

131. Otani M, Shimada T. Efficient embryogenic callus formation in sweet potato (Ipomoea batatas (L.) Lam.). Japanese Journal of Breeding. 1996;46(3):257-260

132. Otani M, Shimada T, Kimura T, Saito A. Transgenic plant production from embryogenic callus of sweet potato (Ipomoea batatas (L.) Lam.) using Agrobacterium tumefaciens. Plant Biotechnology. 1998;15:11-16

133. Lou HR, Maria MS, Benavides J, Zhang DP, Zhang YZ, Ghislain M. Rapid genetic transformation of sweetpotato (Ipomoea batatas (L.) Lam) via organogenesis. African Journal of Biotechnology. 2006;5(20):1851-1857

134. Kim SH, Ahn YO, Ahn MJ, Lee HS, Kwak SS. Down-regulation of β-carotene hydroxylase increases β-carotene and total carotenoids enhancing salt stress tolerance in transgenic cultured cells of sweetpotato. Phytochemistry. 2012;74:69-78

135. Ling YU, Hong ZH, Wei CH, HE SZ, LIU QC. Cloning and functional analysis of lycopene ε-cyclase (IbLCYe) gene from sweetpotato, Ipomoea batatas (L.) Lam. Journal of Integrative Agriculture. 2013;12(5):773-780

136. Lee P, Schmidt-Dannert C. Metabolic engineering towards biotechnological production of carotenoids in microorganisms. Applied Microbiology and Biotechnology. 2002;60(1):1-1

137. Giuliano G, Aquilani R, Dharmapuri S. Metabolic engineering of plant carotenoids. Trends in Plant Science. 2000;5(10):406-409

138. Guerin M, Huntley ME, Olaizola M. Haematococcus astaxanthin: Applications for human health and nutrition. Trends in Biotechnology. 2003;21(5):210-216

139. Misawa N. Pathway engineering of plants toward astaxanthin production. Plant Biotechnology. 2009;26(1):93-99

140. Zhu C, Naqvi S, Capell T, Christou P. Metabolic engineering of ketocarotenoid biosynthesis in higher plants. Archives of Biochemistry and Biophysics. 2009;483(2):182-190

141. Hasunuma T, Miyazawa SI, Yoshimura S, Shinzaki Y, Tomizawa KI, Shindo K, et al. Biosynthesis of astaxanthin in tobacco leaves by transplastomic engineering. The Plant Journal. 2008;55(5):857-868

142. Harada H, Maoka T, Osawa A, Hattan JI, Kanamoto H, Shindo K, et al. Construction of transplastomic lettuce (Lactuca sativa) dominantly producing astaxanthin fatty acid esters and detailed chemical analysis of generated carotenoids. Transgenic Research. 2014;23(2):303-315

143. Jayaraj J, Devlin R, Punja Z. Metabolic engineering of novel ketocarotenoid production in carrot plants. Transgenic Research. 2008;17(4):489-501

144. Fujisawa M, Takita E, Harada H, Sakurai N, Suzuki H, Ohyama K, et al. Pathway engineering of Brassica napus seeds using multiple key enzyme genes involved in ketocarotenoid formation. Journal of Experimental Botany. 2009;60(4):1319-1332

145. Zhu C, Naqvi S, Breitenbach J, Sandmann G, Christou P, Capell T. Combinatorial genetic transformation generates a library of metabolic phenotypes for the carotenoid pathway in maize. Proceedings of the National Academy of Sciences. 2008;105(47):18232-18237

146. Yokoyama A, Shizuri Y, Misawa N. Production of new carotenoids, astaxanthin glucosides, by Escherichia coli transformants carrying carotenoid biosynthesis genes. Tetrahedron Letters. 1998;39(22):3709-3712

147. Nishida Y, Adachi K, Kasai H, Shizuri Y, Shindo K, Sawabe A, et al. Elucidation of a carotenoid biosynthesis gene cluster encoding a novel enzyme, 2, 2⊠-β-hydroxylase, from Brevundimonas sp. strain SD212 and combinatorial biosynthesis of new or rare xanthophylls. Applied and Environmental Microbiology. 2005;71(8):4286-4296

148. Shindo K, Hasunuma T, Asagi E, Sano A, Hotta E, Minemura N, et al. 4-Ketoantheraxanthin, a novel carotenoid produced by the combination of the bacterial enzyme β-carotene ketolase CrtW and endogenous carotenoid biosynthetic enzymes in higher plants. Tetrahedron Letters. 2008;49(20):3294-3296

149. Breitenbach J, Bai C, Rivera SM, Canela R, Capell T, Christou P, et al. A novel carotenoid, 4-keto-α-carotene, as an unexpected by-product during genetic engineering of carotenogenesis in rice callus. Phytochemistry. 2014;98:85-91

150. Maoka T, Takemura M, Tokuda H, Suzuki N, Misawa N. 4-Ketozeinoxanthin, a novel carotenoid produced in Escherichia coli through metabolic engineering using carotenogenic genes of bacterium and liverwort. Tetrahedron Letters. 2014;55(49):6708-6710

151. Umeno D, Arnold FH. Evolution of a pathway to novel long-chain carotenoids. Journal of Bacteriology. 2004;186(5):1531-1536

152. Kang L, Ji CY, Kim SH, Ke Q, Park SC, Kim HS, et al. Suppression of the β-carotene hydroxylase gene increases β-carotene content and tolerance to abiotic stress in transgenic sweetpotato plants. Plant Physiology and Biochemistry. 2017;117:24-33

153. Li R, Kang C, Song X, Yu L, Liu D, He S, et al. A ζ-carotene desaturase gene, IbZDS, increases β-carotene and lutein contents and enhances salt tolerance in transgenic sweetpotato. Plant Science. 2017;262:39-51

154. Ducreux LJ, Morris WL, Hedley PE, Shepherd T, Davies HV, Millam S, et al. Metabolic engineering of high carotenoid potato tubers containing enhanced levels of β-carotene and lutein. Journal of Experimental Botany. 2005;56(409):81-89

155. Ke Q, Kang L, Kim HS, Xie T, Liu C, Ji CY, et al. Down-regulation of lycopene ε-cyclase expression in transgenic sweetpotato plants increases the carotenoid content and tolerance to abiotic stress. Plant Science. 2019;281:52-60

156. Gerjets T, Sandmann G. Ketocarotenoid formation in transgenic potato. Journal of Experimental Botany. 2006;57(14):3639-3645

157. Kim SE, Lee CJ, Park SU, Lim YH, Park WS, Kim HJ, et al. Overexpression of the golden SNP-carrying Orange gene enhances carotenoid accumulation and heat stress tolerance in sweetpotato plants. Antioxidants. 2021;10(1):51

158. Gama MI, Leite RP, Cordeiro AR, Cantliffe DJ. Transgenic sweet potato plants obtained by Agrobacterium tumefaciens-mediated transformation. Plant Cell, Tissue and Organ Culture. 1996;46(3):237-244

159. Takahata Y, Tanaka M, Otani M, Katayama K, Kitahara K, Nakayachi O, et al. Inhibition of the expression of the starch synthase II gene leads to lower pasting temperature in sweetpotato starch. Plant Cell Reports. 2010;29(6):535-543

160. Kim SH, Ahn YO, Ahn MJ, Jeong JC, Lee HS, Kwak SS. Cloning and characterization of an Orange gene that increases carotenoid accumulation and salt stress tolerance in transgenic sweetpotato cultures. Plant Physiology and Biochemistry. 2013a;70:445-454

161. Park SC, Kim SH, Park S, Lee HU, Lee JS, Park WS, et al. Enhanced accumulation of carotenoids in sweetpotato plants overexpressing IbOr-Ins gene in purple-fleshed sweetpotato cultivar. Plant Physiology and Biochemistry. 2015;86:82-90

162. Kim SH, Kim YH, Ahn YO, Ahn MJ, Jeong JC, Lee HS, et al. Downregulation of the lycopene ∈-cyclase gene increases carotenoid synthesis via the β-branch-specific pathway and enhances salt-stress tolerance in sweetpotato transgenic calli. Physiologia Plantarum. Apr 2013;147(4):432-442

163. Ma DFLI. Selection of parents for breeding ed-ible varieties of sweetpotato with high carotene content. Agri-cultural Sciences in China. 2009;8(10):1166-1173

164. Hwang SY, Tseng YT, Lo HF. Application of simple sequence repeats in determining the genetic relationships of cultivars used in sweet potato polycross breeding in Taiwan. Scientia Horticulturae. 2002;93(3-4):215-224

165. Cervantes-Flores JC, Sosinski B, Pecota KV, Mwanga RO, Catignani GL, Truong VD, et al. Identification of quantitative trait loci for dry-matter, starch, and β-carotene content in sweetpotato. Molecular Breeding. 2011;28(2):201-216

166. Laurie SM, Van Jaarsveld PJ, Faber M, Philpott MF, Labuschagne MT. Trans-β-carotene, selected mineral content and potential nutritional contribution of 12 sweetpotato varieties. Journal of Food Composition and Analysis. 2012;27(2):151-159

167. Pillay K, Khanyile N, Siwela M. Acceptance of an orange-fleshed sweet potato complementary food by infant caregivers in KwaZulu-Natal Province—A preliminary study. South African Journal of Child Health. 2018;12(3):100-104

Chapter-11

Prospective Application of Aspergillus Species: Focus on Enzyme Production Strategies, Advances and Challenges

1. Introduction

With the exponential increase in science and knowledge about biochemical processes; it would be fair to say that it is inconceivable to consider any biological process without an enzyme. They are biocatalysts that enhance the rate of reaction in numerous folds. Enzymes usually are reusable. In other words, they are not used up by the reaction and can be reused. Once an enzyme binds to a substrate and catalyzes the reaction, the enzyme is released, unchanged, and can be used for another reaction. This means that for each reaction, it is not necessary to have a ratio of 1:1 between enzymatic molecules and substrate molecules. Enzymes mostly are proteinaceous-based in nature (there are a few RNA-based enzymes) and necessary for all living organisms [1]. A significant number of them have been recognized as safe from a biotechnological perspective. Fungi, as one of the simplest organisms, are often used to produce enzymes. In addition, factors such as low energy, low cost, non-toxic and environmentally friendly nature make them popular in many industrial processes [2]. Also, the need for gentle temperature and pressure for enzymes to function enables them to become a viable alternative to hazardous chemical catalysts [3]. Enzymes are commonly used to make wine, beer, bread, cheese, vinegar, and leather and textiles. However, the pure and clean form of enzymes has found wide applications in industry only a few decades ago [4]. Enzymes produced by the fungal system are commonly used in various sectors including food, chemicals, medicine, agriculture and energy [5]. Today, due to multiple applications, the demand for different kind of enzymes in various food sectors has increased greatly [6, 7, 8, 9, 10, 11, 12, 13, 14, 15, 16, 17, 18], as shown in Table 1. Additionally, the manipulation of strains through recombinant DNA techniques and protein engineering technology has made it possible to meet the growing demand for enzymes [19]. Fungi are metabolically dynamic, simple to ferment and can work on an industrial scale, require simple nutrients and can be used throughout the year and are not subject to seasonal conditions [20]. The genus *Aspergillus* has more than 340 officially known species [21]. These fungi are characterized by an extraordinary capability to produce and secrete large amounts of proteins, metabolites, and organic acids into their growth medium [22]. Notwithstanding the existence of different pathogenic *Aspergillus* strains, a significant number of them have been recognized from a biotechnological perspective [23]. Characteristics such as the existence of a secretory pathway, the eventuality of genetic manipulation, and high productivity using diverse fermentative processes are beneficial and positive for the make use of *Aspergillus* species [23]. Enzymes produced by *Aspergillus* species have been broadly investigated for their potential in the formulation of commercial products [24]. Hence, the main goal of this chapter investigation is to analyze the secreted and cellular proteins from *Aspergillus* species and their utilization in food industries.

Table 1. The main applications of enzymes in different food sectors [6, 7, 8, 9, 10, 11, 12, 13, 14, 15, 16, 17, 18].

Enzymes	Food sectors	Applications
Amylases	Brewing industry	Fermentation of alcohol by converting starch to sugars
	Baking industry	Breakdown of starch into simple sugars; thereby allowing the bread to rise and impart flavor
		Dough conditioning
		Generates additional sugar in the bread, which improves the taste, crust color and toasting quality
		Anti-staling effect during bread making; improves the softness and shelf-life
Cellulases	Fruit industry	Fruit and vegetable juice clarification
		Reducing the viscosity of nectars
		Alteration of fruit sensory properties
	Beverages industry	Concentrating purees
	Health food industry	Carotenoids extraction
	Edible oil extraction industry	Olive oil extraction
	Baking industry	Improvement quality of bakery products
Chitosanases	Seafood industry	The degradation of crustacean chitinous waste
	Agriculture industry	Biological activities such as antifungal effect
Galactosidases	Dairy industry	Production of low lactose/milk free lactose
		Production of prebiotics
		Prevents crystallization of lactose Improves the scoop ability and creaminess of the product
		Production of ice creams, sweetened flavor and condensed milks
		Improves the scoop ability and creaminess of the product
Invertases	Food sweetener market	Invert sugar production
	Confectionery food industry	Production of high fructose syrup
		Manufacturing of soft-centered candies
		Manufacture of artificial honey

Enzymes	Food sectors	Applications
Laccase	Wine industry	Removal of polyphenol, thereby providing stability to wines
		Preparation of cork stoppers of wine bottles
		Reduces cork taint generally imparted to aged wine bottles
	Brewing industry	Removal of oxygen at the end of beer fermentation process
		Prevent the formation of off-flavors (trans 2-nonenal)
	Fruit industry	Juice clarification
	Baking industry	Increase strength, stability and reduce stickiness
		Increase volume, improved crumb structure and softness of the product
Lipases	Fats and oils food industry	Production of mayonnaise and other emulsifiers, Triglycerides synthesis and trans-esterification of triglycerides in non-aqueous media; specially fat production
	Dairy industry	Development of flavoring agent in milk, cheese, and butter
		Hydrolysis of milk, fat, cheese ripening, and modification of butter fats
	Meat industry	Degumming during the refining of vegetable oil
		Flavor development, meat and fish product fat removal
	Baking industry	Flavor development, shelf-life prolongation
Naringinases	Fruit industry	Debittering of citrus fruit juices
	Wine industry	Enhances the aroma in the wine
		Production of pruning, a flavonoid
Pectinases	Fruit industry	Clarification of the fruit juices
		Enhanced levels of fruit juice volume when fruit pulps treated with pectinase
		Soften the peel of citrus fruits
		Enhances the citrus oil extraction such as lemon oil
	Beverages industry	Accelerates tea fermentation
		Reduces foam forming property in instant tea powders
		Remove mucilaginous coat from coffee beans
	Wine industry	Imparts stability of red wine
Phytases	Baking industry	Reduction of phytate content in dough & fresh breads
		Shortening of formulation time without any change in pH
		Increase in bread volume and an improvement in crumb texture
		Softer bread crumbs were obtained
		Other texture parameters like gumminess and chewiness were also decreased

Enzymes	Food sectors	Applications
Proteases	Dairy industry	Prevent coagulation of casein during cheese production
		Flavor development
	Meat industry	Meat tenderization
	Baking industry	Assures dough uniformity
		Improve dough consistency
		Gluten development
		Improve texture and flavor
		Reduce mixing time
Tannase	Brewing industry	Removal of polyphenolic compounds
	Beverages industry	Manufacture of instant tea

2. History and Background

The use of various *Aspergillus* species dates back to almost a century ago, when it was discovered that the fungus *Aspergillus niger* was able to produce citric acid, which is a food and beverage additive and is normally extracted from citrus fruits [25]. Nowadays, *A. niger* is preferable to other microbes for the commercial production of citric acid because of its improved manufacturing yield. *Aspergillus* species are easily manipulated and can ferment different low-priced raw materials and offer high yields [26]. Thus, *Aspergillus* strains can be improved to generate industrial strains for application in the commercial manufacture, and mutagenesis and strain choice have been performed for such improvement. Through the application of different mutagenic agents, including various radiation for example ultraviolet, gamma and X-rays radiation, various chemicals for example, ethyl methane sulphonate and diethyl sulphonate have been commonly used to induce the mutation of *Aspergillus* species [26]. However, the industrial applications of *Aspergillus* species are not limited to the manufacture of citric acid. They have the capability to produce other organic acids such as gluconic, malic and itaconic acids, secondary metabolites and industrially important enzymes [26]. For example, *A. niger* has been used to manufacture different extracellular enzymes such as glucose oxidase, pectinase, α-amylase and glucoamylase, organic acids, and recombinant proteins [26]. *A. oryzae is a* fungus that widely plays an indispensable role in Asian food industrialized, such as soy sauce (sake and shoyu) and soybean paste (miso). Moreover, it has been applicable in the manufacture of industrial enzymes used in food processing [27]. *A. terreus* has attracted interest due to its ability to produce a group of secondary metabolites called statins that are used in the manufacture of cholesterol-lowering drugs [28]. *A. nidulans* is a capable fungal cell factory that can produce different industrial enzymes such as amylase, cellulases, cutinases, glucosidases, etc. [29]. There are many companies (such as Genencor, Novozymes, Pfizer, Amano Enzyme, Verenium, etc.) that used enzymes, proteins and secondary metabolites produced by different species of *Aspergillus* on a large scale to produce commercial products [30].

3. Enzymes Production Process

A general overview of the enzyme production process has been shown in Figure 1. Fermentation has two parts, upstream processes (UsP) and downstream processes (DsP). The UsP for the enzyme manufacture include the selection of *Aspergillus* strain, inoculum and sterilization of media [31]. For primary culture, the frozen culture of the *Aspergillus* strain is inoculated into the medium. The culture media are sterilized before inoculations for primary and secondary culture. Once the strain is inoculated, primary cultivation is acceptable for primary fermentation [31, 32]. The primary culture is used as inoculum during the secondary fermentation process [32]. The DsP of the fermented product depends on the nature of the secretion of an enzyme (intracellular/extracellular) [31]. If the enzymes are intracellular, the cells are first collected with the help of centrifugation and other filtration methods [32]. Disruption of cells by diverse techniques such as sonication, french press, enzymatic lysis and freezing/thawing process is necessary to get the intracellular content [31, 32]. After the cell breaking, the cell residue is excluded by filtration. In the case of extracellular proteins, the enzymes are secreted in the culture medium, so they are easier to purify. The enzymes produce using both intracellular and extracellular fermentation are subsequently concentrated using concentrators. The fermentation is generally followed by cell lysis or cell extraction then elimination of cells-residue from the culture, either using centrifugation, filtration or sedimentation. The cell-free culture is then concentrated using ultra-filtration, ammonium sulphate precipitation or liquid extraction with organic solvents [33]. About 80% of the purification processes employed thus far have used a precipitation step, with 65% of these using ammonium sulphate and 35% using ethanol, acetone or an acid (usually hydrochloric) followed by a combination of ultra-filtration system and various chromatographic techniques (20%) for example gel filtration and affinity chromatography [33]. The purification of an enzyme leads to the procedure of formulation [34]. The formulation may include methods such as drying, vacuum drying, spray drying, and freezedrying [34]. These processes make

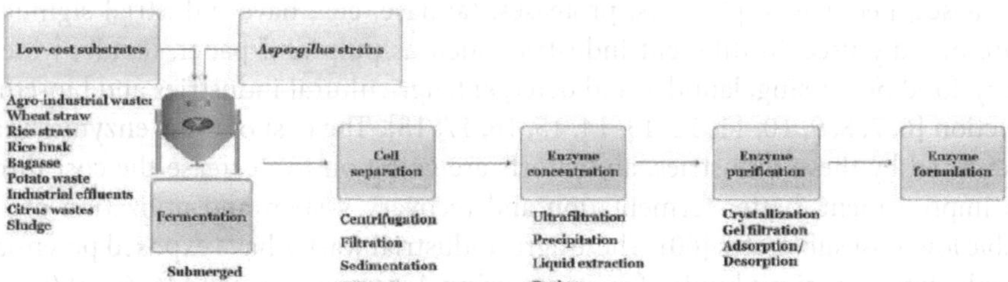

Figure 1. Process steps for conversion of low-cost substrate to various enzymes by Aspergillus strains. The fermentation (submerged or solid-state) is generally followed by cell-lysis/extraction, then elimination of cells-residue from the culture, either using centrifugation, filtration, sedimentation. The cell-free culture is then concentrated using ultra-filtration, ammonium sulphate precipitation or liquid extraction with organic solvents. Then followed by a combination of various chromatographic techniques for example gel filtration and affinity chromatography.

a decision how an enzymatic product will be launched in the market. It can be a mixture of two enzymes or more of them. The formulation is exclusively based upon the research and development of an industry or a lab. Formulation development is a fundamental factor not only in the early stages of enzymes/protein development, but also in the future marketing success of a promising investigational product. This is the operation of discovering novel sciences which can be used to generate novel and valuable products. The development part comes after the study and analysis and is the act of turning the discovered science into a precious product that the company can commercialize and sell. All the UsP and DsP lead to the formation of the final product.

3.1 Fermentation

Both submerged (SmF) and solid-state (SSF) fermentation are used for making different enzymes by fungi [35, 36, 37, 38, 39, 40, 41, 42, 43, 44, 45, 46, 47, 48, 49, 50, 51, 52, 53, 54, 55, 56, 57, 58, 59, 60, 61, 62, 63, 64, 65, 66, 67, 68, 69, 70, 71, 72, 73, 74, 75, 76, 77, 78, 79, 80, 81, 82, 83, 84, 85, 86, 87, 88, 89, 90, 91, 92, 93, 94, 95, 96, 97, 98, 99]. Due to easy measurement and of control fermentation parameters, reduction fermentation time and basic ways for harvesting and refining enzymatic products, more attention has been paid to SmF [62, 63, 64, 65]. In recent time, intensive investigate on SSF has been conducted and has gained reliability due to low water consumption, low energy necessities, less contamination and high manufacture yields [35, 52]. However, *Aspergillus* species have shown potential for producing enzymes under both SSF and SmF (Table 2).

3.2 Use of low-cost/economical substrates for enzyme cost-effective/ commercial production

A large amount of agro-industrial waste like wheat straw, rice straw, rice husk, bagasse, potato waste, industrial effluents, citrus wastes, sludge, etc., are produced annually. They are rich sources of sugars, mineral elements, vitamins, fiber and different phenolic compounds, etc. Consequently, they can be used for the manufacture of commercially important products like enzyme, due to their nutritional potential (Table 2). Enzymes like amylases, celluloses, chitosanases, galactosidases, invertases, laccase, lipases, naringinases, pectinases, phytases, proteases, tannase, etc., have industrial significance and are broadly used in different industries such as pulp and paper, textile, wine and brewery, food processing, laundry and detergent, agricultural industries and bio-ethanol production [6, 7, 8, 9, 10, 11, 12, 13, 14, 15, 16, 17, 18]. The cost of these enzymes is a big subject faced by these industries, and efforts are going on to decrease the cost through strain improvement, better fermentation and recovery system and utilization of easily available low-cost substrates [60]. These agro-industrial wastes have exposed potential for the production of various kinds of enzymes using *Aspergillus* species [46, 60, 100].

3.3 Recovery purification and formulation

Enzymes are recovered from fermentation through chemical engineering operations that are broadly used to produce enzymes [101]. When the enzyme is intracellular, the cells must be broken down to release the enzyme. This can be done using mechanical methods (such as high-pressure press, grinding, or ultrasound) or non-mechanical methods (such as drying or lysis). In the case of extracellular enzymes, an early stage of isolation (centrifugation, filtration or both) is often used to eliminate residue of the cells [101]. Then the dissolved enzyme is concentrated by eliminating the water (cross-flow filtration or evaporation), resulting in an enzyme concentrate. Alternatively, a whole enzyme preparation containing inactivated cells or cell debris may be suitable where the resulting food undergoes further

Table 2. Production of industrial enzymes from various *Aspergillus* species.

Enzymes	*Aspergillus* species	Substrate	Type of fermentation	Fermentation conditions				Yield of enzymes	Ref.
				pH	T (°C)	D	M (%)		
Amylase	*A. niger* JGI 24	Wheat bran	SSF	4.5–9.0	22–40	4	43–81	74 U/mgds	[34]
			SmF				—	58.06 U/ml	
	A. terreus NCFT4269.10	Pearl millet	SSF	7.0	30	4	70	19.19 U/gds	[35]
	A. fumigatus	Maltose	SmF	6.0	30	8–10	—	60–130 U/mgds	[36]
	A. tamarii	Starch or maltose	SmF	4.0–10.0	25–42	6	—	ND	[37]
	A. oryzae LS1	Wheat bran	SSF	6.0	28–30	7	50	14,249 U/gds	[38]
	A. oryzae	Groundnut oil cake, coconut oil cake, sesame oil cake	SSF	4.5	32.5	4–5	64	9868.12 U/gds	[39]
	A. flavus AUMC 11685	Mandarin (*Citrus reticulata*) peel	SmF	4.0–5.5	28–40	4–5	—	26.90 U/ml	[40]
	A. awamori ATCC 22342	Rice flour	SmF	6.5	30	2	—	0.18 U/ml	[41]
	A. niger							0.08 U/ml	
	A. awamori nakazawa MTCC 6652	Wheat bran	SSF	5.5	35	4	85	4528.4 ± 121 U/gds	[42]
	A. carneus SA 1326	Ground millet, starch or carboxymethylcellulose	SmF	5.6	28	4	—	ND	[43]
	A. sydowii IMI 502692	Cassava root fiber	SSF	3.62	ND	2	ND	1.327 U/ml	[44]

Enzymes	*Aspergillus* species	Substrate	Type of fermentation	Fermentation conditions				Yield of enzymes	Ref.
				pH	T (°C)	D	M (%)		
Cellulases	*A. niger* NS-2	Agricultural and kitchen waste residues	SSF	3.0–8.0	20–50	4	57–67	17–310 U/gds	[45]
	A. heteromorphus	Wheat straw	SmF	5.0	30	5	—	3.2 U/ml	[46]
	A. fumigatus	Rice straw and wheat Bran	SSF	5.0–6.0	40	4	75	0.68–42.7 U/gds	[47]
	A. terreus	Lantana leaves	SmF	5.0	25	7	—	213.3 U/ ml	[48]
	A. niger	Wheat bran, rice bran, rice husk, coir waste and saw dust	SSF	6.0	30	4	50	29.11 U/gds	[49]
			SmF					2.04 U/ml	
	A. niger USM AI 1	Corn steep liquor	SSF	7.0	30	5	70	3. 4 U/gds	[50]
Chitosanases	*A. niger*	Wheat bran	SSF	6.6	28	5	65	41.33 U/gds	[51]
	Aspergillus sp. QD-2	Yeast glucose	SmF	4.0	30	1	—	85.816 U/ml	[52]
	A. fumigates ATCC13073	Vogel's medium	SmF	6.0	37	1	—	8.80 U/mg	[53]
	A. oryzae SU-B2	Yeast-peptone glucose	SmF	5.0	30	4	—	352 mg/l	[54]
Galactosidases	*A. niger* ATCC 9142	Rice straw and wheat straw	SSF	7.0	30	6	70	4681 U/mg	[55]
	A. oryzae	Red gram and waste-wheat bran	SSF	5.5	35	6	50	ND	[56]
	A. oryzae ATCC 20423	Lactose and wheat bran	SmF	4.8	30	7	—	ND	[57]
	A. oryzae	Wheat bran and rice husk	SSF	5.0	30	7	90	146.6–386.6 U/ml	[58]
Invertases	*A. caespitosus*	Wheat bran	SSF	4–6	30–40	3	70	117.4 U/gds	[59]
			SmF				—	19.1 U/ml	
	A. niger MTCC 282	Orange fruit peel	SSF	5.0	30	4	80	43 U/ml	[60]
	A. nidulans	Rye flour	SmF	6.0	30	3	—	30–33.6 U/ml	[61]

Enzymes	*Aspergillus* species	Substrate	Type of fermentation	Fermentation conditions				Yield of enzymes	Ref.
				pH	T (°C)	D	M (%)		
Laccase	*A. flavus*	Starch and yeast extract	SmF	7.0	35	14	—	17.39 U/ml	[62]
	A. sydowii NYKA 510	Banana peel and peptone excelled	SmF	5.2	31	7	—	15.1 and 2.60 g/l	[63]
	A. nidulans	Glucose and straw	SmF	5–7	28	2	—	0.052 and 0.0677 U/ml	[64]
Lipases	*A. niger* NCIM 1207	Wheat bran + synthetic oil based	SSF	5.5	30	6	40	630 U/gds	[65]
	A. niger AS-02	Sheanut cake	SSF	7.0	30	7	60	49.37 U/gds	[66]
	A. oryzae RBM4	Sorghum, wheat bran	SmF	5.5	30	3	—	5.66 U/ml	[67]
	A. flavus PW2961	Bran-wood flour-olive oil, bran-soy bean	SSF	5.0	28	3–4	50	37.4 U/gds	[68]
	A. niger MTCC 872	Rice husk, cottonseed cake and red gram husk	SSF	6.0	40	1	75	28.19 U/gds	[69]
Naringinases	*A. niger* van Tieghem MTCC 2425	Citrus wastes	SmF	3–5	26–30	6–8	—	426.4–545.2 U/gds	[70]
	A. foetidus	Orange and grapefruit rind	SSF	5.4	35	8	ND	2.58 U/ml	[71]
	A. aculeatus JMUdb058	Yeast extract, naringin	SmF	6.0	28	7	—	1.16 U/ml	[72]
	A. oryzae 11,250	Orange peel	SmF	5.0	45	4	—	2194.62 U/mgds	[73]
	A. niger MTCC 1344	Rice bran, wheat bran, sugar cane bagasse, citrus peel, and press mud	SSF	4.0	27	4	50	58.1 U/gds	[74]
	A. brasiliensis 1344	Cassava waste	SSF	5.0	27	5	ND	889.91 U/mg	[75]
	A. tubingensis MN589840	Mildew pomelo peel	SSF	4.0	30	5	ND	808.85 U/mg	[76]
	A. sojae	Soybeans	SmF	4.5	28	6	—	1.5 U/mgds	[77]

Enzymes	*Aspergillus* species	Substrate	Type of fermentation	Fermentation conditions				Yield of enzymes	Ref.
				pH	T (°C)	D	M (%)		
Pectinases	*A. niger*	Wheat bran	SSF	4.0	30	3	63	68 U/gds	[78]
	A. carneus NRC1	Orange peels and pulps	SmF	5.0–5.5	30–55	5	—	40 U/ml	[79]
	A. flavipes FP-500	Lemon peel	SmF	4.2	37	6	—	ND	[80]
	A. tamarii	Wheat bran, banana peel, sugarcane bagasse, lemon peel, coffee pulp and orange peel	SSF	6.0	30	4	70	101.05 U/ml	[81]
	A. flavus CECT-2687	Agroindustrial residues and polysaccharides	SmF	3.5–9.0	37	5	—	1.35–7.89 U/ml	[82]
	A. japonicus	Polygalacturonic acid, citrus pectins	SmF	4.0–5.5	30	ND	—	805 and 839 U/mg	[83]
Phytases	*A. niger* CFR 335	Wheat bran, rice bran, and groundnut cake	SSF	2.0–7.5	30	8	10–80	60.6 U/gds	[84]
			SmF			10	—	9.6 U/mL	
	A. ficuum SGA 01		SSF			8	10–80	38 U/gds	
			SmF			10	—	8.2 U/mL	
	A. niger NCIM 563	Chickpea flour	SmF	7.0	35	4	—	164 U/mL	[85]
	A. ficuum	Potato waste	SSF	6.1	27	6	79	12.93 U/gds	[86]
	A. terreus	Rice bran	SmF	4.5	30	4	—	ND	[87]
Proteases	*A. niger*	Wheat bran	SSF	8.0	40	8	3.3	30.21 U/mg	[88]
	A. oryzae LBA 01	Wheat bran	SSF	5–5.5	23	3	50	3961.30 U/gds	[89]
	A. clavatus	Vogel medium with glucose	SmF	9.5	37	10	—	38 U/ml	[90]
	A. flavus IMI 327634	Wheat bran	SSF	7.5–9.5	32	2	63	6.8 U/ml	[91]

Enzymes	*Aspergillus* species	Substrate	Type of fermentation	Fermentation conditions				Yield of enzymes	Ref.
				pH	T (°C)	D	M (%)		
Tannase	*A. tamarii*	Tannic acid, gallic acid and methyl gallate	SmF	5.0	30	2	—	20.6 U/ml	[92]
	A. ruber	Jamun (*Syzygium cumini*) leaves	SSF	5.5	30	4	50	69 U/gds	[93]
	A. niger MTCC 5898	Cashew testa	SSF	3.0–8.0	32–35	3–5	60	97.32–301.7 U/gds	[94]
	A. heteromorphus MTCC 8818	Rosewood (*Dalbergia sissoo*) sawdust—a timber industry waste	SSF	5.5	30	4	70	1.84 U/gds	[95]
	A. ochraceus	Khanna medium	SmF	5.0	40	3	—	0.92 U/mgds	[96]
	A. melleus	Achachairu seed powder	SSF	5.5	40	2	60	452.55 U/ml	[97]
	A. aculeatus DBF9	Wheat bran, rice bran, saw dust, rice straw dust, sugarcane pith	SSF	5.0	30	3	80	1.32–3.95 U/gds	[98]

ND = not determined.

SSF: solid-state fermentation; Smf: submerged fermentation; T: incubation temperature; D: day incubation time of fermentation; M: moisture content in case of SSF.

U/ml = μmoles of fatty acids released per minute per ml of crude enzyme; U/gds = μmoles of fatty acids released per minute per gram of dry solid substrate from which enzyme has been extracted; U/mgds = μmoles of fatty acids released per minute per milligram of dry solid substrate from which enzyme has been extracted; mg/l = milligram per liter is a unit of measurement of mass concentration that shows how many milligram of a certain substance are present in one liter of a mixture; g/l = gram per liter is a unit of measurement of mass concentration that shows how many gram of a certain substance are present in one liter of a mixture.

refinement, for example in potable alcohol production. In all cases, the prepared enzyme is free of viable fungi [101]. The concentrate is then formulated using the correct ingredients to stabilize and standardize the enzyme [101]. The raw materials used for recovery and formulation require to be of suitable purity for the future use and require to be used according to good manufacturing Practices, i.e., in the minimal quantities required to achieve the desired effect [101]. The utilize of potential allergens in the process of producing food enzymes must be addressed and, if necessary, included in the enzyme preparation. At the end of the manufacturing procedure, the last formulated enzyme generate is introduced in the market after testing to verify agreement with qualifications for contaminants (microbial and chemical) established for enzyme preparations by the Food Chemicals Codex and FAO/WHO's JECFA. In other words, enzymes come in three diverse forms. Firstly, there is the enzymatic protein itself, which is a pure substance that is used in labs [102]. Secondly, enzymatic concentrates are products produced following fermentation or extraction [102]. They contain the enzyme produced in much smaller amounts from other substances obtained during the fermentation process, like other (secondary) enzymes or the remainder of the fermentation [102]. These enzymatic concentrates are evaluated for safety prior to being approved for marketing. Finally, there are enzymatic preparations, which are formulations containing one or more enzymatic concentrates with added stabilizers, preservatives, and diluents to stabilize enzymes and maintain activity. These formulas are sold commercially. In general, a proven quality feature of enzymes produced by microorganisms is the lack of viable cells. In addition, other specific features of microbial enzymes include their ability and significant activity under abnormal conditions, mostly temperature and pH. For example, some microbial enzymes are produced in thermophilous, acidophilic or alkalophilic forms.

4. Recombinant DNA (rDNA) Technology

In industries, the rDNA technique will contribute to the manufacture of chemicals of commercial importance, to the advancement of existing fermentation processes and protein/enzyme production from waste materials. For this purpose, more effective strains of microorganisms can be developed. Thus, the technology of rDNA has many useful applications in crop betterment, medication and industry.

The rDNA in microorganisms occurs through three different parasexual processes namely conjugation, transduction, and transformation [103]. Internal genetic rearrangements can also occur via translocatable DNA segments (insertion sequences or transposons) [103]. Conjugation implies DNA transfer through cell-to-cell contact. Transduction occurs from the host cell to the recipient cell through bacteriophaging mediation. Transformation involves the absorption and expression of bare DNA by the appropriate cells. Competence occurs naturally but can also be induced by changes in the physical and chemical environment. In the laboratory, it can be induced by cold calcium chloride treatment, protoplasting, electroporation and heat shock [103]. After 1980, there was a heightened interest in the application of genetic recombination to the production of important microbial products such as antibiotics. The use of rDNA technology has made it possible to produce new enzymes appropriate for specific food-processing conditions [104]. Various substantial enzymes (lipases, pectinases, cellulases, amylases, etc.) are useable for the specific manufactures because of their exclusive roles and utilization in food and feed industries. Manufacture of microbial strains is another vast accomplishment that became feasible with the assist of rDNA technology. Various microbial strains have been expanded to manufacture superior enzymes by particular engineering. Specific strains of fungi have been modified in order that their capability of manufacturing toxic and hazardous materials could be decreased. Wide ranges of recombinant proteins/enzymes have been expressed in different species of fungi to be used as enzymes in industries [105]. *Aspergillus* species are regarded as promising candidates for developing large-scale heterologous protein production platforms. However, production yields of heterologous proteins are usually significantly lower than those detected for native proteins. Failure to achieve the favorable protein amounts in *Aspergillus* cultures can be ascribed to limitations associated to transcription, translation, and the post-translation processing and modifications during protein manufacture. Furthermore, bottlenecks in the fungal secretion system and the problem of extracellular breakdown further impede the effective production of foreign proteins in *Aspergillus* species [106].

Several *Aspergillus* species, in particular *A. niger* and *A. oryzae*, are widely used as protein-production hosts in a variety of biotechnology applications. To improve the expression and secretion of recombinant proteins from these filamentous fungi, several novel genetic engineering strategies have been developed over the past few years. Yu et al. reported construction and application of a novel genetically engineered *A. oryzae* for expressing proteases [107]. Their results showed different degrees of improvement in the protease activity when compared with wild-type *A. oryzae*. A major improvement in the polypeptide yield was achieved when these strains were used in soybean meal fermentation. The polypeptide conversion rate of transformants *A. oryzae* reached 35.9%, which was approximately twofold higher than that exhibited by wild-type *A. oryzae* [107]. Amino acid content analysis showed that the essential amino acid content and amino acid composition of the fermentation product significantly improved when engineered *A. oryzae* strains were used for soybean meal fermentation [107]. Prathumpai et al. reported lipase production by recombinant strains of *A. niger* expressing a lipase-encoding gene from *Thermomyces lanuginosus* [108]. Their heterologous lipase was expressed using the TaKa amylase (the *A. oryzae* amylase is known as Taka-amylase) promoter from *A. oryzae*. Their results showed the transformants strain was found to be the best producer. Record et al. reported expression of the *Pycnoporus cinnabarinus* laccase gene in *A. niger*. *A. niger* preprosequence allowed an 80-fold increase in laccase production [109]. Bohlin et al. reported heterologous expression of *Trametes versicolor* laccase in *A. niger*. Their results showed recombinant laccase from *A. niger* harboring the lcc2 cDNA was purified to homogeneity and it was found to be a 70 kDa homogeneous enzyme with biochemical and catalytic properties similar to those of native *T. versicolor* laccase [110]. Dragosits et al. reported recombination for the production and purification of two previously uncharacterized β-galactosidases from *A. nidulans* as well as one β-galactosidase from *A. niger*. Their results showed that *A. niger* and *A. nidulans* are suitable for various glycobiological and biotechnological applications [111].

5. Enzymes Production By *Aspergillus* Species and their Application

5.1 Amylases

The *Aspergillus* species produce a wide range of intra- and extracellular enzymes, and amylases are one of the main enzymes used in industrial markets (Table 2). Amylases are broadly employed in food manufacturing such as brewing, baking, preparation of digestive aids, manufacture of cakes, fruit juices and starch syrups [112]. They have been extensively used in the baking industry. Amylases can be added to the dough of bread which degrade the starch in the flour into smaller dextrins, and these smaller dextrins are subsequently fermented by the yeast. The addition of these enzymes to the dough results in improving the rate of fermentation and the reduction of the viscosity of dough, resulting in progresses in the volume and texture of the product. Moreover, they produce additional sugar in the dough that improves the flavor, crust color and toasting qualities of the bread [112]. In addition to producing fermentable compounds, they also have an antistaling effect in baking bread which advances the retention of the softness of bakery products, increasing the shelf life of these products. These enzymes are also used for the clarification of fruit and veggie juices or beer or for the pretreatment of animal feed to improve the digestibility of fiber [6, 34, 35, 36, 37, 38, 39, 40, 41, 42, 43, 44].

5.2 Cellulases

Cellulases are groups of enzymes that are secreted using a wide range of *Aspergillus* species (Table 2). According to current enzyme market information, the main areas of the industry where cellulases enzymes are progressively being applied are textile, pulp and paper, laundry detergents, healthcare, food and beverages. Their extensive application in coffee and tea processing, fruit juice production and wine making is associated to food and beverage segment. In other industrial applications, it is widely used to generate laundry detergents and cleaning and washing agents. These enzymes are also being highly known as an effective alternative to available various antibiotics and antifungal. Lemnaru Popa et al. reported bacterial cellulose in skin wound treatment is very attractive due to its unique characteristics. Their results confirmed the drugs' presence in the bacterial-cellulose dressing's structure as well as the antimicrobial efficiency against *Staphylococcus aureus* and *Escherichia coli* [113]. In another study, Limón et al. reported antifungal and chitinase specific activities of *Trichoderma harzianum* CECT 2413 by addition of a cellulose binding domain was increased [114]. Consequently, the potential of these enzymes is a wonderful trend to fight against antibiotic-resistant bacteria which may overcome problems in the healthcare sector [7]. Furthermore, in the food market these enzymes have many applications. Fruit and veggie juice clarification, carotenoid extraction, reducing the viscosity of nectars, alteration of fruit sensory properties, concentrating purees, olive oil extraction, and the quality betterment of bakery products are among the different processes in food industry and biotechnology where cellulases are exploited worldwide. Consequently, these enzymes can cause a huge economic impact. However, there are some significant bottlenecks of employing this enzyme in the industry such as loss of enzyme activity, immobilization of enzyme in undesired conformation and subsequent loss of activity, cost of carrier and additional preparation materials and methods, as well as laborious training strategies, mass transfer limitations, laborious and time-consuming immobilization processes [7, 46, 47, 48, 49, 50].

5.3 Chitosanases

Chitosanases can degrade chitin and it can generate using various *Aspergillus* species (Table 2). These enzymes play a significant role in nutrition and defense. Due to its antimicrobial activity against a wide range of filamentous fungi, yeasts and bacteria, they have attracted much attention as a potential food preservative [8]. These enzymes potentially used in preparation of N-acetyl d-glucosamine and chitooligosaccharides. Chitooligosaccharides and N-acetyl glucosamine are currently of enormous relevance to pharmaceutical, nutraceutical, cosmetics, food, and agriculture industries due to their wide range of biological activities, which include antimicrobial, antitumor, antioxidant, anticoagulant, wound healing, immunoregulatory, and hypocholesterolemic effects. Furthermore, the control of pathogenic fungi in agriculture could be done using chitonases. The degradation of crustacean chitinous waste in seafood industry could be enhanced using them. It is also valuable for the preparation of single-cell protein and also for the isolation of protoplasts from fungi and yeast, etc. [8, 51, 52, 53, 54, 55].

5.4 Galactosidases

Galactosidases can be generated by various *Aspergillus* species and being used in food industry for hydrolysis of lactose in milk and milk by-products (Table 2). They have attracted much attention in view of lactose intolerance in human population and due to importance of milk in human diet. These enzymes can hydrolyze galactopyranosides, that is, lactose, and form a wide range of trans-galactosylation products or galactooligosaccharides capable of providing some health benefits as prebiotics. Furthermore, these enzymes also find applications in production of lactose based sweeteners from high lactose containing effluents of cheese manufacturing industries. At present, these enzymes are mainly obtained from various *A. niger* strains [9, 56, 57, 58, 59].

5.5 Invertase

Invertases are generated using plants, bees, and microorganisms [10], but the filamentous fungi belonging to the *Aspergillus* species are the most prominent organisms used for invertases manufacture (Table 2). They are used to hydrolyze sucrose and polysaccharides, which have the same type of β-d-fructofuranosyl bond, to obtain fructose and glucose as final products [10]. They are significant in the food biotechnology, especially in confectionery and candy manufacturers, as a catalytic factor in obtaining an artificial sweetener. Hence, they are used for the construction of formulations that prevent crystallization of certain sweet preparations, using in the chocolate industry markets. In some syrup, it is also used to growth its sweetening properties such as producing of soft caramel fillings. Honey is the most common form of this inverted sugar that is a supersaturated mixture of fructose and glucose. Furthermore, they are able to synthesize fructooligosaccharides through fructotransferase where sucrose is presented in high concentrations. The fructooligosaccharides are related to improve human health [9, 60, 61, 62].

5.6 Laccases

Fungi such as *Aspergillus* species have been used to generate Laccases (Table 2). Laccases are a broadly studied enzyme because of its potential use in some industries areas such as textile, paper and pulp [16]. They can be used in bioremediation, beverage processing (such as wine, fruit juice and beer), sugar beet pectin gelation, ascorbic acid determination, baking, and as biosensor and to progress food sensory parameters [17]. They can increase the productivity, efficiency and quality of food goods without costly investment, and this is an advantage. Wide areas of the food industry that benefit from processing with these enzymes include juice processing, wine stabilization, baking industry, and bioremediation of waste water [16, 17, 63, 64, 65].

5.7 Lipases

Lipases are one of the most important biocatalysts that perform different reactions in aqueous and non-aqueous media [11]. These enzymes usually catalyze the hydrolysis of long-chain triglycerides. They can operate on a diversity of substrates counting natural oils, artificial triglycerides, and esters of fatty acids. They are manufactured using animals, plants, and microorganisms. Presently, fungal lipases are achieving much consciousness with the rapid development of enzyme technology. Fungi-produced lipases have played an interesting role in industrial biotechnology because many of them are stable in a wide range of pH, high temperatures, and organic solvents. They are signifying one of the most important groups of biocatalysts for industrial applications. *Aspergillus* species is an important chief manufacturer of lipases (Table 2). They are capable to modify the characteristics of lipids by altering the location of fatty acid chains in the glyceride and exchange one or more fatty acid with new ones. Cocoa butter is a crucial ingredient in chocolate that has a high butterfat value because it contains palmitic and stearic acids and has a melting point of almost 37°C. Melting of cocoa butter in the mouth creates a positive favorable cooling sensation in a product such as chocolate [11]. Lipases are used ex-situ to create taste and to change the formation using inter- or transesterification that obtain products of improved nutritional value, or appropriate for feeding [11]. They are also been used in food to modify taste using manufacture of esters of short-chain fatty acids and alcohol, which are known flavor and fragrance compounds. Lipases facilitate the elimination of fat from meat and fish products. They are used for the manufacture of maltose and lactose like sugar fatty acid esters. They have many applications in food and flavor industry and in the production of ice cream [11, 66, 67, 68, 69, 70].

5.8 Naringinases

Various microbial sources of naringinases have been reported worldwide by various investigators [12]. Production of naringinases has been very well studied in fungal sources. Among fungi, *Aspergillus* species (especially *A. niger*) have been reported as major producers of naringinases (Table 2). Naringin is a flavonoid naturally which present in citrus fruits (such as oranges, lemon, and grapefruit). Flavonoids may cause interference during the citrus fruit juice processing and are responsible for the bitter taste. The processing of citrus fruit juice has faced formidable problems in terms of bitterness and delayed onset of bitterness. The bitterness affects its consumer acceptability. In citrus juices two compounds namely flavonoids and limonoids are established responsible for bitterness. The bitterness in grapefruit juice can be reduced by using enzymes such as naringinases which hydrolyzes naringin into relatively non-bitter compounds. They are an enzyme which catalyzes the hydrolysis of naringin into pruning, rhamnose, glucose and then into naringenin, which is non-bitter and tasteless. This enzyme has two different enzyme activities (due to two different subunits). One is α-rhamonosidase which acts on naringin to release prunin and α-rhamnose. Second is β-D-glucosidase which acts on prunin to release naringenin and β-D-glucose. There are only few reports on the commercial manufacture of this enzyme. These enzymes have been used for elimination of bitter flavor in citrus fruit juice (due to naringin) [12, 71, 72, 73, 74, 75, 76, 77, 78].

5.9 Pectinases

Pectinases have the most important role in fruit and veggie juice marketing by breaking the pectin (polysaccharide) structure present in the cell wall of plants. They are mainly manufactured using microorganisms and plants. Among microorganisms, fungi (especially *Aspergillus* species) have a high ability to secrete them (Table 2). Pectinases are a class of enzymes that catalyzes the degradation of pectic substances [13]. They have broad applications in food and agricultural industries. Pectinases can be used and commercially applied for the processing of fruit juices and wines. In food industry, these enzymes are used especially for clarification, maceration, extraction and stabilization of fruit juices. They also enhance fruit juice yield and involved in fermentation of coffee, cocoa and tea and preparation of jams and jellies. They are used in oil extraction from plant but olive oil extraction is the most common. These enzymes are added for easy oil extraction during grinding of olives. They have capability to reduce feed viscosities which directly increase the nutrients absorption capability of animals. These nutrients are released from fibers using hydrolysis process and it also reduces animal defecation [13, 79, 80, 81, 82, 83, 84].

5.10 Phytases

Phytases have a role in food and feed industry. They are synthesized using fungi, mainly from *Aspergillus* species (Table 2). They can reduce the antinutritional effect of phytate and improve the digestibility of phosphorous, calcium, amino acids and energy, as well as minimize the negative impact of inorganic phosphorous excretion on the environment [14]. The benefits of using phytase in animal feed are well recognized [14]. Especially, phytate-degrading enzymes from *Aspergillus* species offer industrial and economical feasibility for their manufacture and application. Phytates have been considered as a threat in human diet due to its antinutrient behavior, which is known as strong chelator of divalent minerals such as Ca^{2+}, Mg^{2+}, Zn^{2+} and Fe^{2+}. There is a high potential for the employ of phytases in processing and developed of food for human consumption. Investigation in this aspect focuses on the progress of the nutritional value of plant-based food and feed as well as on the technical improvement of food processing. A diet rich in phytate leads to a significantly decreased absorption of dietary minerals and the dephosphorylation of phytate during food processing results in the formation of only partially phosphrylated myo-inositol phosphate esters with a lower capability to impair with the intestinal uptake of dietary minerals. Individual myo-inositol phosphate esters have been shown to have some important physiological functions in man. Consequently, phytases may find application in food processing to generate functional foods. Also, phytases were showing to be an excellent bread making improver [14, 85, 86, 87, 88].

5.11 Proteases

Proteases are produce in all organisms, such as plants, animals, and microbes [15]. The peptide bond present in the polypeptide chain is hydrolyzed by proteases. They are degradative enzymes and demonstrate specificity and selectivity in protein modification. They are one of the most important industrial enzymes and their international market is significantly growing annually. Of the 60% of enzymes marketed worldwide, proteases account for 20%. Proteases have been successfully produced by researchers from various microbial sources [15]. Reports suggest that two-thirds of the world's commercial proteases are produced by microorganisms because of their greater yield, reduction in time consumption, reduction in space requirement, lofty genetic manipulation, and cost-effectiveness, which have made them suitable for biotechnological application in the market. Among microbes, *Aspergillus* species have been extensively studied for protease manufacture in a large scale (Table 2). Proteases are used on a large commercial scale in the production of baked goods, bread, crackers and waffles. Proteases produced by *Asperjillus oryzae* have been used to modify wheat gluten using limited proteolysis. They are also used in the dairy industry. Their main application in the dairy industry is in cheese manufacture. In cheese manufacture, the primary function of proteases is to hydrolyze the certain peptide bond to generate para-k-casein and macro peptides [15, 89, 90, 91, 92].

5.12 Tannase

Tannases are a group of enzymes that are employed in multitudinous industries such as food, brewing, and pharmaceutical [18]. They have an expansive range of scattering and are generated form animals, plants, and microbial sources. However, manufactured tannins of microbial origin are favored over other sources for industrial utilization. Fungi such as *Aspergillus* species have been used to generate tannases (Table 2). They operate upon hydrolyzable tannins by cracking the ester and depside bonds so as to release glucose and gallic acid. One of the most significant commercial applications of tannases is gallic acid manufacture. Besides that, they are widely employed in the food industry, especially in the manufacture of instant tea, where it increases the extractability and cold water solubility of key compounds [18]. Another significant applying of tannase is the elimination of haze formation and unflavored phenolic compounds from beer and wine [18]. Moreover, the quality of fruit juices can be enhanced using tannases enzymes. The turbidity and bitterness of fruit and veggie juices are minimized by using these enzymes. While using agro-industrial residues as animal feed, tannins-rich biomass are considered as anti-nutritional factors. De-tannification of feed using tannases enzymes treatment can extensively progress the quality of animal feed [18, 93, 94, 95, 96, 97, 98, 99].

6. Future Perspectives and Conclusions

At present, enzymes have become an important part of various industries [1]. The total enzymes market size in the worldwide is anticipated to reach over $13–14 billion by 2027. However, the manufacture of different enzymes has always been a challenge. They are produced from plants, animals and microorganisms [4]. Microbial enzyme production is generally accepted and occupies approximately 85–90% of the global enzyme market. In microbial enzyme manufacture, the localization of enzyme is a major aspect to be considered. If an enzyme is extracellular, the cost of downstream processing is reduced [100]. However, when it comes to intracellular enzymes, it becomes an expensive process to purify such enzymes. The degree of purification also varies according to the use of enzymes. Among microorganisms, fungi are especially used for the manufacture of various enzymes in a wide range [5]. Out of about 260 commercial enzymes, 60% are sourced from about 25 fungal genera [115]. Fungi can produce a number of industrial enzymes that are used in variety different industrial processes. Owing to their ability to use low-value substrates, their ability to handle and their ability to produce high enzymatic titres, fungi are the subject of extensive studies for industrial enzymes. Enzymes of fungal origin (especially from *Aspergillus* species) are employed in a wide range of industrial applications. The advantages of using *Aspergillus* species in the industry are multiple and they are a rich source of enzymes with valuable properties in industrial processes. A good example is their outstanding high stability, as they naturally evolved to work in a relatively harsh extracellular environment. They can use common "types of waste" as a source of carbon and energy and release the valuable enzymatic product from their cells into the medium. The cost of enzyme manufacture by *Aspergillus* species is another attractive parameter and an essential prerequisite for the exploitation of these catalysts in large-scale industrial settings. In addition, *Aspergillus* species will be able to produce strong, multifunctional (chimeric) enzymes using recombinant DNA technology, high-efficiency screening of the latest, metagenomic screening, silicon enzyme engineering and site-to-site mutagenesis to meet future needs.

References

Patel AK, Singhania RR, Pandey A. Production, purification, and application of microbial enzymes. In: Brahmachari G, editor. Biotechnology of Microbial Enzymes. San Diego, CA: Elsevier; 2017. pp. 13-41. DOI: 10.1016/B978-0-12-803725-6.00002-9

Meyer V, Basenko EY, Benz JP, Braus GH, Caddick MX, Csukai M, et al. Growing a circular economy with fungal biotechnology: A white paper. Fungal Biology and Biotechnology. 2020;7:1-23. DOI: 10.1186/s40694-020-00095-z

Jun H, Kieselbach T, Jönsson LJ. Enzyme production by filamentous fungi: Analysis of the secretome of Trichoderma reesei grown on unconventional carbon source. Microbial Cell Factories. 2011;10:68. DOI: 10.1186/1475-2859-10-68

Singh R, Kumar M, Mittal A, Mehta PK. Microbial enzymes: Industrial progress in 21st century. 3 Biotech. 2016;6:174. DOI: 10.1007/s13205-016-0485-8

Robinson PK. Enzymes: Principles and biotechnological applications. Essays in Biochemistry. 2015;59:1-41. DOI: 10.1042/bse0590001

de Souza PM, de Oliveira Magalhães P. Application of microbial α-amylase in industry—A review. Brazilian Journal of Microbiology. 2010;41(4):850-861. DOI: 10.1590/S1517-83822010000400004

Jayasekara S, Ratnayake R. Microbial cellulases: An overview and applications. In: Pascual AR, Martín MEE, editors. Cellulose. London, England: IntechOpen; 2019. DOI: 10.5772/intechopen.84531

Muzaddadi AU, Devatkal S, Oberoi HS. Seafood enzymes and their application in food processing. In: Dhillon GS, Kaur S, editors. Agro-Industrial Wastes as Feedstock for Enzyme Production. San Diego, CA: Elsevier; 2016. pp. 201-232. DOI: 10.1016/B978-0-12-802392-1.00009-5

Xavier JR, Ramana KV, Sharma RK. β-Galactosidase: Biotechnological applications in food processing. Journal of Food Biochemistry. 2018;42:e12564. DOI: 10.1111/jfbc.12564

Veana F, Flores-Gallegos AC, Gonzalez-Montemayor AM, Michel-Michel M, Lopez-Lopez L, Aguilar-Zarate P, et al. Invertase: An enzyme with importance in confectionery food industry. In: Kuddus M, editor. Enzymes in Food Technology. Singapore: Springer Singapore; 2018. pp. 187-212. DOI: 10.1007/978-981-13-1933-4_10

Mehta A, Guleria S, Sharma R, Gupta R. The lipases and their applications with emphasis on food industry. In: Ray RC, editor. Microbial Biotechnology in Food and Health. San Diego, CA: Elsevier; 2021. pp. 143-164. DOI: 10.1016/B978-0-12-819813-1.00006-2

Yadav M. Naringinase: Microbial sources, production and applications in food processing industry. Journal of Microbiology, Biotechnology and Food Sciences. 2018;8:717-720. DOI: 10.15414/jmbfs.2018.8.1.717-720

Kubra KT, Ali S, Walait M, Sundus H. Potential applications of pectinases in food, agricultural and environmental sectors. Journal of Pharmaceutical, Chemical and Biological Sciences. 2018;6:23-34

Dahiya S. Role of phytate and phytases in human nutrition. International Journal of Food Sciences and Nutrition. 2016;1:39-42

Razzaq A, Shamsi S, Ali A, Ali Q, Sajjad M, Malik A, et al. Microbial proteases applications. Frontiers in Bioengineering and Biotechnology. 2019;7:110. DOI: 10.3389/fbioe.2019.00110

Minussi RC, Pastore GM, Durán N. Potential applications of laccase in the food industry. Trends in Food Science and Technology. 2002;13:205-216. DOI: 10.1016/S0924-2244(02)00155-3

Brijwani K, Rigdon A, Vadlani PV. Fungal laccases: Production, function, and applications in food processing. Enzyme Research. 2010;2010:149748. DOI: 10.4061/2010/149748

Kumar SS, Sreekumar R, Sabu A. Tannase and its applications in food processing. In: Parameswaran B, Varjani S, Raveendran S, editors. Green Bio-Processes: Energy, Environment, and Sustainability. Singapore: Springer Singapore; 2019. pp. 357-381. DOI: 10.1007/978-981-13-3263-0_19

Wang Q, Zhong C, Xiao H. Genetic engineering of filamentous fungi for efficient protein expression and secretion. Frontiers in Bioengineering and Biotechnology. 2020;8:293. DOI: 10.3389/fbioe.2020.00293

Hyde KD, Xu J, Rapior S, Jeewon R, Lumyong S, Niego AGT, et al. The amazing potential of fungi: 50 ways we can exploit fungi industrially. Fungal Diversity. 2019;97:1-136. DOI: 10.1007/s13225-019-00430-9

Park HS, Jun SC, Han KH, Hong SB, Yu JH. Diversity, application, and synthetic biology of industrially important Aspergillus fungi. Advances in Applied Microbiology. 2017;100:161-202. DOI: 10.1016/bs.aambs.2017.03.001

El-Enshasy HA. Filamentous fungal cultures—Process characteristics, products, and applications. In: Yang ST, editor. Bioprocessing for Value-Added Products from Renewable Resources. B.V., The Netherlands: Elsevier; 2007. p. 225–261. Doi:10.1016/B978-044452114-9/50010-4

Luis HSG, Patricia N d CS. Aspergillus biotechnology: An overview on the production of hydrolases and secondary metabolites. Current Biotechnology. 2017;6:283-294. DOI: 10.2174/2211550105666160607071134

Ward OP, Qin WM, Dhanjoon J, Ye J, Singh A. Physiology and biotechnology of Aspergillus. Advances in Applied Microbiology. 2005;58C:1-75. DOI: 10.1016/S0065-2164(05)58001-8

Max B, Salgado JM, Rodríguez N, Cortés S, Converti A, Domínguez JM. Biotechnological production of citric acid. Brazilian Journal of Microbiology. 2010;41:862-875. DOI: 10.1590/S1517-83822010000400005

Show PL, Oladele KO, Siew QY, Aziz Zakry FA, Lan JC-W, Ling TC. Overview of citric acid production from Aspergillus niger. Frontiers in Life Science. 2015;8:271-283. DOI: 10.1080/21553769.2015.1033653

Gomi K. Aspergillus: Aspergillus oryzae. In: Encyclopedia of Food Microbiology. 2nd ed. USA: Elsevier Academic Press; 2014. pp. 92-96. DOI: 10.1016/B978-0-12-384730-0.00011-2

Manzoni M, Rollini M. Biosynthesis and biotechnological production of statins by filamentous fungi and application of these cholesterol-lowering drugs. Applied Microbiology and Biotechnology. 2002;58:555-564. DOI: 10.1007/s00253-002-0932-9

Kumar A. Aspergillus nidulans: A potential resource of the production of the native and heterologous enzymes for industrial applications. International Journal of Microbiology. 2020;2020:8894215. DOI: 10.1155/2020/8894215

Cairns TC, Nai C, Meyer V. How a fungus shapes biotechnology: 100 years of Aspergillus niger research. Fungal Biology and Biotechnology. 2018;5:13. DOI: 10.1186/s40694-018-0054-5

Arnau J, Yaver D, Hjort CM. Strategies and challenges for the development of industrial enzymes using fungal cell factories. In: Grand Challenges in Fungal Biotechnology. Cham: Springer International Publishing; 2020. pp. 179-210. DOI: 10.1007/978-3-030-29541-7_7

Berenjian A, editor. Essentials in Fermentation Technology. 1st ed. Cham, Switzerland: Springer Nature; 2019. DOI: 10.1007/978-3-030-16230-6

Tan CH, Show PL, Ooi CW, Ng EP, Lan JC, Ling TC. Novel lipase purification methods–a review of the latest developments. Biotechnology Journal. 2015;10:31-44

Bhatia S. Introduction to Pharmaceutical Biotechnology. Volume 2: Enzymes, Proteins and Bioinformatics. London, England: Institute of Physics Publishing; 2018. pp. 2-42. DOI: 10.1088/978-0-7503-1302-5

Varalakshmi KN, Kumudini BS, Nandini BN, Solomon J, Suhas R, Mahesh B, et al. Production and characterization of alpha-amylase from Aspergillus niger JGI 24 isolated in Bangalore. Polish Journal of Microbiology. 2009;58:29-36

Sethi BK, Jana A, Nanda PK, DasMohapatra PK, Sahoo SL, Patra JK. Production of α-amylase by Aspergillus terreus NCFT 4269.10 using pearl millet and its structural characterization. Frontiers in Plant Science. 2016;7:639. DOI: 10.3389/fpls.2016.00639

Goto CE, Barbosa EP, Kistner LCL, Moreira FG, Lenartovicz V, Peralta RM. Production of amylase by Aspergillus fumigatus utilizing α-methyl-D-glycoside, a synthetic analogue of maltose, as substrate. FEMS Microbiology Letters. 1998;167(2):139-143. DOI: 10.1111/j.1574-6968.1998.tb13219.x

Moreira FG, Lima FA de, Pedrinho SRF, Lenartovicz V, Souza CGM de, Peralta RM. Production of amylases by Aspergillus tamarii. Revista de Microbiologia 1999;30:157–162. DOI:10.1590/S0001-37141999000200014

Fattah Farid MA, Halim Shata HMA. Amylase production from Aspergillus oryzae LS1 by solid-state fermentation and its use for the hydrolysis of wheat flour. Iranian Journal of Biotechnology. 2011;9:267-274

Balakrishnan M, Jeevarathinam G, Kumar SKS, Muniraj I, Uthandi S. Optimization and scale-up of α-amylase production by Aspergillus oryzae using solid-state fermentation of edible oil cakes. BMC Biotechnology. 2021;21:33. DOI: 10.1186/s12896-021-00686-7

Ali EH, El-Nagdy MA, Al-Garni SM, Ahmed MS, Rawaa AM. Enhancement of alpha amylase production by Aspergillus flavus AUMC 11685 on mandarin (Citrus reticulata) peel using submerged fermentation. European Journal of Biological Research. 2017;7:154-164. DOI: 10.5281/zenodo.818271

Pestana F, Castillo FJ. Glucoamylase production by Aspergillus awamori on rice flour medium and partial characterization of the enzyme. World Journal of Microbiology and Biotechnology. 1985;1:225-237. DOI: 10.1007/BF01742317

Negi S, Banerjee R. Optimization of culture parameters to enhance production of amylase and protease from Aspergillus awamori in a single fermentation. African Journal of Biochemistry Research. 2010;4:73-80. DOI: 10.5897/AJBR.9000178

Shambe T, Ejembi O. Production of amylase and cellulase: Degradation of starch and carboxymethylcellulose by extracellular enzymes from four fungal species. Enzyme and Microbial Technology. 1987;9:308-312. DOI: 10.1016/0141-0229(87)90010-X

Adegoke SA, Odibo FJC. Production, purification and characterization of α-amylase of Aspergillus sydowii IMI 502692. Plant Cell Biotechnology and Molecular Biology. 2019;20:1050-1058

Bansal N, Tewari R, Soni R, Soni SK. Production of cellulases from Aspergillus niger NS-2 in solid state fermentation on agricultural and kitchen waste residues. Waste Management. 2012;32:1341-1346. DOI: 10.1016/j.wasman.2012.03.006

Singh A, Singh N, Bishnoi NR. Production of cellulases by Aspergillus heteromorphus from wheat straw under submerged fermentation. International Journal of Civil Engineering. 2009;1:23-26

Sherief AA, El-Tanash AB, Atia N. Cellulase production by Aspergillus fumigatus grown on mixed substrate of rice straw and wheat bran. Research Journal of Microbiology. 2010;5:199-211. DOI: 10.3923/jm.2010.199.211

Padmavathi T, Nandy V, Agarwal P. Optimization of the medium for the production of cellulases by Aspergillus terreus and Mucor plumbeus. European Journal of Experimental Biology. 2012;2:1161-1170

Mrudula S, Murugammal R. Production of cellulase by Aspergillus niger under submerged and solid state fermentation using coir waste as a substrate. Brazilian Journal of Microbiology. 2011;42:1119-1127. DOI: 10.1590/S1517-83822011000300033

Lee CK, Darah I, Ibrahim CO. Production and optimization of cellulase enzyme using Aspergillus niger USM AI 1 and comparison with Trichoderma reesei via solid state fermentation system. Biotechnology Research International. 2011;2011:658493. DOI: 10.4061/2011/658493

Xu X, Song Z, Yin Y, Zhong F, Song J, Huang J, et al. Solid-state fermentation production of chitosanase by Streptomyces with waste mycelia of Aspergillus niger. Advances in Enzyme Research. 2021;9:10-18. DOI: 10.4236/aer.2021.91002

Zhang H, Sang Q, Zhang W. Statistical optimization of chitosanase production by Aspergillus sp. QD-2 in submerged fermentation. Annales de Microbiologie. 2012;62:193-201. DOI: 10.1007/s13213-011-0246-1

Hirano K, Arayaveerasid S, Seki K, Adams DJ, Mitsutomi M. Characterization of a chitosanase from Aspergillus fumigatus ATCC13073. Bioscience, Biotechnology, and Biochemistry. 2012;76:1523-1528. DOI: 10.1271/bbb.120248

Jebur HA, Abdulateef AA, Thbit ZA. Chitosan production from Aspergillus oryzae SU-B2 by submerged fermentation and studying some of its physiochemical and antibacterial characteristics. Jordan Journal of Pharmaceutical Sciences. 2019;11:609-613

Kazemi S, Khayati G, Faezi-Ghasemi M. B-galactosidase production by Aspergillus niger ATCC 9142 using inexpensive substrates in solid-state fermentation: Optimization by orthogonal arrays design. Iranian Biomedical Journal. 2016;20:287-294. DOI: 10.22045/ibj.2016.06

Kapnoor S, Mulimani VH. Production of α-galactosidase by Aspergillus oryzae through solid-state fermentation and its application in soymilk Galactooligosaccharide hydrolysis. Brazilian Archives of Biology and Technology. 2010;53:211-218. DOI: 10.1590/S1516-89132010000100026

Bailey MJ, Linko M. Production of β-galactosidase by Aspergillus oryzae in submerged bioreactor cultivation. Journal of Biotechnology. 1990;16:57-66. DOI: 10.1016/0168-1656(90)90065-J

Nizamuddin S, Sridevi A, Narasimha G. Production of β-galactosidase by Aspergillus oryzae in solid-state fermentation. African Journal of Biotechnology. 2008;7:1096-10100. DOI: 10.5897/AJB07.325

Alegre ACP, de Lourdes Teixeira de Moraes Polizeli M, Terenzi HF, Jorge JA, Guimarães LHS. Production of thermostable invertases by Aspergillus caespitosus under submerged or solid state fermentation using agroindustrial residues as carbon source. Brazilian Journal of Microbiology. 2009;40:612-622. DOI: 10.1590/S1517-83822009000300025

Raju CHAI, Pulipati K, Jetti A. Production of invertase by Aspergillus niger under solid state fermentation using orange fruit peel as substrate. Advances in Crop Science and Technology. 2016;04:1-4. DOI: 10.4172/2329-8863.1000247

Alves JNO, Jorge JA, Guimarães LHS. Production of invertases by anamorphic (Aspergillus nidulans) and teleomorphic (emericela nidulans) fungi under submerged fermentation using rye flour as carbon source. Advances in Microbiology. 2013;03:421-429. DOI: 10.4236/aim.2013.35057

Kumar R, Kaur J, Jain S, Kumar A. Optimization of laccase production from Aspergillus flavus by design of experiment technique: Partial purification and characterization. Journal, Genetic Engineering & Biotechnology. 2016;14:125-131. DOI: 10.1016/j.jgeb.2016.05.006

Abdallah YK, Estevez AT, Tantawy DEDM, Ibraheem AM, Khalil NM. Employing laccase-producing Aspergillus sydowii NYKA 510 as a cathodic biocatalyst in self-sufficient lighting microbial fuel cell. Journal of Microbiology and Biotechnology. 2019;29:1861-1872. DOI: 10.4014/jmb.1907.07031

Fokina O, Fenchel A, Winandy L, Fischer R. Immobilization of LccC laccase from Aspergillus nidulans on hard surfaces via Fungal hydrophobins. Applied and Environmental Microbiology. 2016;82(21):6395-3402. DOI: 10.1128/AEM.01413-16

Mahadik ND, Puntambekar US, Bastawde KB, Khire JM, Gokhale DV. Production of acidic lipase by Aspergillus niger in solid state fermentation. Process Biochemistry. 2002;38:715-721. DOI: 10.1016/S0032-9592(02)00194-2

Salihu A, Bala M, Alam MZ. Lipase production by Aspergillus niger using sheanut cake: An optimization study. Journal of Taibah University for Science. 2016;10:850-859. DOI: 10.1016/j.jtusci.2015.02.011

Ahmed A, Badar R, Khalique N. Screening and optimization of submerged fermentation of lipolytic Aspergillus oryzae. BioResources. 2019;14:7664-7674. DOI: 10.15376/biores.14.4.7664-7674

Kirsh D. Lipase Production by Penicillium oxalicum and Aspergillus flavus. Botanical Gazette. 1935;97:321-333. DOI: 10.1086/334555

Nema A, Patnala SH, Mandari V, Kota S, Devarai SK. Production and optimization of lipase using Aspergillus niger MTCC 872 by solid-state fermentation. Bulletin of the National Research Centre. 2019;43:1-8. DOI: 10.1186/s42269-019-0125-7

Borkar V, Chakraborty S, Gokhale JS. Fermentative production of naringinase from Aspergillus niger van Tieghem MTCC 2425 using citrus wastes: Process optimization, partial purification, and characterization. Applied Biochemistry and Biotechnology. 2021;193:1321-1337. DOI: 10.1007/s12010-020-03385-9

Mendoza-Cal A, Cuevas-Glory L, Lizama-Uc G, Ortiz-Vázquez E. Naringinase production from filamentous fungi using grapefruit rind in solid state fermentation. African Journal of Microbiology Research. 2010;4:1964-1969

Chen Y, Ni H, Chen F, Cai H, Li L, Su W. Purification and characterization of a naringinase from Aspergillus aculeatus JMUdb058. Journal of Agricultural and Food Chemistry. 2013;61:931-938. DOI: 10.1021/jf303512q

Zhu Y, Jia H, Xi M, Li J, Yang L, Li X. Characterization of a naringinase from Aspergillus oryzae 11250 and its application in the debitterization of orange juice. Process Biochemistry. 2017;62:114-121. DOI: 10.1016/j.procbio.2017.07.012

Shanmugaprakash M, Vinoth Kumar V, Hemalatha M, Melbia V, Karthik P. Solid-state fermentation for the production of debittering enzyme naringinase using Aspergillus niger MTCC 1344. Engineering in Life Sciences. 2011;11:322-325. DOI: 10.1002/elsc.201000128

Shanmugaprakash M, Vinothkumar V, Ragupathy J, Reddy DA. Biochemical characterization of three phase partitioned naringinase from Aspergillus brasiliensis MTCC 1344. International Journal of Biological Macromolecules. 2015;80:418-423. DOI: 10.1016/j.ijbiomac.2015.06.057

Xia X-K, Zhang Y-E, Lei S-J, Hu B, Fu C-X. Identification and iterative combinatorial mutagenesis of a new naringinase-producing strain, Aspergillus tubingensis MN589840. Letters in Applied Microbiology. 2021;72:141-148. DOI: 10.1111/lam.13379

Chang H-Y, Lee Y-B, Bae H-A, Huh J-Y, Nam S-H, Sohn H-S, et al. Purification and characterisation of Aspergillus sojae naringinase: The production of prunin exhibiting markedly enhanced solubility with in vitro inhibition of HMG-CoA reductase. Food Chemistry. 2011;124:234-241. DOI: 10.1016/j.foodchem.2010.06.024

Sandri I, Silveira M. Production and application of pectinases from Aspergillus niger obtained in solid state cultivation. Beverages. 2018;4:48. DOI: 10.3390/beverages4030048

El-Sheekh MM, Ismail A-MS, El-Abd MA, Hegazy EM, El-Diwany AI. Effective technological pectinases by Aspergillus carneus NRC1 utilizing the Egyptian orange juice industry scraps. International Biodeterioration & Biodegradation. 2009;63:12-18. DOI: 10.1016/j.ibiod.2008.06.002

Wolf-Márquez VE, Martínez-Trujillo MA, Aguilar Osorio G, Patiño F, Álvarez MS, Rodríguez A, et al. Scaling-up and ionic liquid-based extraction of pectinases from Aspergillus flavipes cultures. Bioresource Technology. 2017;225:326-335. DOI: 10.1016/j.biortech.2016.11.067

Shanmugavel M, Vasantharaj S, Yazhmozhi A, Bhavsar P, Aswin P, Felshia C, et al. A study on pectinases from Aspergillus tamarii: Toward greener approach for cotton bioscouring and phytopigments processing. Biocatalysis and Agricultural Biotechnology. 2018;15:295-303. DOI: 10.1016/j.bcab.2018.06.013

Pradal-Velázquez M, Martínez-Trujillo MA, Aguilar-Osorio G. Xylanases and pectinases of Aspergillus flavus CECT-2687 on different carbon sources and initial ph values. Revista Mexicana de Ingenieria Quimica. 2018;17:421-431. DOI: 10.24275/10.24275/uam/izt/dcbi/revmexingquim/2018v17n2/Pradal

Semenova MV. Isolation and properties of pectinases from the fungus Aspergillus japonicus. Biochemistry (Moscow). 2003;68:559-569. DOI: 10.1023/A:1023959727067

Shivanna GB, Venkateswaran G. Phytase production by Aspergillus niger CFR 335 and Aspergillus ficuum SGA 01 through submerged and solid-state fermentation. Scientific World Journal. 2014;2014:392615. DOI: 10.1155/2014/392615

Shah PC, Kumar VR, Dastager SG, Khire JM. Phytase production by Aspergillus niger NCIM 563 for a novel application to degrade organophosphorus pesticides. AMB Express. 2017;7:66. DOI: 10.1186/s13568-017-0370-9

Tian M, Yuan Q. Optimization of phytase production from potato waste using Aspergillus ficuum. 3 Biotech. 2016;6:256. DOI: 10.1007/s13205-016-0573-9

Yamada K, Minoda Y, Yamamoto S. Phytase from Aspergillus terreus: Part I. Production, purification and some general properties of the enzyme. Agricultural and Biological Chemistry. 1968;32:1275-1282. DOI: 10.1080/00021369.1968.10859223

Ahmed ME. Extraction and purification of protease from Aspergillus niger isolation. Pharmacy & Pharmacology International Journal. 2018;6:62-65. DOI: 10.15406/ppij.2018.06.00162

de Castro RJS, Sato HH. Protease from Aspergillus oryzae: Biochemical characterization and application as a potential biocatalyst for production of protein hydrolysates with antioxidant activities. Journal of Food Processing. 2014;2014:1-11. DOI: 10.1155/2014/372352

Tremacoldi CR, Carmona EC. Production of extracellular alkaline proteases by Aspergillus clavatus. World Journal of Microbiology and Biotechnology. 2005;21:169-172. DOI: 10.1007/s11274-004-2724-0

Malathi S, Chakraborty R. Production of alkaline protease by a new Aspergillus flavus isolate under solid-substrate fermentation conditions for use as a depilation agent. Applied and Environmental Microbiology. 1991;57:712-716. DOI: 10.1128/AEM.57.3.712-716.1991

Costa AM, Ribeiro WX, Kato E, Monteiro ARG, Peralta RM. Production of tannase by Aspergillus tamarii in submerged cultures. Brazilian Archives of Biology and Technology. 2008;51:399-404. DOI: 10.1590/S1516-89132008000200021

Kumar R, Sharma J, Singh R. Production of tannase from Aspergillus ruber under solid-state fermentation using jamun (Syzygium cumini) leaves. Microbiological Research. 2007;162:384-390. DOI: 10.1016/j.micres.2006.06.012

Viswanath V, Leo VV, Prabha SS, Potty VP, Jisha MS. Optimized production of tannase from cashew testa using Aspergillus niger MTCC 5898. Food Biotechnology. 2016;30:249-262. DOI: 10.1080/08905436.2016.1234392

Beniwal V, Rajesh, Goel G, Kumar A, Chhokar V. Production of tannase through solid state fermentation using Indian rosewood (Dalbergia sissoo) sawdust—A timber industry waste. Annales de Microbiologie. 2013;63:583-590. DOI: 10.1007/s13213-012-0508-6

Gonçalves HB, Riul AJ, Quiapim AC, Jorge JA, Guimarães LHS. Characterization of a thermostable extracellular tannase produced under submerged fermentation by Aspergillus ochraceus. Electronic Journal of Biotechnology. 2012;15:5. DOI: 10.2225/vol15-issue5-fulltext-5

Liu TPSL, Brandão Costa RMP, de Vasconcelos Freitas DJ, Oliveira Nacimento C, de Souza Motta CM, Bezerra RP, et al. Tannase from Aspergillus melleus improves the antioxidant activity of green tea: Purification and biochemical characterisation. International Journal of Food Science and Technology. 2017;52:652-661. DOI: 10.1111/ijfs.13318

Banerjee D, Mondal K, Pati B. Tannase production by Aspergillus aculeatus DBF9 through solid-state fermentation. Acta Microbiologica et Immunologica Hungarica. 2007;54:159-166. DOI: 10.1556/AMicr.54.2007.2.6

Bajar S, Singh A, Bishnoi NR. Exploration of low-cost agro-industrial waste substrate for cellulase and xylanase production using Aspergillus heteromorphus. Applied Water Science. 2020;10:153. DOI: 10.1007/s13201-020-01236-w

Sewalt V, Shanahan D, Gregg L, La Marta J, Carrillo R. The generally recognized as safe (GRAS) process for industrial microbial enzymes. Industrial Biotechnology. 2016;12:295-302. DOI: 10.1089/ind.2016.0011

Deckers M, Deforce D, Fraiture MA, Roosens NH. Genetically modified micro-organisms for industrial food enzyme production: An overview. Food. 2020;9:326. DOI: 10.3390/foods9030326

Adrio JL, Demain AL. Recombinant organisms for production of industrial products. Bioengineered Bugs. 2010;1:116-131

Khan S, Ullah MW, Siddique R, Nabi G, Manan S, Yousaf M, et al. Role of recombinant DNA technology to improve life. International Journal of Genomics. 2016;2016:2405954. DOI: 10.1155/2016/2405954

Archer DB. Enzyme production by recombinant Aspergillus. In: Yoshikatsu M, Tadayuki I, editors. Recombinant Microbes for Industrial and Agricultural Applications. Boca Raton, FL: CRC Press; 2020. pp. 373-393. DOI: 10.1201/9781003067191

Meyer V, Punt PJ, Ram AFJ. Genetics, genetic manipulation, and approaches to strain improvement of filamentous fungi. In: Bull AT, Junker B, Katz L, Lynd LR, Masurekar P, Reeves CD, Zhao H, editors. Manual of Industrial Microbiology and Biotechnology. 3rd ed. USA: American Society of Microbiology; 2010. pp. 318-329. DOI: 10.1128/9781555816827.ch22

Yu X-C, Ma S-L, Xu Y, Fu C-H, Jiang C-Y, Zhou C-Y. Construction and application of a novel genetically engineered Aspergillus oryzae for expressing proteases. Electronic Journal of Biotechnology. 2017;29:32-38. DOI: 10.1016/j.ejbt.2017.07.004

Prathumpai W, Flitter SJ, McIntyre M, Nielsen J. Lipase production by recombinant strains of Aspergillus niger expressing a lipase-encoding gene from Thermomyces lanuginosus. Applied Microbiology and Biotechnology. 2004;65:714-719. DOI: 10.1007/s00253-004-1699-y

Record E, Punt PJ, Chamkha M, Labat M, van Den Hondel CAMJJ, Asther M. Expression of the Pycnoporus cinnabarinus laccase gene in Aspergillus niger and characterization of the recombinant enzyme: P. cinnabarinuslaccase gene expression in A. niger. European Journal of Biochemistry. 2002;269:602-609. DOI: 10.1046/j.0014-2956.2001.02690.x

Bohlin C, Jönsson LJ, Roth R, van Zyl WH. Heterologous expression of Trametes versicolor laccase in Pichia pastoris and Aspergillus niger. Applied Biochemistry and Biotechnology. 2006;129-132:195-214

Dragosits M, Pflügl S, Kurz S, Razzazi-Fazeli E, Wilson IBH, Rendic D. Recombinant Aspergillus β-galactosidases as a robust glycomic and biotechnological tool. Applied Microbiology and Biotechnology. 2014;98(8):3553-3567. DOI: 10.1007/s00253-013-5192-3

Mojsov K. Microbial α-amylases and their industrial applications: A review. International Journal of Management in Education. 2012;2:583-609

Lemnaru Popa G-M, Truşcă RD, Ilie C-I, Ţiplea RE, Ficai D, Oprea O, et al. Antibacterial activity of bacterial cellulose loaded with bacitracin and amoxicillin: In vitro studies. Molecules. 2020;25:4069. DOI: 10.3390/molecules25184069

Limón MC, Chacón MR, Mejías R, Delgado-Jarana J, Rincón AM, Codón AC, et al. Increased antifungal and chitinase specific activities of Trichoderma harzianum CECT 2413 by addition of a cellulose binding domain. Applied Microbiology and Biotechnology. 2004;64:675-685. DOI: 10.1007/s00253-003-1538-6

Kango N, Jana UK, Choukade R. Fungal enzymes: Sources and biotechnological applications. In: Satyanarayana T, Deshmukh S, Deshpande M, editors. Advancing Frontiers in Mycology & Mycotechnology. Singapore: Springer Singapore; 2019. pp. 515-538. DOI: 10.1007/978-981-13-9349-5_21

Index

M

N

O

P

Q

R

S